Adrian Stewart was educated at Rugby School and Cambridge, and is the author of numerous works of military history.

Hurricane

HURRICANE

ADRIAN STEWART

CANELOHISTORY

First published in the United Kingdom in 1982 by William Kimber & Co. Limited

This edition published in the United Kingdom in 2021 by

Canelo
31 Helen Road
Oxford OX2 0DF
United Kingdom

A CIP catalogue record for this book is available from the British Library.

Print ISBN 978 1 80032 532 6
Ebook ISBN 978 1 80032 531 9

Look for more great books at www.canelo.co

Printed and bound in Great Britain by Clays Ltd, Elcograf S.p.A.

1

To My Wife Janet who didn't interrupt

Chapter One

Creation and Early Campaigns

On 6 November 1935, a small, silver-painted monoplane took off from Brooklands aerodrome, by the famous motor-racing track, on its maiden flight. The pilot, P.W.S. 'George' Bulman, chief test pilot of Hawker Aircraft Limited, was delighted by the trial, though he contented himself with reporting merely that the aircraft flew comfortably. It was then known somewhat curtly as the 'Interceptor Monoplane' or more pompously as the 'F36/34 Single Seat Fighter – High Speed Monoplane', registered serial number K5083 – but on 27 June 1936 it received a new official name: 'Hurricane'.

K5083 was designed by Sydney (later Sir Sydney) Camm, who in 1925 became Chief Designer of H.G. Hawker Engineering Company Limited, the privately owned predecessor of Hawker Aircraft Limited, having joined the firm as a draughtsman two years before. Early in 1925, he produced a drawing of a small monoplane fighter to be armed with two Vickers machine-guns and powered by a Bristol Jupiter engine, but the project never proceeded, doubtless because there was a strong belief in high places that the monoplane lacked sufficient structural integrity – a belief that arose in September 1912 when two monoplanes had broken up in the air, resulting the following month in RFC pilots being banned from flying military monoplanes. This was never accepted by the Admiralty and lasted only for five months, but the stigma remained far into the future.

However, if Camm's monoplane never flew, his biplanes did. Of these perhaps the most famous was the Hart light bomber. This first took to the air in 1928, powered by a Rolls-Royce Kestrel engine round which Camm had designed not only for the Hart but also the Hornet fighter. In early 1930, the Hart joined 33 Squadron and, in the air-defence exercises that year, its top speed of 184 mph – some 10 mph in excess of that of any existing fighter – made it quite impossible to intercept.

Understandably, the Air Staff was greatly concerned by the implications of the Hart's invulnerability, so on the principle of 'setting a Hart to catch a Hart', the Demon, a two-seater fighter version of the Hart, was hastily produced. In the air exercises of 1931, the Demons of 23 Squadron successfully intercepted their bomber cousins. However, the Demon had no more than a marginally superior performance, as was almost equally true of the single-seater Hornet, which under the name of Fury also entered squadron service in 1931, with two Vickers machine-guns and a top speed of 207 mph.

In order to obtain clear superiority, Air Ministry Specific-ation F7/30 was issued, offering large production contracts to any firm that could produce a fighter capable of operating by day or night, with a top speed of 250 mph and an armament of four machine-guns. Manufacturers were informed that use of the new Rolls-Royce Goshawk engine would be 'considered sympathetically'. A large number of prototypes was tendered for evaluation, among them a Goshawk-powered improvement of the Fury known as the PV3; the ultimate, worthy winner of the competition was the Gloster Gladiator. Yet the most important result of the study caused by the specification was that it convinced Camm that the biplane had reached its limit, prompting him therefore to look elsewhere for a solution to the problem.

It will be recalled that, in 1925, Camm had projected a small fighter monoplane, and now, encouraged undoubtedly by the success of the Supermarine S6 High-Speed Seaplanes, which in 1931 had won the Schneider Trophy outright, he again directed his attention to a monoplane design. Since the Fury was Hawkers' finest biplane fighter, this naturally formed a basis for his proposals. In August 1933, after a discussion between Camm and Major Buchanan of the Directorate of Technical Development at the Air Ministry, a preliminary scheme for a 'Fury Monoplane' was accepted in principle.

In its initial form, evolved solely as a private venture, the Fury Monoplane had the basic fuselage of its biplane namesake but by definition the top wing had been removed. Both leading and trailing edges of the wings were slightly tapered. The cockpit was enclosed to allow for the greater speed envisaged. A large radiator was located under the centre of the wings. At this stage, it was intended that the aircraft would have a fixed undercarriage, four machine-guns and a Goshawk engine. Because of its derivation from the Fury, it had the standard Hawker fabricated steel tubular structure, with fabric covering throughout except on the fuselage forward of the cockpit.

The project was thus less advanced than a parallel development proceeding under the guidance of Reginald Mitchell of Supermarine, which resulted in the production of the Spitfire, and which had a light metal stressed-skin monocoque structure. On the other hand, the Hawker design, with the central section of the wings built as an integral part of the fuselage, was immensely strong, while its less advanced method of construction made it an easier aircraft to build and repair – both factors of vital importance later.

There would be considerable delay before the aeroplane came to be built but since meanwhile numerous improvements resulted, this may be considered time well spent. In January 1934, the Rolls-Royce PV12 engine – subsequently famous as the Merlin – was substituted for the Goshawk. With the abandonment of the engine went the abandonment of the name 'Fury'. Henceforth, the design was known as the 'Interceptor Monoplane'.

By a lucky chance the radiator had to be moved aft some eighteen inches to balance the increase in weight caused by the installation of the Merlin. This left the wing interspar centre-section bay unobstructed, allowing the inclusion of a retractable undercarriage. On Camm's insistence, the new undercarriage had a wide wheel track, which was later renowned for reliability on every conceivable type of landing surface.

Meanwhile, vital developments were occurring elsewhere. It was realised by the Air Ministry that as the speeds of fighting aeroplanes increased, so the length of time during which an interceptor could align its guns on a target, proportionally decreased. An obvious answer was a further addition to the number of guns but it was feared that this would result in considerable loss of performance even if the additional guns could be made accessible to the pilot – a necessity since those envisaged were Vickers machine-guns, which had a tendency to jam when fired over long periods.

Fortunately for Squadron Leader Ralph Sorley[1], the leading advocate of the eight-gun battery, a solution was presented by the possible introduction of the Browning machine-gun manufactured by the Colt Patent Firearms Company of America, which not only had a faster rate of fire than the Vickers but, being far more reliable, could be housed away from the pilot and fired pneumatically. This used 0.300-inch rimless ammunition, which was unsuitable for use by the Royal Air Force but enquiries quickly confirmed that it could be modified to accept British 0.303-inch rimmed ammunition without loss of its outstanding qualities.

Thereafter negotiations commenced for the grant to the Air Ministry of a licence to manufacture the adapted gun. Final

---

[1]Later Air Marshal Sir Ralph Sorley KCB OBE DSC DFC. For the sake of clarity all RAF officers and non-commissioned officers will be referred to by the ranks they held at the time of the incident described.

terms were ratified in July 1935, whereupon manufacturing contracts were placed with Birmingham Small Arms Company and Vickers Limited. Even prior to this, Air Ministry Specification F5/34 had been issued, calling for an aircraft armed with eight forward-firing Brownings; while Sorley, who greatly favoured mounting such a battery in the Hawker monoplane, since its strong wing would be an ideally steady gun-platform, had visited Hawkers, prompting the design team to consider adapting their project to meet the eight-gun requirement.

The necessity for this had in fact become inevitable on 30 January 1933, when Adolf Hitler became Chancellor of Germany. In October, Germany withdrew from the Geneva Disarmament Conference and the League of Nations, conscription was introduced the following year, and the German aircraft industry, with the aid of massive state loans, began a huge increase in production intended to give the Luftwaffe superiority over any possible rival. During 1933/1934, the designs were commenced, which would lead to the production of the Dornier Do 17, Heinkel He 111 and Junkers Ju 88 bombers, the Junkers Ju 87 'Stuka' dive bomber, the single-engine Messerschmitt Bf[2] 109 fighter and the twin-engine Messerschmitt Bf 110 fighter; Air Minister Hermann Göring was, it will be remembered, a great believer in putting 'guns before butter'.

Since, unfortunately, the only way in which a country can safeguard itself from those who hold such views, is by finding guns of its own, the then National Government, alarmed by increasing international tension, announced its aim, on 19 July 1934, of increasing the Royal Air Force by forty-one squadrons within five years. The first result was a pact between the Labour and Liberal parties to defeat this proposal. A motion of censure

---

[2]Bf was short for 'Bayerische Flugzeugwerke' - Bavarian Aircraft Company. The abbreviation Me, though widely used, was not officially correct until 1944.

was tabled, which on 30 July was resoundingly defeated by 404 votes to 60. It is a terrifying thought that if the verdict had gone the other way, Air Ministry Specifications F36/34 and F37/34, which were issued a month later for what would become the Hurricane and Spitfire prototypes respectively, would probably never have been issued at all.

Henceforward, the Hawker project proceeded with remarkable speed, bearing in mind its radical nature. Scale models had been built previously; now a wooden mock-up was prepared to study the details of the design. In February 1935, work commenced on K5083, the first of the Hurricanes – though, as mentioned earlier, this name was not officially bestowed until June of the following year – at Hawkers' Canbury Park Road plant in Kingston-upon-Thames. The structure was complete by August. In September, the Merlin engine was installed. On 23 October, it was taken by lorry to Brooklands for final assembly and testing. On 6 November, 'George' Bulman took it into the air for the first time.

K5083 was a unique non-standard Hurricane, differing from later production aircraft that had improved performance, due largely to lessons learned from the experiences of the prototype. It had a speed of 318 mph at 15,500 feet, being the first interceptor in Britain, or indeed the world, to exceed 300 mph. It could climb to 15,000 feet in 5.7 minutes; to 20,000 feet in 8.4 minutes. Service ceiling was 34,500 feet and estimated absolute ceiling 35,400 feet. At this time, the machine had small wheel-flaps, soon to be removed after damage when taxying over rough ground, a two-blade Watts fixed-pitch wooden propeller and no guns (or radio), though ballast was placed in the wings to compensate.

Subsequent flights confirmed Bulman's first favourable impressions of the aircraft's steadiness, manoeuvrability and lack of vices. In February 1936, the prototype was delivered for initial service evaluation to the Aircraft and Armament Experimental Establishment (A & AEE) at Martlesham Heath, Suffolk, where it won fervent tributes to its reliability. In July, Squadron Leader Anderson of A & AEE demonstrated the aircraft to the public for the first time at the Royal Air Force Display at Hendon. In August, the fighter at last received its fighting power: eight Browning machine-guns.

In the meantime, on 3 June, Air Ministry approval was indicated by a contract for 600 aircraft. Although this was the largest production order yet placed for a military aircraft in peace-time, the official confidence had already been exceeded by that of Hawkers, for in March, in a superb illustration of private enterprise – in every sense of the word – the Company had instructed its planning department to commence schedules catering for the production of 1,000 aeroplanes – rightly believing that export orders would always cover any excess in numbers over potential RAF orders.[3]

Thus, it will be seen that throughout 1935/1936 Hawkers considered it all-important that they should mass-produce their new aeroplane as soon as possible. This urgency – entirely suitable for an aircraft with the service nickname of 'Hurry' – had certain disadvantages that led to the type having perhaps a shorter life than would otherwise have been the case, but these were outweighed by its benefits. As Francis Mason points out in *Hawker Aircraft since 1920*:

> Because of this anxiety to place the Hurricane in production immediately, the aircraft forfeited the development period which might have bestowed a slightly increased performance in later years. Such was the philosophy which resulted in the Hurricane being somewhat inferior in performance to the Spitfire. But it has been estimated that

---

[3]As a result of this anticipation, the company was able to issue production drawings to the shops only five days after receipt of the Air Ministry contract.

had the Hurricane been delayed for development as long as the Spitfire, Fighter Command would have taken delivery of some 600 fewer Hurricanes than it did by August 1940.

If this estimate is anything like correct – and it might be worth pointing out that by 1940 Hurricane production was running at a total of about 100 machines per month – then had the company not taken these decisions, the result would have been that, following heavy losses in France, there would have been no Hurricanes at all at the start of the Battle of Britain. Bearing in mind that during the battle Hurricanes were responsible for 80 per cent of the interceptions of enemy aircraft, there can be no doubt that without them the Luftwaffe would have been able to swamp the defences by sheer weight of numbers and the country would have been faced with certain invasion and probable defeat.

Indeed, the only real delay in Hurricane production was caused by problems not with the aircraft but with its engine. For Bulman's first flight, K5083 carried Merlin C No 11 engine. On 5 February 1936, this was changed to Merlin C No 15; which in turn was replaced by Merlin C No 17; then by Merlin C No 19; then by a Merlin F. Finally, in September 1936, it was decided that the production Hurricanes would carry the Merlin G or Mark II. Since this necessitated the alteration of the Hurricane's nose, propeller, air-intake, engine cowlings, engine mounting, glycol tank and hand starting system, it caused a four-month delay in production; so that it was not until 12 October 1937 that the first Hurricane Mark I, L1547, took off from Brooklands under the command of Hawker test pilot Philip Lucas.

The old prototype continued to fly, taking part in the Paris Aero Show in November 1938, and in July 1939 giving a breath-taking display at the Brussels Exhibition, being flown by test pilot R. C. 'Dick' Reynell, of whom more later, before vanishing into a maintenance unit shortly before the outbreak of war. Yet it was on the eight-gunned production Hurricane Mark I that the spotlight now fell, four such machines being delivered to No 111 (Fighter) Squadron at Northolt in December 1937.

During the early months of 1938, twelve more Hurricanes joined the squadron, its old Gauntlets disappeared and the pilots experienced the joys and perils of their new monoplanes. Certainly, there were problems with the greater speeds involved, nor was confidence increased by ludicrous stories suggesting that this most forgiving of fighters could only be flown by men of vast strength, fed on special diets. Unhappily, there were a number of accidents during the 'running-in' period, Flying Officer Bocquet being killed on 1 February when he lost control of his aircraft in a dive.

It may therefore have been with an eye on the improvement of morale that, on 10 February, Squadron Leader John Gillan flew a Hurricane the 327 miles from Turnhouse near Edinburgh to Northolt, in 48 minutes – an average speed of 408.75 mph. The Hurricane had the assistance of a wind of considerable velocity, but perhaps inevitably, this received scant mention in reports of the incident. However, it was noted by the squadron who nicknamed their CO 'Downwind Gillan' forthwith; nor was his machine forgotten: Hurricane L1555 became 'State Express 555' thereafter.

However, if Gillan intended his flight to re-assure his pilots, he succeeded beyond question. As Mr Mason expresses it in his book, *The Hawker Hurricane:* 'At once No 111 (Fighter) Squadron became regarded as a *corps d'élite* and the trepidation amongst the pilots vanished.'

In March, No 3 Squadron received Hurricanes, but a few months later it was decided (following a fatal crash) that its base at Kenley was too small for the new fighters, so it re-converted to Gladiators. In April, 56 Squadron gave up its Gladiators for Hurricanes. During the summer, 73 and 87 Squadrons also received Hurricanes but neither was fully operational at the time of the Munich crisis in September. It should be mentioned that, at this time, Fighter Command, which had no Spitfires at all, possessed only ninety-three Hurricanes, which lacked heating for their gun-bays and so could not fight at more than 15,000 feet even in summer. Nor were anti-aircraft guns available in sufficient numbers; nor was there adequate radar cover except over Kent and East Anglia. Germany, by contrast, already possessed a powerful air force with the priceless practical experience gained by the 'Condor Legion' in the Spanish Civil War.

It is therefore worth considering what might have happened had war been declared at this juncture. It is surely unarguable that if German bombers had attacked Britain – even without escorting fighters that lacked sufficient range – then the defences would have been inadequate to prevent them inflicting considerable damage. Yet it might seem that this would have been only a temporary ordeal, for the Allied land forces – which meant, in practice, the French Army – had such superiority that they would have been able to over-run Germany in short order, aided probably by a revolt among the German generals.

However, here the imponderables start – for would the French Army have attacked Germany? A year later, the German Army leaders believed they were saved from disaster by the failure of the Allies to attack them while their forces were engaged in Poland. If in 1939 the Allies watched, apparently unmoved, while Poland went down fighting heroically, would they not perhaps have watched a year earlier while Czechoslovakia went down, even making allowance for the fact that in 1938 Germany's success was unlikely to have been achieved so swiftly, since she was not then so well prepared or equipped?

It does seem at this distance of time that the fundamental fact was that the British people were simply not ready for war. This was the reason for the hysterical praise that greeted Prime Minister Neville Chamberlain's promise of 'Peace for our time'. The immediate result of Munich might be, in Churchill's bitter words, that 'the German dictator instead of snatching the victuals from the table, has been content to have them served to him course by course'; but it had a deeper consequence. Britain had done everything possible to avoid the conflict. Therefore, when this came, no reasonable citizen could doubt the truth of Chamberlain's announcement: 'It is the evil things that we shall be fighting against: brute force, bad faith, injustice, oppression and persecution. Against them I am certain that right will prevail.'

In order to help ensure that desirable result, the Air Ministry, on 1 November, placed a further order for 1,000 Hurricanes. Aircraft were soon pouring off the production lines at Kingston, Brooklands and Hawkers' new factory at Langley, completed in 1939 for the express purpose of building Hurricanes. Also, Hawkers had taken over the Gloster Aircraft Company in 1934, allowing Hurricanes to be built at Gloster's Hucclecote factory, though the first such did not fly until 20 October 1939.

By the end of 1938, ten squadrons flew Hurricanes. In March 1939, 501 'County of Gloucester' Squadron was formed with Hurricanes, being the first Auxiliary Air Force squadron to receive modern eight-gun fighters. By the following September, there were 497 Hurricanes in service with seventeen squadrons (including No 3, which had reverted to the type in May) or in reserve. The hour had come, but so had the aircraft.

While quantity increased, so did quality. As a result of a slight defect noticed in the Hurricane when recovering from a spin, it was decided to enlarge the rudder and incorporate an underfin. Better radios were fitted as were better exhausts. Most important perhaps, the old fixed-pitch two-bladed wooden propeller, which had a tendency to fly into pieces under stress, was replaced by a De Havilland metal three-bladed two-pitch propeller. With this, the pilot could alter the pitch — i.e. the angle — of the blades so as to use fine (low) pitch for take-off and coarse (high) pitch for greater speeds — much as gears are used in a car. This ensured not only fuel economy but an improvement in rate of climb. A further step forward came with the Rotol constant-speed propeller, which adjusted the pitch to the engine speed. To operate this, Rolls-Royce designed the Merlin III engine featuring a standardised propeller-shaft, which could be used with either the De Havilland or Rotol airscrews.

Since the Hurricane was a warplane, its fighting capabilities were improved. Armour plate and bullet-proof windscreens gave added protection to the pilot. Metal wings replaced those covered with fabric. The lines of fire of the guns, originally aligned to converge to a point at 650 yards range, were reduced to 400 yards, then to 250 yards – this at the insistence of Squadron Leader Halahan of No 1 Squadron, which had bitter memories of the report of a previous CO at the time of Munich that the only way their Furys would be likely to destroy enemy bombers was by ramming them. Not all Hurricanes had been fully converted on the outbreak of war – indeed, by August 1940 there were still a handful in service with fabric-covered wings or wooden propellers or both – but the bulk of the changes had taken place by early 1940.

It was also during the period from Munich to early 1940 that the Hurricane began its travels, which were to take it (apart from South America) almost literally throughout the world. With the wisdom of hindsight, it seems unwise to have allowed a single aeroplane to go abroad, but no doubt the British Government thought that a limited export to friendly nations would be beneficial.

Consequently, in December 1938, twelve Hurricanes were flown to Yugoslavia, via France and Italy. These were joined by twelve more in February 1940. In addition, the Yugoslavs were granted a licence to build Hurricanes. Only about twenty such had entered service at the time of the German invasion of 6 April 1941, but one of these deserves special mention as the only Hurricane to fly with an engine other than a Merlin, in this case a 1,050 hp Daimler-Benz DB601A. The Yugoslavs, who were aware of the possibility of being unable to obtain Merlins, appear to have rather preferred the DB powered version – but comparing such a one-off model against the mass-produced standard aircraft was perhaps not a fair test.

Also in December 1938, seven Hurricanes were shipped to South Africa to form No 1 Squadron SAAF. During 1939, twenty aircraft went to Belgium, which also took out a licence to build a further eighty. These Belgian fighters differed from other Hurricanes in being armed with four wing-mounted 12•65 mm Brownings, which gave them a greater weight of fire than those in service with the RAF. Unfortunately, the German invasion prevented more than two being built. Twelve aircraft were sent to Rumania and fifteen to Turkey, but the outbreak of the war prevented the planned supply of machines to Persia and Poland, only two being sent to the former and only one (L2048) to Poland.

Most significant perhaps were the twenty-two aircraft supplied to Canada. Though intended for No 1 Squadron RCAF, they became in addition manufacturing patterns, for – once more on the initiative of Hawkers, though with official approval – every factory drawing of Hurricane components was put on microfilm and flown in duplicate to Canada. A manufacturing contract was entered into with the Canadian Car and Foundry Co Ltd of Montreal. The first production batch of Canadian machines, fitted with Rolls-Royce engines also shipped from Britain, began to appear early in 1940, arriving in Britain in time to take part in the crucial battle for its survival.

The idea behind this scheme was to provide a source of supply outside bombing range – for it was now clear to almost everyone that war had become inevitable. While the Luftwaffe practised attacks on shipping or ports, the Hurricane squadrons took part, in August, in air exercises all over south-east England. French bombers 'attacking' London were intercepted by No 1 Squadron. Pilot Officer Paul Richey, in his superb book *Fighter Pilot*, described these as 'gay' but 'pathetically out of date'.

> Later that day [he added] [Fit Lt] Johnny Walker
> and I intercepted three RAF Blenheims low over
> the Downs under a violent thunderstorm. They
> looked grim and businesslike, with mock German
> crosses on their wings.

Thus, when on 1 September 1939 Hitler set out to re-arrange the world, the Hurricane pilots were ready for their ordeal.

–

Ready; but not yet to be tested. The traditional name for the first seven months of hostilities is the 'Phoney War', which expression varies vastly in accuracy depending on the aspect under discussion. It may be entirely appropriate to the British Expeditionary Force in France (BEF), which did not suffer its first casualty until 9 December, when Corporal Priday was shot dead on patrol. It is nonsense to use it even in the same breath as the vicious month-long campaign that saw the conquest of Poland. Nor can it reasonably refer to the frequently savage fighting at sea. With regard to the Hurricane's war, at least as far as the home-based squadrons were concerned, it was largely true to begin with, but became progressively less so as time passed.

On 8 September, the Hurricane fired its first shots. A barrage balloon that had broken loose from its moorings, threatening damage from its trailing steel cable, was brought down by Flying Officer Kilmartin of 43, destined to be one of the most famous of Hurricane squadrons, who required 1,200 rounds for the purpose. It seems that in the early days of the war, balloons were not secured adequately, for over the next fortnight five more of 43's pilots had to destroy one or more; 32 and 79 Squadrons at Biggin Hill also indulged in this activity, as did the first Hurricane squadron, 111, which, on 4 October, became the reigning champions with a score of eleven in a day.

Then, on 21 October, the Hurricane began its attacks on German aircraft. No 46 Squadron on convoy patrol near the east coast, intercepted a formation of Heinkel He 115 seaplanes. The Hurricanes promptly fell on these, shooting down three and damaging a fourth, which landed on the sea.

This action may have been a trifle inglorious but, exactly a month later, the home-based Hurricanes destroyed its first enemy bomber. Flying Officer Davies and Flight Sergeant Brown of 79 Squadron intercepted a Dornier Do 17 on weather reconnaissance over the Channel. Attacking from above, they sent this diving into the sea, pouring smoke from one engine.

Eight days later, Squadron Leader Harry Broadhurst, commanding 111 Squadron, shot down a reconnaissance Heinkel He 111 into the North Sea. Thereafter, a steady succession of enemy bombers fell victim to Hurricane pilots. Like 111, 43 Squadron, whose first victory was another He 111 shot down by Flight Lieutenant Hull and Sergeant Carey on 30 January 1940, was particularly successful, which seems to indicate that the 'balloon-busting' had provided useful practice. Among the enemy aircraft destroyed, special mention might be made of a Heinkel He 111 attacked by Flight Lieutenant Peter Townsend of 43 on 3 February. This crashed near Whitby, Yorkshire, with two dead crew-members on board – the first hostile aeroplane to be brought down on the soil of England as distinct from in Scotland or the sea.

Probably as a result of their achievements, in February 1940 43 and 111, together with another Hurricane squadron, 605 of the Auxiliary Air Force, were sent to Wick to protect the naval base at Scapa Flow. On 8 March, Flying Officer Dutton of 111 Squadron shot down a He 111 and later German attacks also found the Hurricanes too much for them. The most satisfactory engagement came on 8 April, the day before the 'Phoney War' became only too real, during which 43 destroyed three Heinkel He 111s, while a fourth, damaged by Sergeant Hallowes, decided to 'ditch' and crash-landed on Wick airfield, having mistaken the flare path for lights on the sea. Some amusement was then provided by its crew who threw out a dinghy, removed their flying boots to be able to swim more easily, then dived out – onto dry land!

Yet for the home-based Hurricanes, this early period was one of preparation, during which more pilots learned to fly Hurricanes, more aircraft received their metal wings or variable-pitch propellers or very high frequency radios, which permitted a far greater control from the ground, and indeed more Hurricanes arrived, for six further squadrons were converted or formed during this time. This was perhaps as well for already the Hurricane had begun its widespread service overseas – a tendency that would shortly have alarming consequences.

–

Mention has already been made of the dispatch of the British Expeditionary Force to France. Clearly it would require fighter cover but, unfortunately, the French airfields were not only grass strips but had quite inadequate drainage. The obvious fighter to send, therefore, was the Hurricane, with its wide, strong undercarriage. In consequence, four squadrons flew to France to protect the disembarkation, after which 85 and 87 were stationed in north-eastern France to form the fighter portion of the Air Component in support of the BEF, while 1 and 73 moved to Lorraine behind the Maginot Line to give protection to the Advanced Air Striking Force, a body of twelve Battle or Blenheim bomber squadrons detailed for attacks or reconnaissance across the German frontier.

The Air Component Hurricanes, like their comrades in Britain, saw combat with enemy reconnaissance aircraft, Flight Lieutenant Voase-Jeff achieving 87's first victory – a Heinkel He 111 shot down over Merville, on 2 November – while Flight Lieutenant Lee claimed the first for 85 by sending another He 111 into the sea off Boulogne on the 21st. However, the greatest peril facing the pilots in the early months of the war seems to have been the risk of landing in Belgium by mistake. This happened on several occasions but fortunately the Belgian authorities, though neutral, were sympathetic, allowing the interned pilots to escape within a few weeks.

The Hurricanes in the Advanced Air Striking Force saw considerably more action. On 30 October – i.e. before either the Air Component or the home-based Hurricanes had destroyed an enemy bomber – Pilot Officer Mould of No 1 Squadron, flying Hurricane L1 842, took off from Vassincourt in pursuit of three unidentified aircraft. Overtaking one of these, which proved to be a Dornier Do 17, Mould attacked from astern, firing a long burst from point-blank range. The Dornier burst into flames, to spiral to the ground where it exploded – the first enemy to fall to the Hurricane in France.

When the squadrons reached France, they were equipped with elderly aircraft with wooden propellers, but more modern machines with de Havilland variable-pitch propellers were soon arriving, and, in one such, Flying Officer Edgar Kain scored 73's first victory on 8 November, by shooting down a Do 17 at the then unheard-of combat height of 27,000 feet. 'Cobber' Kain, who was not an Australian as might be supposed from his nickname but a New Zealander, destroyed another Do 17 on 23 November, and by the end of March 1940 had become the first RAF pilot to have downed five enemy aircraft. His remaining three victims, however, deserve more notice since all were enemy fighters.

The first encounter between the Hurricane and the Messerschmitt Bf 109 came on 22 December. Three aircraft from 73 Squadron were on patrol near Vassincourt when they were ambushed from cloud cover by four 109s, which shot down two, killing the pilots, both of whom had just come out from Britain, but losing only one of their own number destroyed and one damaged in the process. Air Vice-Marshal Peter Wykeham in *Fighter Command* calls this 'not a very inspiring beginning', but since the Hurricane pilots were novices, outnumbered, taken by surprise and attacked from above, it cannot really be considered too discouraging either. It does seem, moreover, somewhat misleading that the Air Vice-Marshal makes no mention of the other combats with the 109 in the 'Phoney War' since these were very much in the Hurricane's favour.

By the end of March 1940, five more 109s had been credited to 73 Squadron or rather to two pilots who, unlike the victims of the action just described, were far from inexperienced. On 2 March 'Cobber' Kain destroyed one Messerschmitt and damaged two more before his Hurricane was badly hit, in spite of which he was able to glide with a dead engine to crash-land on a French airfield. On 26 March, he shot down two more 109s but this time was forced to bale out of his own blazing aircraft with a wounded leg. He came down safely, receiving compensation in the award of the DFC.[4] Also on the 26th, Flying Officer Orton, who ranked second only to Kain among 73's pilots, shot down a further two Messerschmitts, after which he returned safely to base to receive another DFC.

No 1 Squadron gained its first fighter-against-fighter victory on 29 March, when Paul Richey shot down a 109 near Metz; he was credited only with a 'probable', but German records show that the Messerschmitt crash-landed with a wounded pilot at the controls. During April, the squadron was credited with the destruction of five more 109s for the loss of one Hurricane whose pilot escaped by parachute. Finally, on 22 April, five Hurricanes met fifteen 109s, which instead of using their height advantage to attack, retreated rapidly to Germany. It is believed that the Germans were a group of trainee pilots, but the incident

---

[4] By the end of the Battle of France, Kain was credited with seventeen enemy aircraft destroyed which made him one of the then top-scoring RAF pilots. He survived the fighting, only, ironically enough, to have a fatal crash while 'beating up' the squadron's airfield prior to what would have been his return flight to Britain.

shows that neither side felt that the Hurricane was inferior to its opponent – despite later statements to this effect.

There was no doubt that the Hurricane was superior to the twin-engine Messerschmitt Bf 110. This was manned by the finest of the Luftwaffe pilots, and Göring believed that it would prove equally deadly against fighters and bombers. The RAF also had an exaggerated respect for the *Zerstörer* (destroyer) in consequence of which Air Marshal Barratt, C-in-C British Air Forces in France, had offered dinner with him in Paris to the first pilot to down one. On 29 March, the prize was won by Flight Lieutenant Walker of No 1 Squadron. He duly dined with Barratt at Maxims but insisted that the other members of his section, Flying Officer Stratton and Sergeant Clowes, should be invited also. It is reported that the wide variety of ranks at the Air Marshal's table caused considerable amazement.

As if to confirm the Hurricane's superiority, three other pilots from No 1 attacked nine 110s on 1 April, although these had the advantage of height as well as numbers. They shot down three at the cost of damage to one Hurricane, which landed safely. All three victories were later confirmed by wreckage on the ground. It was clear that the Hurricane could out-manoeuvre the 110 so comprehensively as to put it into a quite different class; a matter of great encouragement to the RAF pilots.

In all, before the German invasion of the West on 10 May, No 1 Squadron claimed twenty-six enemy aircraft destroyed while 73 claimed thirty. It was a creditable enough performance even if allowance is made for the slight exaggerations of over-enthusiasm, but what gave cause for greater satisfaction was the point that although aircraft had been lost, in landing accidents in the poor conditions as well as in combat, 1 Squadron had had only one pilot killed and 73 only three, including the two ambushed by the Messerschmitts.

The lack of casualties was clearly due in part to the skill of the pilots but also in part to the ability of the Hurricane to withstand murderous punishment and yet return home, or at least stay airborne long enough for its pilot to escape by parachute. Perhaps the most outstanding example of this trait came on 23 November 1939. Sergeant Clowes of 1 Squadron, flying Hurricane L1842, the same aircraft in which Mould had scored No 1's first victory, had just shot down a Heinkel He III, when a French fighter, usually reported as a Morane but apparently a Curtiss Hawk acquired from the United States, eager for a share in the combat, collided with the Hurricane's tail, knocking off one elevator and most of the rudder. The Frenchman was forced to take to his parachute, but Clowes managed to fly his battered machine back to Vassincourt, where he had to land at 120 mph to maintain control. The Hurricane tipped up on its nose but such was its strength that the Sergeant escaped unscathed – physically unscathed at least:

'I saw him straight after this little effort,' remarks Richey, 'and, though he was laughing, he was trembling violently and couldn't talk coherently' This description should be remembered when similar incidents are coldly recited.

Later on the same day, Flying Officer Palmer attacked a Dornier Do 17. By the time he had finished, the Dornier was losing height, with one engine on fire. The rear-gunner and navigator had baled out while the pilot was slumped over the controls, apparently dead. As Palmer flew alongside, however, he ceased feigning death, swerved onto Palmer's tail and put thirty-four bullets into the Hurricane, one of which came through the locker behind the pilot's head to smash his windscreen – in spite of which the Hurricane force-landed in a field without further damage. Two other Hurricanes then downed the Dornier but the pilot survived to be entertained by No 1 Squadron in their Mess, a pleasant gesture which may seem out of place in total war but which one cannot regret having been made.

As a result of Palmer's narrow escape, 1 Squadron decided more back-armour was needed. Higher Authority opposed the request, fearing this would upset the Hurricane's centre of gravity. The squadron, unconvinced, 'borrowed' some armour plate from a wrecked Battle, fitted it behind the seat in a Hurricane and sent Flying Officer Brown to Farnborough to demonstrate this to the Air Ministry experts. His display was sufficient to cause back-armour to be fitted as standard equipment throughout Fighter Command, thus saving countless lives.

By its easy acceptance of back-armour, the Hurricane showed another of its main characteristics: its remarkable adaptability. This merit would be illustrated once again before the end of the 'Phoney War' in unusual circumstances.

During January and February 1940, twelve Hurricanes were sent to Finland, then suffering an attack from Russia as brutally unprovoked as, but fortunately far less efficient, than the German onslaught on Poland. Unhappily, two of them were written off en route. In the somewhat different climatic conditions in which they found themselves, the remaining aircraft abandoned their tail-wheels for snowskids. Only six weeks later, the so-called Winter War ended on 12 March, but two more Hurricanes were subsequently destroyed in accidents.

Then on 22 June 1941, Germany attacked Russia, and Finland, understandably wishing to recover lost territory, joined in the hostilities three days later. So did the Finnish Hurricanes that shot down five Russian aircraft during July. One Hurricane was later brought down by AA, fire but the other seven survived the fighting and one survives still, though in a non-flying condition.

The Winter War may have ended but this was not to be the case with the larger conflict. On 4 April 1940, Mr Chamberlain, as sincere but short-sighted as ever, proclaimed in the House of Commons that Hitler had 'missed the bus'. Five days later, the expression 'Phoney War' became ridiculous in every respect. A new word, the significance of which was already too well-known to the hapless Poles, replaced it as the appropriate term for military events: Blitzkrieg!

Chapter Two

Blitzkrieg

As the good citizens of the Danish capital, Copenhagen, made their way to work on the morning of 9 April, they were surprised to encounter a column of German soldiers marching towards the Royal Palace – but not alarmed for they believed a film was being made. It was no film. The troops really were German. So were the warships in harbour, the motorised units forcing their way into Jutland, the paratroopers seizing the Aalborg airfields. Within twenty-four hours, this charming little country that had asked only to be left alone had been swallowed up by the Third Reich.

For Hitler, the occupation of Denmark meant the gain of advanced air-bases for his planned attack on Norway, the capture of which would open the North Atlantic to the German Navy, provide air and naval bases for flanking attacks on Britain, secure Germany's supply of Swedish iron-ore, which passed through the ice-free port of Narvik, and, in the long-term, threaten communications between Britain and Russia – for Hitler's ultimate aim always was the destruction of his then ally in order to achieve *Lebensraum*.

April 9 therefore saw also audacious German naval attacks on all the main Norwegian ports. At Oslo, the coastal defences temporarily beat back the assault, sinking heavy cruiser *Blücher*, but airborne troops seized the Oslo airport, Fornebu, reinforcements were rushed in by air from newly captured Aalborg and the Norwegian capital was taken from the rear. By midday, paratroopers had also taken Oslo/Kjeller airfield as well as the even more vital one at Sola, Stavanger. Air-transports were soon pouring in supplies at all the captured bases, while the Luftwaffe moved in to support their ground forces.

Apart from the initial failure at Oslo, the German seaborne attacks were equally successful, capturing Kristiansand, Bergen and Trondheim. The boldest thrust came at Narvik, more than 600 miles from Oslo as the crow flies. Ten destroyers raced into Ofotfjord, where, sinking two old Norwegian coastal defence vessels, they landed General Dietl's III Mountain Division, 2,000 Austrians specially trained to operate in snow-conditions. Meanwhile, the main forces under General von Falkenhorst began pushing northwards to link up with their isolated beach-heads.

Anglo-French counter-moves were sadly ineffective. On 15 April, a landing was made at Harstad, near Narvik, but although the Royal Navy in two brilliant actions, on 10 and 13 April, had already annihilated the enemy's destroyer force, no attempt was made for over a month to attack Dietl's men who were for a time – but only briefly – demoralised by the destruction of their naval support.

Meanwhile, plans had been laid for the recapture of Trondheim. On 16 April, Allied troops landed at Namsos, some 125 miles by road to the north. On the 18th, another landing took place at Aandalsnes, some 200 miles south of Trondheim by road. It had been intended that the main assault would be a direct seaborne landing at the town. However, this was now abandoned, largely, it seems, because of the risks to the Navy from air-raids. Yet without Trondheim, the Allies would have no port adequate for unloading supplies – no airfield from which fighter-protection could be given. Thus, the two diversionary forces were doomed. Viciously attacked from the air, both were evacuated by the night of 2 May.

It will be clear from this brief account that the most vital ingredient of the German success was complete control of the air. All suitable landing-grounds were in enemy hands. Fighter Command possessed no long-range interceptors and a scheme to rectify this by fitting Blackburn Roc floats to Hurricanes never materialised owing to lack of time.

In consequence, although some protection was afforded by aircraft from the carriers *Ark Royal* and *Glorious*, the only land-based fighter cover in the central Norwegian campaign was that by eighteen Gladiators of 263 Squadron, which on 24 April flew off *Glorious* to frozen Lake Lesjaskog, near Aandalsnes. Tragically, in spite of splendid performances in the air, their supporting administration was so inadequate that within seventy-two hours every machine was destroyed on the ground.

In the Narvik area also, the Luftwaffe had begun to dominate, sinking anti-aircraft cruiser *Curlew* on 26 May. Thus, next day, when the attack on Narvik finally took place in the bright light of the Arctic 'midnight', there was considerable fear of enemy air-bombardment; but, says Captain Donald Macintyre in his book *Narvik*:

> Anxious eyes, red-rimmed from day after day of scanning the skies for enemy aircraft, saw with incredulous relief the patrol of Hurricanes circling overhead. Cheers, not unmixed with ribaldry, rose from ships' companies which had prayed so long, so fervently and so vainly for such a sight.

It is perhaps as well that the remarks made are left to the imagination but the absence of enemy bombers abundantly demonstrated the Hurricanes' deterrent value.

They came from 46 Squadron, commanded by Squadron Leader Kenneth Cross. They had been hoisted on board *Glorious* from lighters in somewhat undignified fashion, but on 26 May they flew off under their own power to their proposed base at Skaanland: this unfortunately proved quite unsuitable. After the first three aircraft had been wrecked on landing

– happily, without injury to the pilots – the remaining fifteen were diverted to Bardufoss, whither 263 had already proceeded with a fresh gaggle of Gladiators.

Even here, conditions were far from ideal. Despite the devoted efforts of the ground-crews – and any pilot would admit that its fitter and rigger are just as important to a fighter as the man who flies it – two more Hurricanes were written off on the ground during the squadron's stay in Norway. In addition, Bardufoss was some fifty miles north of the bases of the expeditionary force. There was no radar, indeed virtually no warning system of any sort, so that the pilots were forced to fly wasteful standing patrols.

Yet for all these disadvantages, the protection afforded by the Hurricanes, as already noticed, ensured the success of the initial stage of the Narvik landing – which indeed had been postponed pending their arrival – but then, early on the morning of 28 May, a sea-fog descended on the Hurricanes' base forcing their return thither just in time. At once, the value of fighter-cover became obvious. The Luftwaffe raided the bombarding warships, damaging the flagship, light cruiser *Cairo*. The German land forces, freed from the threat of the Navy's shells, mounted a critical counter-attack. Luckily, the fog lifted. The Hurricanes re-appeared; the enemy bombers fled; reinforcements were landed. By evening, the little town was firmly in Allied hands, while Dietl, with dogged determination, fell back into the mountains.

Meanwhile, a Hurricane flown by Flying Officer Lydall had brought down a Junkers Ju 88 over Tjelboten. By the end of May, 46 and 263 had destroyed half-a-dozen enemy aircraft, though three Hurricanes were lost on the 29th, Lydall and Pilot Officer Banks being killed but Pilot Officer Drummond baling out. The culmination of the fighting came on 2 June, when the Luftwaffe launched a series of raids by bombers, escorted by Bf 110s – the range was too great for the 109s. The Hurricanes and Gladiators frustrated every attack, claiming to have shot down or damaged nine enemy aircraft without loss, and earning a glowing tribute from the Army commander, General Auchinleck. Nor, incidentally, had the Hurricanes' efforts been restricted to the enemy in the air. On their first day at Narvik, they had destroyed two Dornier Do 26 flying-boats disembarking reinforcements in Rombaksfjord; a fine piece of flying, for these were moored under the edge of a cliff at a very narrow part of the fjord.

However, for all these successes the ultimate captor of Narvik would be not Auchinleck but Dietl. The only use the Allies made of their occupation was to demolish the port install-ations and industrial plant, though ironically their efforts were less thorough than those already carried out by the Germans prior to their retirement. Already on 31 May, the Hurricanes had covered the disembarkation from Bodö of a small force sent south to delay German troops advancing from Trondheim. Now they had to cover the evacuation of Narvik also.

This took place over five days, commencing on 3 June. During it, the Hurricanes and Gladiators gave constant protec-tion to such an extent that the enemy did not even realise until too late that a withdrawal was taking place. On 7 June, the Hurricanes fought their last action, destroying a Heinkel He 111, and damaging another that crash-landed at its base. The last men to leave the shore next day were the ground staff and spare pilots from Bardufoss who sailed home on the merchantman, *Arandora Star*.

As for the aircraft, the Gladiators flew off to *Glorious*. They could land on her without too much difficulty, since although they lacked arrester hooks to catch the wires stretched across her deck, their speed was so low that if the carrier steamed at maximum power, she could produce enough wind down the desk to reduce the length of their landing run sufficiently. For the Hurricanes, with their far greater landing speed and far longer landing run, this was thought to be too risky. Orders were given that they be destroyed at their airfield.

The orders were not carried out. Squadron Leader Cross, knowing that Fighter Command needed every available Hurricane, begged that the squadron be allowed to attempt a landing on *Glorious*. The senior RAF officer in Norway, Group Captain Moore, with considerable moral courage, took the responsibility of granting permission.

But how could the Hurricanes' landing run be reduced? The problem was that the pilots would have to brake so hard that this, combined with the strong wind over the deck, would make it impossible for them to keep their aircrafts' tails down. If the Hurricanes did not nose over completely, they would never come to a halt but would keep going until they ran off the end of the deck. To try to solve this difficulty therefore, the airmen added what the squadron's diary called 'extra weight in the tails of the Hurricanes' – which consisted of bags of sand strapped to the rear of the fuselages.

This by no means ended the dangers of the operation. It would not be easy to land the unbalanced Hurricanes on a small flight deck – the *Glorious* had been designed as a huge cruiser, then converted – particularly for pilots who, though otherwise very experienced, had never before attempted this. Yet when volunteers were called for, every man stepped forward.

As a slight concession to the risks involved, it was decided to make a limited trial run. The senior flight lieutenant, New Zealander Patrick Jameson, would fly to *Glorious* accompanied by Flying Officer Knight and Sergeant Taylor. The remaining seven aircraft would await news from them before following. Led by a Swordfish from *Glorious*, the first three took off on the evening of 7th June. They reached the carrier after an hour's flight. Jameson, throttled back and flying at just above stalling speed, went in to land first. There was no bounce from the weighted tail when he touched down and the Hurricane rolled safely to a halt.

Knight and Taylor then also made safe landings. Confirmation of the success was sent at once to Cross but, in the high, iron-bound Norwegian mountains, radio communications were very poor and it seems the signal was not received. It made no difference; one feels that having gone this far, nothing was going to stop Cross from making his attempt. In the early hours of 8 June, daylight in the Arctic, a Swordfish led seven more Hurricanes from Bardufoss to the carrier. Every one of them got down safely as well.

Sheer tragedy struck that afternoon. The German battle-cruisers *Scharnhorst* and *Gneisenau*, hunting for ships retiring from Norway, came upon *Glorious*. Their first, hurried salvos, though fired at long range, struck home. Two hours later, burning furiously, the carrier went to the bottom. With her went more than 1,100 men. With her also went her escorting destroyers *Ardent* and *Acasta*, which defended her nobly, the latter before she sank getting a torpedo into *Scharnhorst*. With her went all ten Hurricanes. With her went eight Hurricane pilots. Only two managed to reach a Carley float on which they survived the cold for three days before being rescued by a Norwegian fishing-boat. They were Squadron Leader Kenneth Cross and Flight Lieutenant Patrick Jameson.

–

The reason why the Allies had occupied Narvik only to abandon it was that far greater calamities had befallen them elsewhere. For 73 Squadron, 10 May was not the day Germany invaded the Low Countries, nor the day Churchill became Prime Minister, but the 'day of ceaseless activity' when, at last encountering the full strength of the Luftwaffe striking deep into France, its Hurricanes were forced out of their base at Rouvres, though they destroyed four enemy aircraft before retreating.

The other Advanced Air Striking Force fighter squadron, No 1, enjoyed its most spectacular success on 11 May, when five pilots, Flight Lieutenant Walker, Flying Officers Brown, Kilmartin and Richey and Sergeant Soper, sighted thirty

Dormer Do 17s escorted by fifteen Messerschmitt Bf110s near Sedan. The odds were thus exactly nine to one against. The Hurricane pilots attacked immediately. Much to their delight, the Dorniers retired to Germany but the 110s, which it will be recalled were manned by crack crews, promptly took up the challenge, still with a three to one advantage.

Since this was the first major Hurricane versus 110 clash, it is fortunate that among the pilots engaged was Paul Richey who in *Fighter Pilot* gives the following graphic description:

We went in fast in a tight bunch, each picking a 110 and manoeuvring to get on his tail. I selected the rear one of two in line-astern who were turning tightly to the left. He broke away from his No 1 when he had done a half-circle and steepened his turn, but I easily turned inside him, holding my fire until I was within fifty yards and then firing a shortish burst at three-quarters deflection. To my surprise a mass of bits flew off him – pieces of engine-cowling and lumps of his glasshouse (hood) – and as I passed just over the top of him, still in a left-hand turn, I watched with a kind of fascinated horror as he went into a spin, smoke pouring out of him. I remember saying 'My God, how ghastly!' as his tail suddenly swivelled sideways and tore off, while flames streamed over the fuselage. Then I saw a little white parachute open beside it. Good!

Scarcely half a minute had passed, yet as I looked quickly around me I saw four more 110s go down – one with its tail off, a second in a spin, a third vertically in flames, and a fourth going up at forty-five degrees in a left-hand stall-turn with a little Hurricane on its tail firing into its side, from which burst a series of flashes and long shooting red flames. I shall never forget it.

All the 110s at my level were hotly engaged, so I searched above. 'Yes – those buggers up there will be a nuisance soon!' Three cunning chaps were out of the fight, climbing like mad in line-astern to get above us to pounce. I had plenty of ammunition left, so I climbed after them with the boost-override pulled. They were in a slight right-hand turn, and as I climbed I looked around. There were three others over on the right coming towards me, but they were below. I reached the rear 110 of the three above me. He caught fire after a couple of bursts and dived in flames. Then I dived at the trinity coming up from the right and fired a quick burst at the leader head-on.

During this combat, No 1 destroyed ten of the enemy, all of which, it is worth remarking, were confirmed by the French on the ground. One Hurricane was lost; that flown by Richey. Since he was able to write the account just recorded, it is obvious that he survived his descent by parachute without serious bodily injury – but the word 'bodily' should be stressed. When Richey went to sleep:

> Scarcely had I dropped off when I was in my Hurricane rushing head-on at a 110. Just as we were about to collide I woke up with a jerk that nearly threw me out of bed. I was in a cold sweat, my heart banging wildly. I dropped off again – but the nightmare returned. This went on at intervals of about ten minutes all night. I shall never forget how I clung to the bed-rail in a dead funk.

That description also should be remembered.[5]

---

[5]Even in his misfortunes Richey provided a good example of the resilience of the Hurricane pilots as well as the strength of their aircraft, which helped

For 501, County of Gloucester Squadron of the Auxiliary Air Force, 10 May was the day it took its Hurricanes to France to join 1 and 73 with the AASF It was not a happy arrival, for when 501 landed at Betheniville (near Rheims) one of its Bombay transports crashed, killing three pilots and six ground crew. However, later that day, Flying Officer Pickup shot down a Domier Do 17.

Thereafter, 501 quickly showed its worth. Next day, its Hurricanes claimed to have destroyed or damaged six enemy aircraft without loss. The day after they claimed to have destroyed or damaged another twelve but lost two pilots, killed. In retrospect, the most surprising aspect of these first days is that none of the squadron's victims fell to the man destined to be its finest pilot, Sergeant James 'Ginger' Lacey. However, he made up for this on 13 May, when he earned a Croix de Guerre by destroying a Bf 109, a Heinkel He 111 and a Bf 110, all confirmed by the French on the ground, all before breakfast. He was then attacked by four more 110s, but escaped by some violent manoeuvres.

The Air Component Hurricanes were also receiving reinforcements. 85 and 87 Squadrons had already been joined by 607 and 615, former Gladiator squadrons converted to Hurricanes. Now, on 10 May, 3 and 79 Squadrons flew out from Britain, followed by 504 shortly afterwards.

All these were quickly in action. Pilot Officer Carey gained 3 Squadron's first success, a Heinkel He 111, on 10 May. Two days later, the squadron destroyed or damaged eight enemy aircraft. 87 Squadron destroyed or damaged six on 10 May. The same day, 85 Squadron claimed seventeen enemy raiders destroyed

---

them to survive. On 15 May, he shot down two more 110s before having to bale out again. On 19 May, he destroyed or damaged three Heinkel He 111s but was then badly wounded in the neck, in spite of which he was able to force-land his Hurricane without further injury.

(undoubtedly an exaggeration); followed by eight the next day; seven the day after.

Unfortunately, while the Hurricanes dealt faithfully with their opponents, elsewhere events followed the usual depressing pattern of the Blitzkrieg. Paratroops landed on airfields and bridges throughout Holland; Panzer forces raced through the crumbling defences to link up with these; Rotterdam was savagely bombed. Within five days, the Dutch had capitulated.

Meanwhile, the new air weapon had proved equally successful against Belgium. Airborne troops took the vital Veld-wezelt and Vroenhoven bridges over the Albert Canal intact. A few miles to the south, Fort Eben-Emael was captured by nine troop-carrying gliders, which landed literally on top of it. The handful of Belgian Hurricanes could do nothing in the face of this storm. Nine of them were destroyed on the ground by the first bombing attack on Schaffen aerodrome, early on the morning of 10 May. The remainder were quickly finished off either at their bases or in the air, for the pilots' lack of experience made them easy victims.

As German forces flooded westward, Belgian and French bombers made desperate attempts to destroy the captured bridges. These having failed with heavy losses, the RAF was called upon on 12 May. Five Battles of 12 Squadron were entrusted with the task, while eight Hurricanes of 1 Squadron, led by Squadron Leader Halahan, were ordered to arrive just before the bombers to keep enemy fighters out of the way.

This part of the plan succeeded. Not only did the Hurricanes draw the Bf 109s away from the Battles, but in the tremendous dog-fight that followed they claimed to have destroyed eleven of them, including two which collided when engaged by Flying Officer Brown. Halahan's Hurricane was hit in the engine by a cannon shell but he managed to glide to a safe crash-landing in Allied territory. Flying Officer Lewis, a Canadian, had to bale out of his blazing aircraft behind enemy lines. He was captured, but liberated almost at once by a Belgian counter-attack.

Under cover of the Hurricanes' fight, the Battles attacked the bridges through vicious anti-aircraft fire on a scale previously unknown. All were shot down or crash-landed though the Veldwezelt bridge was shattered, almost certainly by the section leader, Flying Officer Garland. Posthumous VCs were awarded to Garland and his observer Sergeant Gray, the navigator for the strike, but as the *RAF Official History*[6] remarks sadly:

> Since Victoria Crosses are distributed sparingly, and there was no other appropriate honour which might be conferred on a dead man, the third member of the crew, Leading Aircraftman L.R. Reynolds, the wireless operator/gunner, received no award. Let his name then at least stand recorded with those of his companions.

In the meantime, the northern Allied armies, pivoting on Sedan, wheeled into Belgium. While German aircraft continued to pound airfields, railway junctions and towns – though not, significantly, road-bridges – they made no attempt to interrupt this advance. This was 'strange, and I feel, very suspicious,' noted the anonymous author of *The Diary of a Staff Officer*, who was, in fact, Colonel Philip Gribble, an incredibly adaptable man, perhaps best known in the racing world, but then an Army Intelligence liaison officer on the Staff of Air Marshal Barratt. 'It looks almost', he added, 'as if the Germans want us where we are going.'

Originally, the intended German attack on France, apart from the inclusion of Holland as well as Belgium, was a Schlieffen Plan with modern weapons. While Army Group

---

[6] *Royal Air Force 1939-1945* by Denis Richards and Hilary St George Saunders. HMSO 3 Volumes: *The Fight at Odds*; *The Fight Avails*; *The Fight is Won*. For simplicity's sake this indispensable work will henceforth be referred to simply as *RAF Official History*.

C, under General von Leeb, watched the Maginot Line, the decisive thrust would be given by General von Bock's Army Group B with the aid of Luftflotte (Air Fleet) 2 under General Kesselring. General von Rundstedt's Army Group A, supported by General Sperrle's Luftflotte 3, had only the secondary role of covering von Bock's southern flank.

It seemed to von Rundstedt's Chief of Staff, General von Manstein, that this plan, being unlikely to achieve surprise, would end in stalemate. He urged that the bulk of the armoured forces should be transferred to Army Group A, that von Bock should push into Holland and Belgium as arranged, but that when the Allies advanced to meet his threat, von Rundstedt should cut them off by breaking out of the Ardennes, difficult country but therefore likely to be less well defended.

Von Rundstedt eagerly approved these ideas, as did the Panzer expert, General Guderian, who was convinced they would work if sufficient weight of armour was provided, but when they were put to the Army High Command – Oberkommando des Heeres (OKH) – the Commander in Chief, General von Brauchitsch, rejected them; nor did he deign to forward them to Hitler.

Then, on 10 January, a German aeroplane carrying an Airborne Division courier, force-landed in Belgium. Documents were captured that revealed much of the existing operational plans. Hitler, whose famous intuition was by no means always wrong, had never liked the original proposals. He now began to consider turning this mishap to advantage by moving the main attack further south, while encouraging the Allies to think that Belgium would still be the primary target.

By this time, the persistent von Manstein had become such a nuisance that OKH had promoted him out of the way to a home command. Before taking up his post, however, von Manstein, as was customary, called to pay his respects to Hitler, upon whom he urged the virtues of his plan. To his delight, Hitler adopted it with enthusiasm. The main weight of the attack was transferred to von Rundstedt's Army Group, to which was allotted General von Kleist's Panzer Group, consisting of 19th Corps (three Panzer divisions under Guderian) and 41st Corps (two Panzer divisions under General Reinhardt). Also under Army Group A came a separate Panzer Corps, the 15th, with two divisions, led by General Hoth.

By the afternoon of 13 May, all these forces, having pushed through the 'impassable' Ardennes, were preparing to cross the Meuse. They had been unable to bring up their supporting artillery, but one of the features of Blitzkrieg was the use of aircraft as flying field-guns, especially the Junkers Ju 87 Stuka dive-bombers, hideous machines fitted with screaming sirens to heighten their already terrifying effect. Some 200 of these horrors now fell on the French defenders, who were largely elderly, untrained reservists, totally demoralising them. Under cover of the Stukas, the Panzers crossed the river near Sedan. Next day, counter-attacks by Allied bombers failed with ruinous losses. By nightfall on the 15th, all the Panzer divisions were pouring into France while the Luftwaffe ranged ahead to pounce on any opposition.

In the face of such catastrophes, the cries for more Hurricanes grew in intensity. On the 13th, thirty-two of them flew out as reinforcements – the equivalent of two extra squadrons. Shortly afterwards, eight flights from different Hurricane units, amounting in practice to four more squadrons, also crossed the Channel.

This dispersal of his forces horrified Air Marshal Sir Hugh Dowding, Commander in Chief of Fighter Command, who, in his own words, 'knew that if this wastage, this great flood of Hurricane exports to France, was not checked it would mean the loss of the war'. So firm were Dowding's arguments that no other Hurricanes were committed to the crumbling French front although in order to bring more into action without too great risk, squadrons were detached to France for a morning or afternoon, flying home by evening. Moreover, as the Germans neared the Channel coast, Hurricanes from British bases also took part in the conflict.

Among the squadrons thus engaged was No 32 from Biggin Hill, which hitherto had seen virtually no combat. Nor did its first trip to France alter the pattern to judge from the squadron diary, which recorded this in the facetious fashion, traditionally cultivated by the RAF though subsequently much mocked by critics who fail to realise that such things help to keep men sane: 'All set off to France, land at Abbeville, refuel, hear dreadful stories, get very frightened, do a patrol, see nothing, feel better, do another, see nothing, feel much better, return to Biggin Hill, feel grand'. It should be remarked in passing that the irrepressible diarist was Flight Lieutenant Michael Crossley, not only the finest pilot in the squadron but one of those rare, fortunate beings who never appear very frightened of anything.

On 19 May, the war ceased to be a joke. Over Cambrai, 32 Squadron sighted a Dornier Do 17 with a strong escort of Messerschmitt Bf 109s and Bf 110s. This line-up usually meant either that there were important personnel in the bomber or that it was on an important 'recce' mission. The Hurricanes attacked, losing one of their number whose pilot became a prisoner of war, but destroying the Dornier and destroying or damaging six of the escorting fighters. Three days later, they destroyed or damaged another six 109s, this time without loss.

The first Hurricane Squadron, No 111, had already seen action. On 18 May it destroyed or damaged seven enemy aircraft, of which Flying Officer Ferris (in Hurricane L1822) shot down no fewer than four Bf 110s, all of which were confirmed beyond doubt.

As Army Group A scythed across France and Army Group B pressed on steadily in Belgium, the Hurricanes saw every type of activity. 1, 73 and 501 Squadrons escorted British bombers attacking the enemy advancing from Sedan – a largely fruitless task since the Luftwaffe so outnumbered the RAF that one force of its fighters could always engage the Hurricanes while others attacked the bombers. 213 and 615 guarded Blenheims raiding the German troops in Belgium; 87 protected the Army co-operation Lysanders – not without heavy losses since in such a role the Hurricanes were at their most vulnerable.

However, in the main, the Hurricanes were engaged in intercepting enemy formations. During this period, with due allowance made for the exaggerations inevitable in the sheer speed of air-combat, they probably destroyed or damaged about 350 enemy aircraft. 85 Squadron claimed to have shot down eighty-nine (though, with respect, the correct figure was probably about sixty). Of this total, three pilots, Pilot Officer Lewis, a South African, Sergeant Howes, an Englishman, and Flying Officer Count Czernin, a naturalised Pole, claimed thirteen between them in the two days of the 19-20 May. 607 claimed seventy-two enemy aircraft (though again, perhaps the real figure was nearer forty). 79 Squadron is said to have accounted for twenty-five of the enemy; 111 for sixteen; 56 for fourteen.

Such figures soon become meaningless. Doubtless they were even more meaningless to the pilots. For them, the days were a kaleidoscope of action: of aircraft climbing, turning, or diving into the ground in flames; of parachutes opening, or not opening, or opening only to catch fire; of escapes through tree-tops when out of ammunition, perhaps seeing a comrade whom it was impossible to assist left battling against half-a-dozen enemy fighters. There was never any time for reflection. Nothing could illustrate the pace of the fighting better than the career of Flight Lieutenant Soden. On 17 May, he led a detachment of 56 Squadron to France. Two days later he was dead, having in the meantime destroyed six enemy aircraft and won a DSO for attacking single-handed some fifty Bf 109s, one of which he shot down.

The pressure on the pilots was immense. There was the strain of constant moves to quite inadequate bases such as that of 87 Squadron at Lille Marque, where, in an emergency, the Hurricanes had to take off towards each other with a hump in the middle of the field ensuring that it was impossible to judge whether any were on collision courses until too late to do much about it. There was the strain of taking off amid the crash of bombs only to be attacked at low level by 109s or 110s – as happened to 607. There was the strain of landing a Hurricane with an aileron hinge severed, all controls damaged to some extent, and so riddled with cannon and machine-gun fire that it was destroyed by the ground crews without further ado – as did Sergeant Soper of No 1. There was the strain of taking off six times in one day, each sortie resulting in combat, and destroying or damaging six enemy aircraft – as did Squadron Leader More of 73.

The exhaustion caused by such activities can perhaps be better imagined than described. One evening, every pilot of No 1 fell asleep at the supper-table. Sergeant Allard, a lynx-eyed marksman from 85 Squadron, on one occasion fell asleep in the cockpit as his Hurricane rolled to a standstill after his fifth flight of the day.

On 21 May, Guderian's tanks reached the Channel near Abbeville, cutting off the northern Allied armies. By this time, the bases of the Air Component had been over-run or were seriously threatened. In any event, the fatigue of the pilots was now such that they could not have continued much longer even had airfields been available. The decision was taken to evacuate the Hurricanes, which by the following evening had returned to Southern England, whence they continued to provide such cover as was possible.

Since the Blitzkrieg began, the Air Component had lost seventy-five Hurricanes to enemy action, of which a large proportion had been destroyed on the ground. Yet no fewer than 120 more had to be burned on French airfields. These were damaged aircraft that could not be repaired in time before their bases fell. 504 Squadron flew home only four Hurricanes; 85 only three; 79 none at all. This loss represented about a quarter of Britain's modern fighters, which gives an idea of the importance of Dowding's refusal to send further reinforcements to France where they would inevitably have been swallowed up in the holocaust. However, there was one great consolation: the now tremendous rate of Hurricane production: in June alone, 309 would be delivered.

While the remaining Air Component Hurricanes retired across the Channel, the squadrons of the AASF, with the aid of commandeered transport, moved to bases further south around Troyes. However, in practice, one of them also returned to Britain at this time. Almost all No 1's pilots were now taken off operations. While twelve new pilots reached the squadron, these men went home, where, after a period of rest, they split up, some to add stiffening to other squadrons, many to become instructors to pass on their hard-won experience to others.

During the ten days from the German attack to the capture of Abbeville, No 1 claimed the destruction of 114 enemy machines. Of course, over-optimism resulted in some exaggeration, but careful research by later compilers of the squadron's history indicates that its pilots brought down eighty-seven at the very least. For this, they paid the price of two dead, one a prisoner of war and two more in hospital, both of whom subsequently recovered to fight with distinction. Bearing in mind these slight losses, the achievement was never to be equalled. There seems little doubt that the squadron at this time was the finest fighter unit in the RAF. Its successes were fittingly acknowledged in the unique mass award of ten DFCs and three DFMs – abundantly justifying the squadron's proud motto, 'First in all things'.

–

By now the isolation of the northern armies, including the British Expeditionary Force, was complete. A week later the surrender of the exhausted Belgian Army made their position even more hopeless. Yet from the trap the bulk of the BEF with large numbers of French troops – for it should never be forgotten that to their eternal credit the rearguards holding the perimeter during the evacuation were almost all French – escaped over the sea, though with the loss of most of their equipment.

The 'miracle' had its origins in an action south of Arras on 21 May, when a British force under Major-General Franklyn achieved a complete surprise against the advancing Germans. The opposing general, one Rommel, always apt to exaggerate when attacked resolutely, magnified Franklyn's seventy-four tanks into 'hundreds' and his three battalions of infantry into five divisions. From this he concluded he was facing a major counter-attack.

Rommel's fears were communicated to von Rundstedt, who was alarmed lest his armoured formations be cut off from the supporting infantry, which was being outstripped – an anxiety shared by several of his subordinate generals, notably von Kluge, commanding Fourth Army, and von Kleist. As a result, checks were imposed on the ever-eager Guderian, who furthermore had now to divert his panzers to deal with garrisons such as Calais, which though isolated, were still resisting valiantly.

On the evening of 23 May, in response to a plea from von Kluge, von Rundstedt issued instructions for a halt on the line of the Aa Canal to allow his infantry to close up. This was then intended as a purely temporary move as is shown by the fact that both Guderian and Reinhardt later resumed their advance, gaining bridgeheads across the canal.

However, in the mid-morning of the 24th, Hitler visited von Rundstedt's headquarters where he stated emphatically that the armour should be conserved for future operations. That evening, he instructed OKH to forbid the panzers to advance beyond the canal. Although undoubtedly influenced by the caution prevailing in Army Group A, the decision was entirely Hitler's. It is surely significant that he did not once attempt to answer protests at OKH by claiming the support of the highly respected von Rundstedt.

Hitler's motives have been the subject of innumerable arguments. Captain Liddell Hart has made much of the theory that Hitler did not wish to defeat the British too heavily in the hope that they would agree to peace proposals, but aside from the fact that they would surely be more eager for peace after a disastrous defeat, this view is expressly contradicted by a directive from Hitler on the same day specifically calling for the 'annihilation' of the trapped armies, while he also ordered the Luftwaffe to prevent the escape of the BEF over the Channel. Nor is there any need for such fantastic suggestions since there were perfectly adequate military reasons for Hitler's decision.

It should always be remembered that Hitler's aim was the defeat of France, after which he fondly believed that Britain also would abandon the struggle. The destruction of the northern armies was merely an important milestone on the road to this objective. After it would come an attack across the Somme, leading in due course to the conquest of the remainder of France. For this, tanks were vital, yet already 50 per cent of von Kleist's and 30 per cent of Hoth's were out of action. Hardly surprisingly therefore, Hitler viewed with much concern the likelihood of further losses in the area around Dunkirk to which the trapped forces were retreating, since, being criss-crossed with canals, it was unsuitable for tank warfare.

However, it seemed to him there was a better way in which the BEF could be wiped out. On the previous evening – the timing is surely important – Göring had telephoned to claim that he could destroy the isolated Allies.

'Leave it to me and the Luftwaffe,' he bragged. 'I guarantee unconditionally that not a British soldier will escape.'

That this promise was in Hitler's mind appears confirmed by a comment in Group A's war diary that further pressure in the Dunkirk area would only cause 'restriction of the activity of the Luftwaffe', as also by a note in the diary of General Halder, the German Chief of the General Staff: 'Finishing off the encircled enemy army is to be left to the Luftwaffe!'

The fighting on land was by no means ended. Army Group A continued to enlarge its bridgeheads while to the south in terrain favourable for tanks, the Germans pressed on to Lille, where they cut off the major part of the French First Army. The main attack came from Army Group B's infantry, but von Bock's handling of his troops was uninspired. By 26 May, Hitler, who had become concerned by von Bock's slow progress, ordered the panzers to renew their advance, but by this time resistance had so stiffened that two days later Guderian, after touring the forward positions, considered that 'further tank attacks would involve useless sacrifice of our best troops'. Commenting that infantry were 'more suitable than tanks for fighting in this kind of country', he requested his armoured divisions be withdrawn to refit. Since he was then only eight miles from Dunkirk harbour, his decision suggests that Hitler's original appreciation was not so very wide of the mark.[7]

Of course, none of this would have mattered greatly if Göring's men had lived up to their leader's boast. That they did not was due in part to bad weather, especially on 30 May, but more to bad tactics. Far too much effort was concentrated against troops on the Dunkirk beaches. Not until 1 June was an all-out assault launched on the rescuing vessels − on this day eleven ships were sunk, including one French and three British destroyers[8], forcing Admiral Ramsay, in charge of the operation, to prohibit future evacuation in daylight, though it continued at night until 4 June. Still more was the Luftwaffe's

_____

[7]It is odd to find that whereas Captain Liddell Hart is always ready to query Allied statements, he is equally prepared to accept enemy accounts at face value though one would have thought it was only human for defeated men to try to shift the blame for their errors onto the shoulders of others. In Guderian's memoirs, he constantly criticises Hitler, emphasises how he stood up to him, but carefully makes no mention of several of his orders of the day that display a revolting sycophancy.

[8]During the evacuation, a total of six British destroyers were lost together with three French. Nineteen other British destroyers were badly damaged.

failure due to the unbelievable tenacity of the Royal Navy backed by the French Navy, the Merchant Navy and the civilian volunteers – the famous 'little ships' – in defying such losses. But the main reason that the German airmen did not attain their objective was that they had other problems to consider.

The RAF was not popular during or immediately following the Dunkirk evacuation. Both Army and Navy felt that their sister-service was not 'pulling its weight'. Detached reflection might indicate that much of the effort in the air would inevitably take place out of sight of the beaches, but there was scarcely leisure for such at Dunkirk.

It was also natural, if somewhat unjust, that the soldiers and sailors should assume automatically that every aeroplane sighted was hostile. Of several stories to illustrate this point the most delightfully ironic is recounted by Tom Moulson in *The Flying Sword: The Story of 601 Squadron*. Flight Lieutenant Sir Archibald Hope, a pilot of 601 making his way home via Dunkirk after being shot down, had occasion to inform a high-ranking Army officer that the aircraft he was reviling were in fact Hurricanes. The officer refused to believe him, protesting indignantly: 'But we've seen lots of those.'

Lots of Hurricanes might well have been seen, for in the Battle of France all available Hurricane squadrons saw combat, the majority during or just before the evacuation. One of the Hurricanes that flew over Dunkirk on several occasions was piloted by Air Vice-Marshal Keith Park, the tall, lean New Zealander commanding No 11 Group, responsible for most of the protective patrols – though in passing it may be remarked that Hurricanes also served by escorting Blenheims on raids against the enemy on the ground.

Statistically, Dunkirk was not one of the Hurricane's greatest successes, but the reasons for this are not far to seek. Having occupied the airfields of northern France, the Luftwaffe was able to muster very large numbers at selected moments, whereas Fighter Command, operating well away from its bases, had to adopt continuous patrols with the inevitable result that some encountered no enemy at all while others faced overwhelming odds. If the size of the protecting forces was increased, then there would not be enough fighters for continuous cover, with the consequent danger of disastrous attacks developing between patrols. This indeed is what happened on 1 June. Though with the wisdom of hindsight, one can perhaps suggest that the fullest use was not made of the available forces, these difficulties were largely insoluble.

An even greater problem was combat training. The Germans had experience stretching back, in some cases, to the Spanish Civil War. The Hurricane squadrons already in France had learned the tricks of the trade but those now fighting over Dunkirk, in most cases, had not. They were still trying to carry out standard 'copybook' attacks, which were quite unsuitable to high-speed combat. They were still flying in the inflexible 'Vic' or 'Vee' formations where more time was spent in keeping station than in looking for hostile aircraft.[9]

It is interesting that one of the most resounding successes was scored by 43 Squadron, which had gained considerable experience from actions around Scapa Flow. On the otherwise grim 1 June, it claimed the destruction or probable destruction of seven Messerschmitt Bf 109s and two Bf 110s with six more enemy aircraft damaged (probably an exaggeration); losing two Hurricanes destroyed, two more damaged, one pilot dead and one wounded. This was achieved against odds of about seven to one. One pilot, Sergeant Hallowes, alone claimed two 109s and one 110 destroyed, and a third 109 probably destroyed – while Sergeant Ottewill claimed two 109s destroyed and one 110 probably destroyed.

---

[9] A similar rigidity greatly handicapped the RAF in its early encounters with the Luftwaffe in the Mediterranean theatre. Here also those squadrons, which were the first to appreciate the deficiencies of the recommended pre-war tactics, gained the most successes as well as suffering the least casualties.

Less knowledgeable squadrons had grimmer encounters. Even before the evacuation, No 17 in its first action had shown the perils of rigid tactics. Attacked by Bf 109s, it downed three of these but lost five of its own number including Squadron Leader Tomlinson. 605 also lost its CO (Squadron Leader Perry) in its first combat. 242, meeting 109s, shot down or damaged three of them but lost two - Hurricanes. 213, attacking a formation of Heinkels, was hit from above by 109s. It brought down one but lost two Hurricanes, though one of the pilots escaped by parachute.

Fortunately, those Hurricane pilots who survived gained the necessary skills very quickly. 213's action, just described, took place in the early morning of 28 May. After breakfast, it made a second sortie over Dunkirk, again meeting an escorted bomber formation. This time it destroyed or damaged seven of the enemy. Three Hurricanes were lost but two of the pilots baled out, one of whom, Sergeant Butterfield, no doubt felt consoled for his misfortune by the belief that he had previously shot down two 109s, a 110 and a Junkers Ju 88. Another pilot of 213 Squadron who deserves mention here was Flight Lieutenant Wight, who during the Dunkirk fighting destroyed or damaged six of the finest German fighters, the Bf 109s.

Throughout the period of the evacuation, the Luftwaffe lost 132 machines, a few of which were claimed by AA fire from Allied warships. Fighter Command lost 106. This may not seem a tremendous victory.

Also, by a grim irony, the most famous success story held the seeds of future disaster. 264 Squadron flying Defiants – two-seater monoplanes intended as successors to the Hawker Demons but, unlike them, lacking forward–firing guns – claimed to have destroyed fifty-seven enemy aircraft in three days. Though the real score was probably as low as fifteen, the Defiants did prove valuable against 109s which, mistaking them for Hurricanes, dived straight into the fire of the rear guns. Amid the jubilation, it was forgotten that 111 Squadron, after comparative trials in October 1939, had reported that any average Hurricane pilot could master a Defiant, as could any 109 pilot who met it a second time. Later, however, this judgement would prove only too accurate.

As a result, many critics have claimed that Churchill's statement that 'there was a victory inside the deliverance ... gained by the Air Force' was incorrect – mere propaganda to counter the complaints against the RAF. Such comments miss the point. The importance of Fighter Command's efforts appeared not in the number of enemy machines shot down, though in view of the odds faced this was not discreditable. It lay in the number of formations broken up, the number of raiders turned back or forced to jettison their bombs blindly in hurried evasive action – in short, in the way the Luftwaffe was prevented from concentrating on its main objective. There was a victory inside the deliverance – gained by the Royal Air Force.

The completion of the Dunkirk evacuation did not end the Hurricane's fight in France. The home-based Hurricane squadrons acted as escorts to Blenheims on bombing or reconnaissance missions, or covered other lesser evacuations chiefly at Le Havre and Le Treport, the fighters refuelling on French airfields if necessary. It was over the latter port on 8 June that 32 Squadron, encountering an escorted formation of Heinkel He 111s, claimed nine enemy aircraft destroyed – six Heinkels, three Bf 109s – plus four 'probables' (though no doubt they exaggerated slightly). Three Hurricanes were brought down. Two pilots were killed but Pilot Officer Grice crash-landed, subsequently making his way home via Jersey.

Operating at long range, these sorties frequently proved expensive. On 7 June for instance, 43, ambushed by 109s, lost four Hurricanes and two pilots over Abbeville, shooting down only two of the attackers. That afternoon it again had trouble over France, claiming three 109s and a 110 but at a cost of three more Hurricanes. All the pilots escaped but Sergeant Ottewill, who had destroyed a 109 on each sortie, was so badly burned that he never rejoined the squadron, though eventually he was able to reach Allied territory.

By this time, several squadrons were becoming demoralised by sheer exhaustion. Thus 79, which had already lost six pilots killed during May, now re-entered the fray with sad results. On 7 June, the squadron claimed to have destroyed or damaged five 109s but that was its last success and it now suffered savage casualties. Before the end of the month, it was attacked by 109s while on escort duty, losing two pilots, killed. Early in July, another pilot died in action on convoy patrol. On 7 July, Squadron Leader Joslin perished when his aircraft crashed for reasons unknown. Next day, the squadron was 'bounced' by 'free-chasing' 109s, two more pilots being killed. Mercifully Park then ordered 79 north to rest, but on the 12th it lost two more Hurricanes, though not this time the pilots, in collisions with training aircraft – accidents that show clearly the squadron's utter fatigue.

However, the last phase of the Hurricane's war in France belonged mainly to the AASF Squadrons. While the Dunkirk saga continued, these were already battling with hostile raiders in circumstances even more difficult than usual. On 27 May, 501 moved to a grass strip near Rouen with the promising name of Boos. Scarcely had the Hurricanes landed than they were bombed by Heinkel He 111s, though miraculously they escaped damage. That afternoon, they gained revenge. Sighting twenty-four Heinkels escorted by twenty Bf 110s, they attacked at once, whereupon the Messerschmitts headed for Germany, though not before one had been severely damaged. 501 then set about the bombers, destroying eleven, plus three 'probables', without loss – but again, it should be recorded that there were no doubt a number of duplicated claims in that total.

Of course, the pilots of the re-vitalized No 1 Squadron would not be left out of the reckoning. On 4 June, they also refuelled at Boos, after which they took off to tackle an escorted bomber formation, being credited with destroying or damaging three Messerschmitt Bf 110s, three Heinkel He 111s and a Dornier Do 17. Two Hurricanes were badly mauled but force-landed without injury to the pilots.

These raids were preliminaries to the Germans' main assault. On 5 June, they broke across the Somme, after which Blitzkrieg proceeded with its usual unstoppable rapidity. On 10 June, Italy entered the war. Paris fell on the 14th. On the 17th, Petain asked for an armistice, while de Gaulle flew to London whence he vainly urged his country's leaders to fight on. Four days later, the French delegation heard Hitler's terms in the same railway coach in which had been signed the Armistice of 1918.

During these catastrophes 1, 73 and 501 Squadrons, moving from one base to another, attempted to guard the remaining British troops struggling to withdraw through north-western France; a task in which they were joined on 7 June by the Hurricanes of 17 and 242 Squadrons. For the new arrivals, thus hurled unprepared into the holocaust, the situation was, if possible, even worse. The pilots of 242, a mainly Canadian unit, finally returned home in a state approaching collective shock, leaving behind seven killed, three in hospital, most of their machines, all their spare equipment and almost all their personal effects. This grim tale again showed how vital had been Dowding's refusal to commit his command as a whole in France.

By 15 June, the fighters had moved to the west coast. 17 and 501, based first at Dinard, later in the Channel Islands, covered evacuations from St Malo and Cherbourg. The other three squadrons, stationed at Nantes, covered that port, Brest and St Nazaire.

As the German bases were now at a greater distance, the withdrawals proceeded successfully despite inevitably sparse fighter cover. There was one frightful exception: the sinking, on 17 June, of the troopship *Lancastria* with some 5,000 men on board. Heavy clouds enabled the attacking Heinkel He 111s to evade No 1's Hurricanes, but these at least gained some revenge when Sergeant Berry shot down the bomber responsible.

By next afternoon, the troops had made good their escape. 1 and 242 followed them. Lieutenant Demozay, No 1's French liaison officer, flew an old Bombay packed with groundcrew to Britain, where under the nom-de-guerre of Moses Morlaix (to avoid reprisals against his family), he was later to rejoin No 1 as a highly successful fighter pilot. 73 mounted a final patrol, then set off after the others. Three days later, the Hurricanes in the Channel Islands also left. The AASF Squadrons had lost sixty-six Hurricanes, the majority unserviceable machines destroyed on the ground, since 10 May.

In all, during the Battle of France, Fighter Command lost 477 aircraft, the great majority of which were Hurricanes. One-fifth of the professional peacetime pilots were already dead or disabled. About 50 per cent of those remaining in Britain were newly trained and quite inexperienced. These losses had brought the country to a perilously weak condition at the very moment it was faced by its greatest threat.

Yet one great advantage had been gained. Most Hurricane squadrons at least had gained vital combat experience. The rigid formations, the inflexible tactics were being discarded. While some of the men who had won laurels in France were passing on their knowledge in training establishments, most remained on the front line for the crucial fighting that was to follow – pilots such as Allard, Clowes, Crossley, Hallowes, Lacey, Czernin, who would provide the steel heart of the defence of the homeland.

As the Hurricanes withdrew, the Luftwaffe moved in to their old bases in northern France, north-western France and Norway. The Germans too had not emerged unscathed. Of their 1,284 aircraft destroyed in battle (nearly 200 more were lost in accidents), it will never be known how many fell to the Hurricanes for detailed enemy records have been lost, but as they encountered far more Hurricanes than all other types of Allied fighters, it must have been a high proportion of this figure.

Nor was German morale as high as before. Though normally the Luftwaffe had been beyond reproach, a few less creditable incidents had occurred: the 110s that left the Heinkels to the mercy of 501; the 109 that retired before a Hurricane pilot who was out of ammunition; the bomber whose crew baled out before the attacking Hurricane could open fire. Some 400 Luftwaffe prisoners were freed after the fall of France. They would fight again against Britain, but they would not have been human if they had not had memories of their previous mishaps. And of course, many of the finest enemy airmen would never fight again; the Luftwaffe's own Flying Personnel Casualty List shows 129 officers and nearly 600 other ranks killed in the last ten days of May alone.

It was just as well that the Luftwaffe had had its striking power blunted, for by 2 July, after initial hesitations, Hitler announced that 'a landing in England is possible'. He added a significant rider: 'Provided that air superiority can be attained'.

Chapter Three

The Bomber-Butcher

The story of the Battle of Britain, dramatic in its details, vital in its importance, momentous in its results, has been told and re-told. Unfortunately, in the process, several myths have been created which may be disposed of before the Hurricane's role in the drama is related.

The first myth is that Hitler did not really intend to invade – it was all some monstrous bluff. Certainly Hitler, whose ultimate objective always was the destruction of Russia, would have preferred to avoid a hard, costly campaign if he could bring the British to accept his terms, but he intended with equal certainty to break them if they refused to yield.

As already mentioned, Hitler's first statement of his intentions, though he stressed this was 'only a plan and has not been decided upon', came on 2 July. By the same day, Halder was writing that 'a landing operation in England was, on the part of the Army General Staff, (now) considered feasible'. The Luftwaffe began preparatory steps to establish air superiority over the Channel, which was to be closed to British shipping. This task was given to Oberst (Colonel) Fink to whom was given the impressive title of Kanalkampfführer or Channel Battle Leader. Fink began without delay; the Battle of Britain officially started on 10 July but, while there was a definite increase in activity on that day, Fighter Command earlier in that month had already lost eighteen aircraft, the equivalent of a full squadron, together with thirteen pilots killed and six injured.

On 16 July, Hitler issued Directive No 16: 'As England, in spite of her hopeless military situation, still shows no sign of willingness to come to terms, I have decided to prepare, and if necessary to carry out, a landing operation against her.' This operation was given the code-name 'Sealion'.

The qualification indicated that Hitler still hoped the invasion would prove unnecessary. Three days later, encouraged by heavy British fighter losses, made even more impressive by German exaggeration, he made a 'last appeal to reason'. This was formally rejected by the Foreign Secretary, Lord Halifax on the 22nd, though by then it had already received a more violent rejection by Fighter Command in general and the Hurricane in particular.

Thereafter, detailed plans for the invasion proceeded. Barges, motor vessels and tugs assembled in the Channel ports to carry the attacking troops – causing, incidentally, considerable disruption of Germany's vitally important inland waterway transport. The divisions for the assault were allocated; their commanders named; their objectives detailed. Overall control was entrusted to von Rundstedt, who on 19 July was encouraged by promotion to Field-Marshal along with eight other Army officers and three Luftwaffe Generals. There was even a deception plan: a dispatch of escorted but empty merchant vessels to feint towards the north-east coast two days before the landing.

Nor was the treatment of the conquered forgotten. Von Brauchitsch, as Army Commander in Chief, issued 'Orders concerning the Organization and Function of Military Government in England', which although designed to spread terror, by, among other methods, the deportation of all able-bodied men between seventeen and forty-five to the Continent, appear mild compared with the plans of Heydrich's Reich Central Security Office.

SS Colonel-Professor Six, one of those foul intellectuals who can always be found on the fringes of dictatorships – he later demonstrated his peculiar abilities in Russia – was appointed director of this body's activities. He was ordered to form 'action groups' – in reality murder squads – in London, Edinburgh, Birmingham, Bristol, Liverpool and Manchester, which would destroy all civic or cultural leaders, all Jewish communities and all organisations that might offer a common ground for resistance – from the public schools to the trade unions, from the established Church to the Freemasons; even the Boy Scouts were not forgotten.

Hitler resolutely defied pressure from his Naval chief, Admiral Raeder, to postpone the invasion to the following May, realising that by then the British Army and Air Force would have mustered a far more formidable strength. He did not invade in September as planned only because he did not achieve the 'conditions necessary', in particular the attainment of air superiority. Until this was clear to him, Sealion was a genuine threat; the fact it was amended from time to time, a proposed attack at Lyme Bay being abandoned in mid-August, proves this, for there is no point in amending a bluff.

The next myth is that even if Fighter Command had been annihilated, the Royal Navy would still have been able to thwart the invasion. Certainly, the Navy's morale, as always, was high. One destroyer captain wrote to a friend: 'I shall not go near his (Hitler's) ruddy warships... nor shall I waste my time crumpling my bows on any old barge. When I have expended my very considerable stock of ammunition and my torpedoes I am by no means at the end of my resources...'

To which Captain S.W. Roskill, in *The Navy at War 1939-1945*, adds: 'As there were about half a hundred British destroyers ready to hand, and all were imbued with the spirit of the writer of that letter, the reluctance of the Germans to hazard their troops on the short sea crossing can readily be understood'.

However, it is noticeable that no mention is made of anti-aircraft guns, yet, from Captain Macintyre's *Narvik*, it appears that British warships were 'equipped with high-angle guns in inadequate numbers and with fire-control systems which could not cope with the high-performance aircraft brought against them'; that though cruisers and above could deal with high-flying aircraft, 'against the dive-bombing technique favoured by the Luftwaffe which could only be countered by massed machine-gun fire, they were singularly ill-equipped'; and that destroyers 'had only the clumsy unreliable 2-pounder pom-pom guns and 0•5-inch machine-guns to bring against air attack of any sort' since their main armament could not be elevated sufficiently. Nor did the position improve before 1942.

Therefore, a more realistic assessment would seem to be that of Colonel Fink, who reported that, always provided air supremacy could be attained, his airmen could dominate the Channel in daylight though not at night and keep British warships from the invasion area again by day only. By the end of July, even without such supremacy, the Luftwaffe had sunk four destroyers, whereupon the Admiralty, acknowledging that command of the Straits in daylight had been lost, abandoned Dover as a base, with the result that the destroyers, whatever the spirit of their captains, would have been too far from the invasion area to have prevented it.

On the other hand, the Navy, by the distasteful but necessary method of attacking ports already in enemy hands, could greatly have hampered the Germans' consolidation of their bridgehead. What would have followed may be judged from the fighting around Crete, which Admiral Cunningham called 'nothing short of a trial of strength between the Mediterranean Fleet and the German Air Force'. 'I am afraid,' he added, 'that in the coastal area we have to admit defeat' – this although the Luftwaffe then had available only about a quarter of the bombers packed around the Channel coast in 1940.

It is also worth mentioning that the conquest of Crete was achieved solely by enemy airborne troops. It was planned to use these against Britain in 1940. While they alone could not have conquered the country, in the absence of fighter interception they would probably have been able to keep building up an existing bridgehead.

That the British Army, aided by the spirited but woefully ill-equipped volunteer organisations, would have fought desperately in defence of the homeland cannot be doubted, but it was inexperienced in mobile warfare, and, with command of the air, the Germans could have broken up its concentrations. Even if the Navy had cut off later reinforcements by sea, this would in some ways merely have made matters worse. The German advance would not have been another Blitzkrieg but a slow, relentless push forward, causing far more destruction, while in front of their armies the German airmen launched terror-raids as at Warsaw or Rotterdam and the SS murder-squads embarked enthusiastically upon their loathsome tasks.

It is worthwhile at this point to consider a report made to Churchill by the Chiefs of Staff at the time when the surrender of France was imminent. Their conclusions were not designed to establish any point or prove any theory. They were the judgements of capable, responsible men facing grim realities. They deserve therefore to be quoted in detail:

> While our Air Force is in being, our Navy and Air Force together should be able to prevent Germany carrying out a serious sea-borne invasion of this country.
>
> Supposing Germany gained complete air superiority, we consider that the Navy could hold up an invasion for a time, but not for an indefinite period.
>
> If with our Navy unable to prevent it, and our Air Force gone, Germany attempted an invasion, our coast and beach defences could not prevent German tanks and infantry getting a firm footing on our shores. In the circumstances envisaged

above our land forces would be insufficient to deal
with a serious invasion.

The crux of the matter is air superiority.

Thus, everything depended on the Luftwaffe, whose chief had
been consoled for having ceased to be Germany's only Field-
Marshal on 19 July, by promotion to the newly created rank
of Reichsmarschall, with a special king-sized baton to prove it,
and by the grant of the Grand Cross of the Iron Cross, the only
one awarded during the war. Whatever doubts existed in the
minds of others, Göring was certain his air force could carry
out its allotted task.

To fulfil his dreams, the new Reichsmarschall could muster
three Luftflotten (Air Fleets) grouped strategically around the
British Isles: – Luftflotte 2 under Field Marshal Kesselring
based in Holland, Belgium and north-east France; Luftflotte
3 under Field Marshal Sperrle in north-west France; Luftflotte
5 commanded by General Stumpff in Denmark and Norway.
Excluding reconnaissance aircraft, these contained more than
2,600 aeroplanes of which the proportion serviceable was 50
per cent at the beginning of July but 80 per cent by the end.

Since mention will be made of various German formations, it will be convenient to explain that their Air Fleets
were divided into Fliegerkorps. These in turn contained
Geschwader, the main tactical unit of ninety to 120 aircraft,
which although having no real British equivalent may be called
super-wings. They were designated according to their function: Kampfgeschwader (KG) – bombers; Stukageschwader
(StG) – dive-bombers; Jagdgeschwader (JG) – single-engine
fighters; Zerstörergeschwader (ZG) – twin-engined fighters,
etc. The Geschwader were split up into Gruppen, then into
Staffeln of ten to twelve machines each. There were also inde-
pendent Gruppen not attached to Geschwader. By far the most
important was Erprobungsgruppe (Test Group) 210. This was
a fighter-bomber unit of two Staffeln of Bf 110s and one of

Bf 109s, all fitted with bomb-racks. Under the inspiration of their brilliant Swiss-born leader, Hauptmann (Captain) Walter Rubensdörffer, their daring, pin-point attacks represented the height of German skill during the Battle of Britain.

The enemy commanders enjoyed not only superior numbers but the ability to choose time, place and strength of attack. By combined raids they could hope to confuse the defenders, while by achieving surprise they could catch British fighters on the ground or still climbing to engage.

Mercifully, Dowding had one vital asset: radar. Without this, Fighter Command, having to rely on the wasteful standing patrols, would have been unable to achieve concentration, so would have been even more outnumbered. It would have received much less warning so have been even more vulnerable. In short, it would not have been able to intercept effectively.

Yet the importance of radar has led to the myth that it solved all problems. In reality, it gave, in 1940, a rough idea of where, when and in what numbers the enemy would attack but it was vastly removed from perfection. Also, like any other system, it depended on expert administration by highly practised operators.

For example, on 15 August the radar picture was so confused that a strike on Martlesham Heath by Erprobungsgruppe 210 was incorrectly identified until far too late. The Hurricanes of 17 Squadron were all but caught on the ground. The Hurricanes of 1 Squadron, unable to gain sufficient height in time, were attacked from above by 109s, which shot down three of them, killing two pilots. The airfield was not fully operational again for forty-eight hours.

In other instances, enemy aircraft were not located at all. On 25 July, a convoy in the Channel suffered three separate raids by fighters flying too low to be recorded on the screens. On 1 August, thirty Heinkel He 111s raided Norwich quite undetected, causing considerable damage, then escaping without interference.

Even when proper warning was given, radar reports were frequently inaccurate. The height of intruders was underestimated so notoriously often that many controllers, when ordering squadrons to intercept, came almost as a matter of course to add a few thousand feet to the altitude specified, while formation leaders sometimes made still further additions. Estimates of numbers were equally inadequate; on 25 August, for instance, a raid on Weymouth reported as a hundred-plus, materialised nearly three times as strong.

Yet the greatest defect of radar was so obvious that it is often overlooked: it could not read the enemy's mind. However accurately it reported a Luftwaffe formation, it could not tell whether this was an attack or a diversion – the main strike or a fore-runner of a still bigger raid timed to catch British fighters re-fuelling; a free-chase' by 109s from which the controllers would attempt to divert fighters to avoid useless attrition or Erprobungsgruppe 210 on a precision attack, which must be stopped at all costs.

In addition, radar did not operate inland. Here, reporting was the responsibility of the Observer Corps, rightly honoured in 1941 by the award of the prefix 'Royal'. No praise could be too high for the efforts of these volunteers, who risked the dangers and discomforts of service life without its pay or amenities. However, where the electronic 'eye' of radar was faulty, so inevitably was human vision. By sharp changes of course or splitting up formations, the Luftwaffe achieved surprise all too often.

For this reason, a vital card in the hands of the defenders was the high-frequency radio that made possible clear, direct speech with the men in the air. This was further aided by Pip Squeak, which automatically reported the aircraft's position to the controllers. The British system was flexible above all, while the German commanders, who could not communicate adequately with their fliers, were unable to adapt pre-arranged plans.

The controllers gave directions, but although anti-aircraft gunfire proved surprisingly effective during the battle, the fighting was mainly in the hands of the young pilots. They often saw action against heavy odds four times a day. The sheer physical exhaustion can only be imagined. A very few broke under the pressure; those who have endured similar strain are alone entitled to pass censure. The vast majority remained unconquered and unconquerable.

The final link in the defence was the interceptor fighter. If radar had been perfect, the controllers infallible, the pilots super-men, the RAF would still have failed if its aircraft had been the biplanes of the time of Munich. Nor were only biplanes inadequate. On 19 July, nine Defiants of 141 Squadron were attacked by Bf 109s of JG2, which, having recognised them, would not repeat the errors of Dunkirk. Six Defiants were shot down. Of those that returned, one was not only so badly damaged that it had to be written-off but was minus its gunner, who baled out never to be seen again. Only the timely arrival of the Hurricanes of 111 prevented the Defiants' complete annihilation.

Thus, the defence had perforce to be entrusted to two makes of fighter: the Hurricane and the Spitfire; which is where there appears the greatest myth of all.

Messrs Derek Wood and Derek Dempster in their account of the battle, *The Narrow Margin*, write:

> Since the Battle the importance of the Hurricane
> to victory has been slowly undermined. The Spit-
> fire tends to hold pride of place to the extent that a
> fallacy runs the risk of being accepted as historical
> fact.

Similarly Wing Commander Roland Beamont in *Phoenix into Ashes*[10] complains that:

> Twenty years later, books and magazine articles began to appear decrying the merits of the Hurricane in the Battle of Britain, suggesting that it was inferior, outmoded and of such low performance as to somehow have contributed to the hazards of the time. How different were the true facts, for the Hurricane not only bore the brunt of the fighting but was exceptionally well fitted to do so.

One reason for the exaggerated emphasis on the role of the Spitfire was that it was a remarkably beautiful aircraft with an arresting name. Another was that, owing to its superior speed, the Luftwaffe respected it more than the Hurricane. They rather preferred to be shot down by Spitfires. Group Captain Peter Townsend in *Duel of Eagles*, delightfully calls this 'Spitfire snobbery'. He points out that in any case the Luftwaffe

> often mistook Hurricanes for Spitfires … Both Kesselring and Osterkamp (his fighter commander) fell into the trap. Uncle Theo 'saw' Spitfires on the ground in the Battle of France; Kesselring said, 'Only the Spitfires worried us'. Both were wrong. There were no Spitfires in France and in the Battle of Britain they shot down, in the aggregate, fewer than the Hurricanes.

As an example of the Germans' bad aircraft recognition, it suffices to mention that on 12 August, they claimed to have

---

[10]This rather odd title derives from the fact that it is mainly an account of the decline of the British aircraft industry after the War.

downed forty-six Spitfires, twenty-three Hurricanes and one non-existent French Morane. Not only were more Hurricanes in action but they suffered heavier losses than the Spitfires (twelve to eight). Clearly, the Luftwaffe could not always distinguish between them.

The reality is that Fighter Command's main weapon was inevitably the Hurricane, if only because there were so many more available. The Command was divided into four groups: Park's No 11 Group in the south-east of England; 12 Group in the Midlands, whose leader was Air Vice-Marshal Trafford Leigh-Mallory, a brother of the Mallory of Everest fame; 13 Group under the command of Air Vice-Marshal Richard Saul protecting a vast area from northern England to northern Scotland; and (after mid-July) 10 Group in the south-west led by the South African Sir Christopher Quintin Brand.

These had some 700 aircraft on strength of which well over half were Hurricanes. There were two squadrons of Defiants; seven of Blenheim fighters (including one operating solely at night); nineteen of Spitfires; twenty-seven of Hurricanes. In the vital No 11 Group three squadrons were equipped with Blenheims, six with Spitfires, fifteen with Hurricanes.

The Hurricane squadrons available at the beginning of July were Nos 1, 3, 17, 32, 43, 46, 56, 73, 79, 85, 87, 111, 145, 151, 213, 229, 238, 242, 249, 253, 257, 501, 504, 601, 605, 607, and 615. Another, 245, was stationed in Northern Ireland, where it provided a useful source of fresh pilots for the squadrons more actively involved in the fighting. In addition, 263, which was replacing the Gladiators lost in Norway with twin-engine Westland Whirlwinds, had a Hurricane flight for airfield defence. Not all these squadrons were fully operational, still less fully equipped, but the ease with which Hurricanes could be produced had enabled Hawkers, in contrast to other manufacturers, to pass every monthly target since the outbreak of war. In July, thanks mainly to the dynamic impulse of Lord Beaverbrook, the aircraft industry produced 488 fighters. Of these, well over half were Hurricanes. This not only made good losses in the Hurricane squadrons but allowed a new one, 232, to be formed on 16 July.

It is also important to note that the Hurricane's combat experience had brought about the solution of any teething troubles before the battle started. Messrs Wood and Dempster describe it as 'a fine fighting aircraft, an excellent gun platform and it was magnificently manoeuvrable up to 20,000 feet. It was extremely strong and could take an extraordinary amount of punishment.'

Francis Mason, in his most splendid work *Battle over Britain*, concurs, calling the Hurricane 'a robust, manoeuvrable aircraft capable of sustaining fearsome combat damage before write-off'; adding significantly: 'and unlike the Spitfire, it was a wholly operational, go-anywhere, do-anything fighter by July 1940.'

The Hurricane Mark I, which bore the brunt of the battle, had a top speed of 320-328 mph, varying according to the degree of skill in the finishing shops in the different factories.[11] It climbed to 20,000 feet in 8.2 minutes; service ceiling was 34,200 feet; range 505 miles. This performance admittedly was inferior to that of the Spitfire or the Messerschmitt Bf 109E. One of the latter had been forced down undamaged near Amiens on 2 May. It was flown by pilots of No 1 against their Hurricanes in mock combats sufficiently realistic to unnerve most of the spectators and, since the pilots might well owe their lives to the knowledge gained, their assessment may be considered the ultimate impartial judgement on the comparative values of the two fighters.

It was agreed that the 109 was faster at most altitudes. It could outclimb and out-dive the Hurricane. It also had a superior field

---

[11]Dowding is widely quoted as having checked the average maximum speed of six Hurricanes at 305 mph, but it is clear that the aircraft in question were battle-scarred veterans, which had undergone considerable repairs.

of view – which was not true of those taking part in the Battle of Britain since, by then, armour plate had been fitted behind the pilot's head, which restricted his visibility.

The Hurricane, by contrast, was considered 'infinitely more manoeuvrable at all altitudes'. It could out-turn the 109 with ease. Curiously, no mention was made of armament but the 109, with two 20 mm cannon and two 7•9 mm machine-guns, carried a bigger punch than the Hurricane's eight machine-guns. However, the Hurricane was such a steady gun–platform that it could make the best possible use of its weapons. Also, as will be now have become obvious, it was better able to endure combat damage.

However, the 109 did have one priceless advantage – as the aggressor it could usually attack with a height advantage. Since the Hurricane lacked the speed or climb of the Spitfire, it was vulnerable longer. Accordingly, it was the aim of the defenders in the Battle of Britain that Spitfires should attack the fighters, while Hurricanes engaged the lower-flying bombers with their escorting Bf 110s, which usually operated below 18,000 feet. This worked better on paper than in practice when either the Spitfires fought the 109s while the bombers carried on unmolested, or the Hurricanes got among the bombers but were attacked by enemy fighters from above. However, it should be recorded that once Hurricanes had gained the necessary height, they achieved numerous successes against the 109s as they had done already in France, and would do again in Malta, the Balkans, North Africa and Russia.

It might also be noted that enemy fighters could not greatly harm Britain. The bombers could. In consequence, the RAF pilots were ordered to concentrate on them, avoiding the fighters if possible. Actions against 109s were basically defensive – to keep them from protecting their bombers. The main role of killing the latter, as already mentioned, was allocated to the Hawker fighter. The Hurricane that came back with blood all over the engine cowling and bits of splintered bone in its radiator summed up only too grimly the success achieved by the type.

Exactly how great was the Hurricane's success will never be known. In the first place it is impossible to reach an exact account of German losses. As is well known, the figures issued at the time were greatly exaggerated. Although it has been suggested that this was done deliberately to raise morale, it should be stated that on many of the quieter days of the battle, RAF claims were less than the true figures. Inevitably on the most active days, the sheer speed of air-fighting resulted in numerous duplicated claims; as many as four pilots might well inflict damage on the same enemy, each afterwards swearing in perfect good faith that he had made a kill. Indeed, if 'shared' successes were credited to each pilot involved as in the French Air Force, the scores claimed by the RAF aces would probably prove not inaccurate – which is more than can be said for many of those credited to the leading pilots of the Luftwaffe during or after the Battle.[12]

The 'official' figures of German losses were taken after the war from the Luftwaffe Quartermaster General's returns, which were used as a basis for getting replacements. Since the last thing any commander would do in such a case is underestimate his losses, these figures were accepted without question for about twenty-five years. Then, however, Francis Mason, to whose monumental *Battle over Britain* a vast debt is owed by all interested in this problem, completely shattered existing preconceptions by pointing out two basic flaws in the above argument.

---

[12] Air Vice-Marshal Peter Wykeham in *Fighter Command*, dealing in particular with the case of Oberst Galland, officially credited with 103 victories, remarks: 'It is no disparagement of a brave opponent to insist that this and other impressive German personal scores were largely created in the propaganda departments.'

The first flaw is that the official British researchers into German losses misinterpreted the Quartermaster General's returns. This they did in two ways. The Germans divided their losses between those on 'war flights', those on 'war support flights' and those on non-combat flights. The assessors quite rightly omitted the last class but quite wrongly also omitted the second class, which, as the name implies, were essential collaterals to the enemy's efforts. As a result the large number of aircraft shot down for instance, on air-sea rescue support have not been included in the official figures.

In addition, the Luftwaffe stated its losses on a percentage basis. An aeroplane that went down in Britain or the sea was a 100 per cent loss but one that crashed in France, becoming a total write-off, was assessed at a figure varying from 60 per cent to 100 per cent depending on what parts could be salvaged for further use. The British official figures, while including written-off machines that came down in enemy territory, took the figure of a destroyed aircraft at 80 per cent. This did not reflect German practice. Thus, those machines with 60 per cent to 80 per cent damage should properly be added to the loss returns. Any figures quoted hereafter will include them.

Even more significant, Mr Mason discovered that there was no mention in the loss returns of several enemy aircraft that had been destroyed beyond doubt – the usual case being where prisoners had been taken. The obvious solution is that there were other channels than the Quartermaster General's returns by which losses could be made good. This was a theory confirmed by the fact that where units that suffered such unreported losses can be identified, they often incurred them in more than one instance.[13]

---

[13]It has also been suggested that these were units in the process of re-equipping, and the aircraft that were lost but did not appear in the Quartermaster General's returns were older types for which replacements would not be required. In any case the fact remains that the returns are incomplete.

Taking examples more or less at random, on 11 July, Flying Officer Davis of 601 Squadron shot down a Bf 110, one crew member baling out to be taken prisoner. On 19 July, Pilot Officer Simpson of 111 destroyed one of the Bf 109s that had been slaughtering Defiants, the pilot of which was rescued wounded from the sea. Next day, 56 Squadron routed some Ju 88s, bringing down one, from which four prisoners were taken. Not one of these aircraft is mentioned in the Quartermaster General's returns.

It may also be pointed out that these losses came about in isolated actions that could be pinpointed. What happened on more violent days can best be guessed from Mr Mason's detailed comments. On 20 August, the Hurricanes of 615 broke up a raid destroying two Do 17s and one Bf 109 confirmed in the German records, but 'it is clear from prisoners that two other Bf 109s and a Dormer Do 17 also fell at this time'. Of 1 September, Mr Mason remarks that: 'It must be assumed that the loss returns of I and II Fliegerkorps were submitted separately (if they were in fact submitted at all) and have not been traced—' because although no Domier Do 17s are reported as missing, 'at least seven ... are known to have fallen over southern England on this date'. Of 11 September, he notes: 'At least three other Bf 109s of JG3 (not mentioned in the German loss returns) are known to have been destroyed. Wreckage was examined and prisoners taken...'

Even if the true loss-total could be discovered, it would not be possible to tell what proportion fell to the guns of the Hurricanes. Repeatedly in German reports appear the words 'shot down by British fighters' – type not specified. However, bearing in mind 'Spitfire snobbery', which led enemy pilots to claim they had encountered Spitfires when this is known to be untrue, as well as the simple fact that there were more Hurricanes about, it seems that the great majority of such casualties were, in fact, inflicted by the Hawker fighters. It is worthy of notice that very often they belonged to the same Staffel as others undeniably shot down by Hurricanes.

The final words before embarking on the details of the fighting must belong to the Hurricane pilots. Roland Beamont, who as a test pilot flew practically every aircraft known, was a young pilot officer with 87 Squadron during the battle. He summed up the virtues of his aircraft thus:

> The Hurricane had an altogether exceptional combination of manoeuvrability, rugged strength, stability, ease of control and gun aiming, and viceless landing characteristics, which went far towards offsetting the fact that its climb, level and altitude performance were slightly lower than the Spitfire and 109. Hurricane pilots knew that once in combat they could outmanoeuvre any enemy and that with their eight Browning machine-guns and the aiming accuracy of their aircraft, hitting the target was no problem.
>
> Furthermore they knew that if in a tight spot the Hurricane could be pushed into a violent full power corkscrew dive which no 109 could follow; and that it could do this without serious consequences which was more than could be said either for the 109 or the Spitfire, both of which developed unhealthy reputations for structural failure in high-speed dives.
>
> In late September with the battle at its height, there was a rumour to the effect that 87 were to be re-equipped with Spitfires, and there was such concern over this that the pilots asked the CO to intervene and prevent such a happening.
>
> While this was going on Michael Lister Robinson, an old friend and CO of 609 at Warmwell and an acknowledged virtuoso of the Spitfire, came to see us in a new Spitfire 2. It was a beautiful aeroplane and much admired, and when

he left, [Fit Lt] Ian Gleed and I took off in our Hurricanes to escort him. Michael decided that the opportunity to have a go at us was too good to miss, but within one and a half turns we were firmly on his tail and after five or six exhilarating minutes at all levels and altitudes around Charmy Down, Michael gave it up and straightened out for Warmwell with a Hurricane firmly stationed on each side at two or three feet from his tailplane for a short time until with his superior speed he began to draw away.

Even more convincing perhaps, is the verdict of ex-Spitfire pilots. When Squadron Leader Robert Stanford Tuck took over 257 Squadron, equipped with Hurricanes,

My first reaction wasn't good. But after the first few minutes I began to realize the Hurri had virtues of her own. She was solid, obviously able to stand up to an awful lot of punishment... steady as a rock – a wonderful gun platform ... The pilot's visibility was considerably better than in the Spit... The undercart was wider, and, I think, stronger than the Spit's. This made landing a lot less tricky, particularly on rough ground.

Another ex-Spitfire pilot who commanded a Hurricane squadron, 242, in the battle was Squadron Leader Douglas Bader. He reported:

Like all pilots who flew and fought in the Hurricane I, I grew to love it. It was strong, highly manoeuvrable, could turn inside the Spitfire and of course the 109. Best of all, it was a marvellous gun platform. The sloping nose gave you a splendid forward view... The aeroplane remained

rock steady when you fired... The Spitfire was less
steady when the guns were firing.

In the last resort, the final answer must be that given by
85's CO, Squadron Leader Peter Townsend: 'We thought they
(the Hurricanes) were great,' he remarked disarmingly, 'and
we would prove it by shooting down around 1,000 Luftwaffe
aircraft in the battle.'

—

The Hurricanes had begun shooting down Luftwaffe aircraft in
large numbers on 10 July, the day the battle began officially – a
somewhat arbitrary date but justified by the first really massive
dogfight involving more than 100 aircraft. This took place over a
convoy in the Straits of Dover, which was attacked by Dornier
Do 17s of KG2 escorted by Bf 110s of ZG26 and Bf 109s of
JG3. That this force sank only one small ship was due to some
fine interceptions, particularly by 111 Squadron, which attacked
the Dorniers head-on, scattering them in all directions, though
at the cost of Flying Officer Higgs, who was killed when his
Hurricane crashed into a bomber (the crew of which also died).
A second Do 17 was brought down in more orthodox fashion,
the crew baling out.

When the Hurricanes attacked, the 110s formed a defensive
circle to protect themselves rather than their charges. The 109s
showed more spirit. Flying Officer Ferris downed one but was
then assailed by three more, which shot away an aileron control.
Despite this he was able to evade them, returning safely to
Croydon where he 'picked up a fresh Hurricane'.

In all, eleven enemy aircraft were lost on this date. Of these,
one fell by unknown means, one was shot down by anti-aircraft
guns, two by Spitfires, seven by Hurricanes. The distribution of
successes should be noted in these early combats, for they give
some indication of the probable state of affairs on later days

when the fighting was too confused for such detailed accounts to be possible. The only Hurricane lost was that of Higgs.

The ratio of Hurricane successes was the same on 11 July, which cost the Luftwaffe seventeen machines. The fate of one is not known, two were claimed by Blenheims, no fewer than fourteen by Hurricanes, including a Do 17 that fell to 242's legendary CO Squadron Leader Douglas Bader. Four Hurricanes were lost but only one pilot – even he baled out but was drowned. Two Spitfires also fell, both pilots being killed.

As an example of the pressure on the Hurricane pilots even this early in the battle, the exploits of 601 Squadron should be mentioned. Its pilots first saw action at about 1015, when Flying Officer Rhodes-Moorhouse and Pilot Officer Bland shot down a reconnaissance Do 17. About an hour later, they met a raid on Portland, bringing down two Ju 87s and two Bf 110s. Finally, at about 1730 they routed an attack on Portsmouth, destroying a Bf 110 and three Heinkel He 111s. Two of these were shot down by Flight Sergeant Pond, who attacked a Heinkel so savagely that it blew up, dragging its neighbour down with it. However, it was in this fight that 601 suffered its only loss when 'friendly' AA fire brought down Sergeant Woolley's Hurricane – happily, though burned and wounded, he escaped.

Throughout the first weeks of the battle, the Luftwaffe mainly attacked convoys, as much in the hope of inflicting serious shipping losses – though in this respect their mine-laying aircraft achieved similar results at far less cost – as of bringing Fighter Command to action in disadvantageous conditions. On 20 July, another fifteen German aircraft fell, one to causes unknown, one to AA guns, one to Blenheims, three to Spitfires, nine to Hurricanes. The RAF lost four Hurricanes (with all their pilots), two Spitfires and one Blenheim.

The day's real importance lay in a clash over a convoy with the interesting code name of 'Bosom'. This was raided by Ju 87s from StG 1 escorted by 110s and 109s, which were in turn engaged by the Hurricanes of 32 and 615 Squadrons. The resulting dogfight lasting almost half an hour, saw the beginning of the end of the Stuka legend.

The pilots of 32 attacked out of the sun, bursting clean through JG51's 109s, one of which they shot down, then falling on the Junkers, which they routed: two were brought down, two others force-landed in France and two more were damaged. One Hurricane was lost as was its pilot, who baled out but was drowned. In the meantime, 615 took on the 109s of JG27, shooting down three without loss. The action gave a clear indication that, provided it had equality of altitude, the Hurricane could more than hold its own with the 109 – a factor that did much for morale. As for the 110s, it is sufficient to say that they preferred to circle rather than take part in the encounter.

By 6 August, the Luftwaffe was ready for its major offensive. That day, Göring summoned his Luftflotten chiefs to his sumptuous home, Karinhall, to hear the operations orders. The assaults would commence on 10 August codenamed 'Adlertag' or 'Eagle Day' by the theatrical Reichsmarschall. Four days, he considered, would suffice to crush the RAF like the Polish air force before it. The end of that period would see all fighter airfields south of a line between Chelmsford and Gloucester wiped out or untenable. A further four weeks should see the complete elimination of all British defences against invasion; just right for the launching of Sealion in mid-September.

As a kind of dress-rehearsal, on 8 August, major attacks of unprecedented intensity were flung against convoys in the Channel. These not only marked the culmination of the anti-shipping operations but were on such a scale that they are generally considered to have opened the main battle. This day was also the last on which the story of the efforts of individual squadrons can be disentangled. For all these reasons therefore, it deserves to be recounted in some detail.

8 August saw twenty-seven Luftwaffe machines destroyed or damaged beyond repair. The fate of a reconnaissance Heinkel He 111 is unknown. Spitfires of 41, 64, 65, 152 and 609 saw

action, downing four of the enemy at the cost of three of their own number (all the pilots being killed). A Blenheim of 600 was also lost.

The Hurricanes, as usual, bore the brunt. First into combat was 145 Squadron, which at about 0900 encountered a force of Ju 87s, escorted by Bf 109s, attacking a Channel convoy. Squadron Leader John Peel had the honour of firing the first shots in the main part of the battle. His squadron destroyed three Messerschmitts and two Stukas, damaged a third Stuka and broke up the raid, but lost two pilots, killed. It was the start of a desperate day for Peel's men.

At about 1230, the convoy attacks were resumed by 57 Stukas. They sank four merchantmen, while their escort of 110s and 109s strove to fend off the intercepting Hurricanes. Those of 257 Squadron were the first to make contact. Attacked by 109s, they shot down one without loss but were not able to interfere with the dive-bombers. 145, with its previous experience, was able to get among the Stukas, shooting down two, damaging two more and this time suffering no casualties. 238 was less successful. It attacked both 110s and 109s, shooting down two of the latter and damaging a third as well as two 110s, but two RAF pilots were killed. On hearing they were missing, Squadron Leader Fenton took off again from Middle Wallop to search for them. At about 1350, he met a Heinkel He 59 seaplane, which he promptly brought down, but was hit in the engine by return fire so badly that he had to ditch. He was picked up safely by a trawler, one of only two Hurricane pilots to be shot down but survive on 8 August.

Finally, at 1630, eighty-two more Ju 87s, again heavily escorted, struck at the same convoy, this time scoring no fatal hits, thanks mainly to a pair of Hurricane squadrons. 43 Squadron engaged the 109s, destroying three of these but losing three Hurricanes. Two pilots were killed but Pilot Officer Upton escaped unhurt.

While 43 occupied the escorts' attention, Squadron Leader Peel led 145 out of the sun to gain complete surprise. Three more of 145's pilots died in the fighting but German records confirm their astonishing achievements: five Stukas, two 110s, one 109 all destroyed. During the encounter, Flight Lieutenant Dutton suffered engine failure. While gliding down, he came across a Stuka, which he promptly set on fire with a quick burst. His engine then re-started so he engaged a second Stuka intent on attacking a ship. This one Dutton shot straight into the sea. His engine then packed up for good, so he glided back to base where he landed safely.

The day's clashes had seen the end of eleven Hurricanes, nine of whose pilots had died. They had destroyed ten Bf 109s, nine Ju 87s, two Bf 110s and one He 59 – twenty-two in all. This was not the way for the Luftwaffe to achieve air supremacy. Fortunately for their peace of mind, the German commanders, believing their pilots' inflated claims, thought they had had the better of the fighting. Preparations for Eagle Day went ahead. Although bad weather forced a postponement to the 13th, important preliminaries took place on the two previous days.

On 11 August came attacks on Portland and Dover, in which large numbers of enemy fighters stayed high above their bombers to attack RAF machines, climbing to engage, from above. 87 Squadron, at the cost of one pilot dead and another in hospital, shot down two Junkers Ju 88s and four Bf 109s; but it had managed to gain adequate height. Several other Hurricane squadrons were less fortunate. 213 lost both its Dunkirk champions, Flight Lieutenant Wight and Sergeant Butterfield, killed. 601, attacking Ju 88s, of which it shot down three, was engaged from above losing four aircraft, all the pilots being killed. 111 and 238, attacked by Messerschmitts while climbing, each lost five Hurricanes and four pilots. Each was able to account for only one of the enemy in return.

Saddest perhaps was the fate of 145 Squadron, which now paid the price for the absence of the five experienced men who had died on its day of triumph on the 8th. It shot down a Bf

109 and a Bf 110, but lost four of its own machines. Two pilots perished. Three more died the next day. Two days later, the remnants were withdrawn to the north. The word 'attrition' began to be mentioned by Fighter Command's anxious leaders.

Strikes on the 12th concentrated on radar stations. That at Ventnor was put out of action temporarily while four others were damaged. However, so flexible was the defence that this was not apparent to the enemy, with the result that Göring, believing that such raids were unlikely to succeed without very heavy losses, largely abandoned them – though on the 16th the Ventnor station, which had just been restored, was again put out of action for a further seven days, while two days later that at Poling was knocked out for over a week. Göring's decision was a fatal one for had such attacks continued they might have blinded Fighter Command. On the other hand, in view of the emphasis placed on the radar cover, it is rather ironic that four of the Command's five finest achievements took place when that cover was least effective.

The first of the 'great days' was the vaunted Adlertag, thanks in large part to German inefficiency. Unexpected bad weather caused the cancellation of an attack on the Thames Estuary but although the order reached the escorting Bf 110s, so bad were enemy ground-to-air communications that it was not received by the bombers, seventy-four Dornier Do 17s of KG2. Nor did highly dangerous manoeuvres by their escort across the bombers' noses bring enlightenment to their leader, the redoubtable Oberst Fink, who continued on his mission unescorted.

As KG2 crossed the coast it was engaged by 74 Squadron, led by the famous South African Squadron Leader 'Sailor' Malan. However, Fink's crews were worthy opponents – they pushed on with only minor damage while their gunners fired on the Spitfires so resolutely that they forced the pilot of one to take to his parachute.

Over the Isle of Sheppey, the raid divided. Half made for the aerodrome at Eastchurch, which it bombed to some effect, though since this was a Coastal Command station its complete obliteration would in no way have harmed Dowding's defences. The other section made for the naval base at Sheerness, but before reaching it met the Hurricanes of 151 Squadron. Only one Dornier was brought down but the others split up, dropping their bombs well short of the target. At this point, Squadron Leader John Thompson led 111 into its famous head-on attack, which completed the rout, shooting down four more Dorniers without loss.

Away to the west, the Luftwaffe reversed its error. Twenty-eight Bf 110s arrived over Portland minus the bombers they were supposed to be guarding. They stayed only long enough to lose one of their number to the guns of a Hurricane of 238 Squadron, flown by Flying Officer Hughes.

By the middle of the afternoon, some order had been restored. Several heavy strikes were launched causing considerable damage to Southampton docks, while even more was done at the airfield near Maidstone – though this also was not a Fighter Command base. Everywhere the raiders met determined opposition. During the day, the Luftwaffe lost forty-five aircraft to fighters, AA guns or accidents on take-off or landing. The Hurricanes' share of the total was at least twenty-five.

Of the thirteen RAF fighters brought down, all except one were Hurricanes but only three pilots died. Four more Hurricanes were badly damaged but returned safely. 56 Squadron lost four aircraft but all the pilots escaped with their lives. It destroyed four 110s and damaged two more so badly that they crash-landed at their bases. 601 again lived up to its badge of a flying sword by shooting down four 110s and two Junkers Ju 88s without loss.

The Hurricanes did even better on 15 August, the day on which the Luftwaffe made its greatest efforts, flying more than 2,000 sorties. The first raids came in against 11 Group airfields in the south-east. Lympne was heavily hit but a strike by Stukas on Hawkinge was disrupted by 501 Squadron, although bombs that fell short severed power cables putting three radar stations out of action. 501 shot down two Stukas. Two Hurricanes were lost but both pilots baled out – an important point this, for airmen were less easy to replace than aeroplanes and they would thus be able to return to action quickly, whereas even if the Junkers' crewmen had survived, which none did, they would have become prisoners of war.

While the enemy retired in the south-east, Air Vice-Marshal Saul, commanding 13 Group in the north-east, had picked up his first major radar plot of the battle as Heinkel He 111s escorted by Bf 110s from Luftflotte 5 came in over the North Sea. When the Spitfires of 72, led by Flight Lieutenant Graham, sighted these, they were somewhat shocked to find that the thirty-plus reported had grown to about 100 in number. After a moment's hesitation 'wondering what to do', Graham dived out of the sun onto the hostile formation with immense success, not so much in the number of 'kills', for 72 despite its claims in fact shot down only three of its foes, as in splitting the enemy up into sections.

So dispersed, the Germans proved easy targets for other squadrons, losing twelve more of their number, of which nine were destroyed by Hurricanes, including those from 605 Squadron, which flew eighty miles from the Firth of Forth to engage. Minimal damage was caused by these attackers.

As these combats were in progress, fifty Junkers Ju 88s from Aalborg in Denmark struck Luftflotte 5's next blow. They arrived on 12 Group's front but since radar again grossly underestimated their numbers, Air Vice-Marshal Leigh-Mallory did not put enough fighters up to deal with them adequately. Pressing on boldly, they inflicted heavy damage on the aerodrome at Driffield – yet unfortunately for their scheme to knock out Fighter Command, this happened to be a Bomber Command base. They were also so unlucky as to encounter that crack Hurricane Squadron No 73, which, without loss, shot down seven of the eight Ju 88s that did not return.

The fighting now again switched south as 11 Group, with 10 Group supporting its flank, grappled with major strikes from Luftflotten 2 and 3 against Kent, Sussex, Hampshire and Dorset. Again, 601 did exceptionally well, bringing down five Ju 88s at the cost of two Hurricanes, whose pilots baled out.

One final raid came in at about 1830, when the brilliant Hauptmann Rubensdörffer led fifteen Bf 110s and eight Bf 109s of Erprobungsgruppe 210 (which had made a number of highly successful strikes over the past few days) against Croydon aerodrome. Squadron Leader Thompson was just able to get 111 Squadron away in time as the bombs crashed down with murderous efficiency. As the Germans pulled clear, Squadron Leader John Worrall led 32 Squadron into a flank attack. Moments later, 111 charged head-on. Erprobungsgruppe 210 split into small bunches, trying to escape into cloud-cover, but a 109 and a 110 fell to 32, while no fewer than five 110s were shot down by 111. Near Rotherfield, a falling 110 swerved aside from a cottage to crash on the far side, the blazing remains marking the funeral pyre of the valiant Rubensdörffer and his gunner, Feldwebel (Sergeant-Major) Ehekercher.

The death of their finest bomber pilot must have completed the misery of the German airmen as they contemplated the day's results. Their biggest efforts had been rewarded by their heaviest losses; seventy-five machines brought down by fighters, flak or accident. Of the sixty-seven that fell to fighters, it is impossible to disentangle the proportions scored by Hurricanes, Spitfires or Blenheims. However, it happens that a number of enemy formations that reported losses to unnamed victors are known to have been attacked only by Hurricanes, which for reasons already given almost certainly made up the great majority of such 'unknown fighters'. It can be deduced from this that Hurricanes shot down at least thirty-eight enemy aircraft, possibly more than fifty.

The RAF lost twenty-eight fighters in combat. Of these, seventeen were Hurricanes, ten of whose pilots survived; – the official loss figure of thirty-four includes badly damaged machines that were repaired later. Most of these were also Hurricanes, which again showed great resilience. Pilot Officer Truran of 615 had his Hurricane hit in the rear fuselage by cannon shells, which caused a small fire, while the wings were riddled by machine-gun bullets. Although wounded, he was able to land safely at Kenley. Squadron Leader Dewar of 213, his cockpit filled by smoke, managed to fly home by raising himself enough to put his head clear of the fumes. He also landed safely which, he remarked, 'was very fortunate as the aircraft is not seriously hit except in the engine and wings'. Pilot Officer Law of 605, badly damaged over the North Sea, still got back over the coast, his Hurricane crash-landing near a railway station without injury to the pilot. Pilot Officer Wlasnowalski, a Polish member of 32, with his Hurricane's wings damaged by machine-gun bullets, shot down a Bf 109, then force-landed unhurt in an Essex cornfield.

Mr Mason – this time in *The Hawker Hurricane* – calls 15 August 'the day that justified the Hurricane. It was then that almost every Hurricane squadron – built up and nurtured over the past three years – went into action to defend this country against the heaviest attack that Germany could mount'. To this it need only be added that rarely can any aircraft have been justified so abundantly or so gloriously.

The Luftwaffe had its own name for 15 August: 'Black Thursday'.

–

Amid the gloom, Kesselring and Sperrle were consoled by one hope: that the scale of combat on the 15th had placed such a strain on the smaller Fighter Command that another day of maximum pressure would find it too exhausted to intervene. The 16th therefore saw further massive raids which, aided by cloud-cover, did break through in several instances, destroying on the ground more than sixty aircraft – a misleading figure in that forty-six of them were Oxford trainers caught in their hangers at Brize Norton. Seven Hurricanes were damaged beyond repair at Tangmere. The attackers also shot down twenty British fighters. Eleven of these were Hurricanes, seven of whose pilots survived. One who did not was Pilot Officer William Fiske of 601 Squadron, the first American volunteer to die in the defence of Britain.

Yet everywhere the defenders resisted tenaciously, costing the Luftwaffe forty-eight aircraft. Of the forty-three claimed by fighters, at least twenty, possibly twenty-five, fell to Hurricanes. 16 August indeed was another of the Hurricane's greatest days.

Kesselring's main contribution was a heavy raid by Dornier Do 17s, escorted by Bf 109s, which crossed the Kent coast about midday only to meet 111 Squadron, once again perfectly positioned for their head-on attack. The pilot of the leading bomber, Oberleutnant (Lieutenant) Lamberty, was immensely impressed by the Hurricanes' performance, which he thought put that of the escorting 109s in the shade, as it probably did in these particular dog-fight conditions where the Messerschmitts were tied down guarding the bombers. His judgement as to the capabilities of the defenders was confirmed two days later when his Dornier was shot down, leaving him a prisoner of war.

However, it was on the 16th that 111 Squadron lost one of its finest pilots when Flight Lieutenant Ferris collided with a Dornier both aircraft disintegrating with 'very little left, except a few pieces floating down'. A second Hurricane went down in flames but the pilot baled out. One 109 and two more bombers fell, after which the formation retired across the Channel. Determined to score, Sergeant Wallace pursued one group to the French coast where he was attacked by six 109s. In turn, he roared low over the sea taking violent evasive action. Five of his tormentors turned back but the remaining one persisted. But whatever may have been said later, no Hurricane pilot feared the 109 in an even duel. The sergeant attacked head-on, shot his foe into the sea, then returned to base where he discovered his machine was practically untouched.

Further west, an action of great future importance took place as Luftflotte 3 made its main effort. A formation of Junkers Ju 87s of StG2, attacking Tangmere, was opposed ferociously by the Hurricanes of 43 Squadron to whom this airfield belonged. Hopelessly outclassed, the Stukas fought with the courage of despair, to their great credit downing two of their attackers, though both pilots escaped. 43 shot down seven of them, damaged three more, then landed skilfully between the bomb-craters, finally completing the discomfiture of the surviving enemy airmen by making them help to repair the damage they had caused.

Yet the exploit that, in a sense, symbolised the Hurricane's fight against the Luftwaffe was strictly an individual one. On 16 August, Flight Lieutenant James Brindley Nicolson, a 23-year-old Yorkshireman from 249 Squadron, who had never previously fired at an enemy aircraft, was about to attack an enemy formation when he was 'bounced' by Bf 109s. Four cannon shells struck home. As he recounted later:

> The first shell tore through the hood over my cockpit and sent splinters into my left eye. One splinter, I discovered later, nearly severed my eyelid. I couldn't see through that eye for blood. The second cannon shell struck my spare petrol tank and set it on fire. The third shell crashed into the cockpit and tore off my right trouser leg. The fourth shell struck the back of my left shoe and made quite a mess of my left foot.

With any other aeroplane, that would probably have been the end of the story. Hurricane P3576 continued to fly under control. Nicolson instinctively dived to the right out of the line of fire. He then prepared to bale out but, as he did so, a 110 loomed into his sights. Furious at having been mauled in this manner, he decided to remain in the cockpit.

Diving after the Messerschmitt, Nicolson pumped bullets into it.

By this time, fire had engulfed the floor of the cockpit and the lower part of the control panel, from which melting paint was dripping. Nicolson drew his feet up under the seat while his right thumb blistered on the firing button and the skin of his left hand peeled off in the flames. The 110 dodged desperately but he pursued it, firing continuously until it plunged down to crash, according to eye-witnesses on the ground, straight into the sea.

And still the Hurricane flew. Nicolson struggled to free himself from his cockpit. This was no easy matter with his injured hands but he finally succeeded. His parachute opened but as he floated down a 109 twice flew close to him. Fearing it would attack him, he feigned death, lolling limply in his harness. He never did know if the enemy fired – if it did, it missed him.

In this position, Nicolson was able to have a look at his injuries, which were not pleasant.

> I could see the bones of my left hand showing through the knuckles. Then for the first time I discovered I'd been wounded in the foot. Blood was oozing out of the lace holes in my left boot. My right hand was pretty badly burned too.

As if he had not suffered enough, just before he landed he was shot in the buttocks by a trigger-happy Local Defence Volunteer. He was rushed to the Royal Southampton Hospital, where he remained on the danger list for forty-eight hours before making a complete recovery. It was found that the glass of his watch had melted in the heat, the strap had been burned to a thread; but it still worked. It was working three months later when he received the only Victoria Cross ever awarded to a pilot of Fighter Command.

Ironically, it was the Luftwaffe that was now exhausted – so much so that although the 17th was a fine day, only a few reconnaissance aircraft crossed the coast; but on the 18th more massed attacks were made, which on this occasion were directed with far greater foresight. Biggin Hill, Croydon and Kenley were all savaged, some fifteen fighters including eleven Hurricanes being destroyed on the ground. Kenley in particular was so hard hit that the Sector Operations Room had to be transferred to a butcher's shop in Caterham.

Inevitably, those Hurricane squadrons stationed at the smitten bases suffered heavily. 32 shot down two enemy aircraft but lost two Hurricanes, though the pilots escaped with wounds – later in the day three more were destroyed but again all the pilots were saved. 111 also claimed two enemy aircraft but at a cost of three of its own machines (plus another on the ground), one pilot being killed. 615, based at Kenley, brought down three Dornier Do 17s and a Heinkel He 111, but it was then attacked from above by 109s, losing four aircraft though only one pilot – it also lost six more machines on the airfield. 501 shot down two 110s but the day cost it seven of its Hurricanes, though mercifully only two of its pilots.

Indeed, 18 August marked the heaviest RAF casualties to date. For reasons unexplained, whereas on the 15th the official losses quite wrongly include several machines that were damaged but later repaired, on the 18th the official figure of twenty-seven apparently omits aircraft that returned home but were written off. It should be about thirty-five. Twenty-nine of these were Hurricanes but only nine of their pilots died.

Again, the ruggedness of the Hurricanes saved many lives. Hurricane P3649 carried Pilot Officer Marshall of 85 Squadron back to base with its starboard wingtip missing. It was later repaired to see more service, this time with 32 Squadron. Hurricane R4109, flown by Flight Lieutenant Carey of 43, had a hole blown in the port wing 'big enough for a man to crawl through', while one elevator and the rudder were shot off. The wounded pilot, 'having', as he remarked later, 'only about three-quarters of an aircraft to control', returned to his airfield where he came under fortunately inaccurate fire from AA gunners, who did not recognise the Hurricane's 'odd' silhouette. It crashed on landing, but Carey walked away without further injury.

Again also, the Hurricanes took their toll. Of the sixty-five enemies brought down by fighters (ground defences or accidents claimed five more) at least thirty-two, possibly thirty-seven, fell to Hurricanes. Once more, 601 did well, having two pilots killed but bringing down a Stuka and three 109s. 43 was lucky enough to catch another Stuka formation, destroying five. The only Hurricane damaged was Carey's, as already mentioned. However, pride of place went to 56, which utterly routed the 110s of ZG26, shooting down six without loss.

These combats had immense repercussions. Luftflotte 5 had already withdrawn from the daylight battles following the disasters of 15 August. Now the vulnerable Stukas were also pulled out of the front line. The almost equally ineffective Bf 110s were now to be given cover by 109s whenever possible – escorts for escorts, a ludicrous situation. Their new duties placed an intolerable burden on the single-engine fighters, for they were forced to be ever closer guardians of their bombers, so that the 'free chases', which had previously claimed by far the greatest number of RAF victims, had virtually to be abandoned.

Göring, it will be remembered, had considered four days ample to smash Fighter Command. The four days of 13, 15, 16 and 18 August showed, on the contrary, that the German airmen would never achieve a swift success. If they were to conquer at all it would be only after a long struggle; their attempt was not yet over, but it was quite clear that they had suffered a disastrous set-back.

It was therefore entirely appropriate that, two days later, Churchill rose in the House of Commons to pay his famous tribute to the RAF. 'All hearts', he added, 'go out to the fighter pilots.'

Chapter Four

The Crisis

There would soon be more fighter pilots: before August was over, four new Hurricane squadrons had seen combat. No 1 Squadron Royal Canadian Air Force became operational on the 17th, but, two days earlier, Squadron Leader McNab, flying with 111 for experience of Fighter Command's tactics, had destroyed a Dornier Do 17.

The Polish squadron No 303 also claimed a 'kill' before it became operational. Officially this was on 31 August but the previous day, when the Poles had been on a training exercise making dummy attacks on Blenheim bombers, they had sighted an enemy raid. The English CO, Squadron Leader Kellett, ordered them to escort the Blenheims to safety, but young Flying Officer Paszkiewicz could not bear to forego the chance of action. Diving upon the enemy, he opened fire. As a Do 17 burst into flames, 303 was operational. By the end of the battle, it was the most successful squadron in Fighter Command. Its greatest pilot, oddly enough, was a Czech, Sergeant Josef Frantisek, a lone wolf with a speciality of destroying 109s, whose seventeen victims in September made him the battle's top-scoring pilot.

The other Polish squadron, 302, had shot down its first enemy aircraft, a Junkers Ju 88, on 20 August. 310, a Czech squadron, first saw action on the 26th, destroying three enemy aircraft but losing three of its own machines, though all the pilots were unhurt. All the squadrons mentioned were formed with Hurricanes because, despite losses suffered, there were more than 600 of these available by the evening of 19 August.

This was perhaps as well for, on 24 August, after a pause dictated by the weather, the Luftwaffe resumed its massed assaults, with far greater effect. When Luftflotte 5 was withdrawn from daylight attacks it sent its 109s to strengthen Luftflotte 2. After 26 August, Luftflotte 3 also concentrated on night raids, transferring most of its 109s to Kesselring. Thus, the main attacks came almost continuously in the area of Park's No 11 Group, against which Luftflotte 2 could now send bombers with huge fighter escorts.

However, the main reason for the seriousness of the next fortnight's raids was that at last, instead of wasting time on training aircraft or Coastal Command bases, the Luftwaffe was striking at the heart of Fighter Command. Although German Intelligence was still abysmal, German pilots by using their eyes and brains had located — though they did not know the full significance of – the most vital RAF bases – the sector stations.

These were the stations with operational command of the fighters in the air, each usually controlling three squadrons. In 11 Group there were seven of them: Biggin Hill and Kenley just south of London, Hornchurch, North Weald and Northolt just north of London, Debden in Essex, Tangmere in Hampshire. If the sector stations were knocked out, 11 Group would be paralysed.

On 24 August the Luftwaffe did knock out a forward airfield – Manston. Here the luckless Defiants of 264 Squadron, which had done so well at Dunkirk, were caught on the ground. They managed to get airborne only to be attacked from above, losing three of their number. The remainder retired to the sector station at Hornchurch, where another attack again all but took them by surprise. Why the slow, clumsy Defiants had been brought into the battle area is inexplicable. In five days, the squadron lost twelve aircraft, fourteen air crew. Thereafter, the Defiant was withdrawn from daylight operations.

Also on the 24th, the Luftwaffe heavily damaged the sector station at North Weald even though the Hurricanes of 111 and 151 prevented more than half the bombers reaching their target. Two days later, Debden was hit. On the 30th, Biggin Hill was reduced to wreckage to such an extent that temporary control of the sector had to be passed to Hornchurch. Further heavy damage was done next day. Hornchurch, Debden and Croydon were also struck, but 111 Squadron saved a sector station in No 12 Group by a head-on attack that broke up a mixed formation of Do 17s and Bf 110s heading for Duxford. During the first week of September, Biggin Hill and North Weald were hit badly – Hornchurch and Kenley (already mauled on 18 August) less severely. By 6 September, all but one of the seven stations were working at greatly diminished efficiency – particularly Biggin Hill.

As the bombers pounded the vital bases, the Hurricanes strove to break through the packed fighter protection to engage them. Sometimes they were able to disperse the bombers or inflict heavy losses. 87 Squadron shot down two Junkers Ju 88s and three Bf 109s on 25 August, though losing a pilot, killed. 43 brought down three Heinkel He 111s and a 109 next day, losing an aircraft, but not the airman, who was wounded but baled out.

Usually, however, the enemy fighter escort was too strong, which meant that the Hurricanes would be battling against 109s or 110s. In combat with the latter, whether acting as a fighter or in the increasingly common role of a fighter-bomber, the British fighters were almost always superior. On the 25th, No 17 shot down five 110s (plus a 109) losing two aircraft and one pilot. Flying Officer Count Czernin destroyed three of the 110s, all of which are confirmed in the German records. On 2 September, 249 destroyed four 110s (plus a Dornier Do 17). Three Hurricanes were lost but all the pilots survived. Next day it was the turn of 310 to bring down four 110s. This time only one Hurricane fell from which the pilot escaped. The day after, 43 brought down yet another four, without loss.

It was also on 4 September, that the Hurricanes of 601 caught Erprobungsgruppe 210 attacking the radar station at Poling. The Germans retired fighting fiercely. They lost only one 110 but this carried to his death Hauptmann von Boltenstern, Rubensdörffer's successor. It gives no pleasure to record the end of a brave, honourable opponent – unfortunately, it was essential for the Hurricanes to kill such men in order to help to save their country from the horrors of invasion.

Against the Bf 109s, of course, the Hurricanes had more problems but it is unarguable that frequently they proved greatly superior to these also. The most fantastic individual success was that of Squadron Leader Gleave of 253 who, on 30 August, engaged a mass formation of 109s from JG27 of which he shot down no fewer than four. Those who believe that RAF claims were inflated deliberately for propaganda purposes may like to know that officially Gleave was credited only with 'four probables'. Later research among enemy records, as well as examination of wreckage on the ground, shows conclusively that all four Messerschmitts did, in fact, crash.

The champion 109 specialist squadron of the period was No 85, one of whose best efforts had come two days earlier. 85 had timed its achievement well, for among the spectators was Winston Churchill who was paying a visit to the south-east coastal defences. The squadron log tells the story thus:

> Croydon, 28/8/40. Ten Hurricanes took off at 1602 hours to patrol Tenterden and were then ordered to intercept Raid 15. About 20 Me (sic) 109s were sighted at 1625 hours flying at 18,000 ft. in the Dungeness area. One Me (sic) 110 was seen at the same time in the same vicinity.
>
> The squadron approached from the sun, but were spotted at the last moment by E/A (Enemy Aircraft) who appeared not to be anxious to engage and broke formation in all directions. Six Me 109s were destroyed and 1 Me 110 damaged.

Sqn Ldr Townsend gave a two-three second burst and E/A rolled over and dived steeply, seen going down by Flt Lt Hamilton and Fg Off Gowers, both of whom saw black and white smoke coming out of E/A. Sqn Ldr Townsend estimated his position at about 12 miles NW of Lympne. This confirmed by Maidstone Observer Corps who received a report at 1646 that a Me 109 had crash-landed at R5167, the pilot being taken prisoner. Plt Off Allard attacked E/A at 200 yds. closing to 20 yards and it caught fire and dived vertically into the sea, two-three miles outside Folkestone harbour. Witnessed by Plt Off English. He then fired several short bursts at another Me 109 which was making for France. E/A dived and flew at 20 feet from sea but engine failed with black smoke coming out, position then was about 5 miles N of St Inglevert, which was confirmed by 11 Group signal 11G/174 of 29/8. Plt Off Hodgson chased E/A from 17,000 ft down across the Channel to 20 ft. above sea level. Fired several bursts and saw pieces falling off and only 1/3 of rudder left. E/A was going very slowly when last seen and emitting much black smoke.

Plt Off Hodgson had to turn back when 5 miles NW of Cap Gris Nez owing to lack of ammunition but he was certain E/A was finished. This was confirmed by 11 Group signal 11G/174 of 29/8.

Fg Off Woods-Scawen attacked E/A from quarter following to astern and gave two long bursts. Black smoke and what appeared to be petrol from the wing tanks poured out of E/A and it dived down vertically. He followed it down for several thousand feet and left it when it was obviously out of control. It was believed to have crashed near Dungeness and this was confirmed by Maidstone Observer Corps who reported a Me 109 in the sea Dymchurch at 1640 hrs.

Sgt Walker–Smith attacked Me 109 and the port petrol tank was seen to explode and then (E/A) went into steep dive. At that moment Sgt Walker-Smith was fired at and had to take evasive action but immediately afterwards dived and saw a large explosion on the sea and black smoke. Pit Off Hodgson confirmed having seen a Me 109 dive into the sea at this point. Fg Off Gowers attacked Me 110 from astern and saw bullets entering but it went into a shallow dive towards French coast and got away. Fg Off Gowers used all his ammunition and definitely damaged E/A.

The Me 109s all had yellow wing tips.

Enemy Casualties: 6 Me 109s destroyed, 1 Me 110 damaged.

Our Casualties: Nil.

The squadron's own account deserves to be recalled in detail for three reasons. First: it makes it quite clear that far from indulging in reckless claims, 85 did its utmost to verify its successes. Second: it establishes beyond reasonable doubt that six enemy machines fell on this occasion; the fact that only four can be located in German records suggests that this episode provides another instance of one of the enemy units involved having other channels of making good losses, apart from the Quartermaster General's returns. Third: it shows that whatever the actual figures and whatever may have been stated to the contrary later, the Hurricanes were capable of dealing more than adequately with the finest fighters the Luftwaffe could put into the sky.

In fact, the Hurricane pilots' most dangerous foe at this time was not the 109 but their own fatigue. The problem can be illustrated by the experiences of 56 Squadron, which shot down three 109s on 24, 26 and 28 August – but on the first of these days suffered no casualties, on the next lost two aircraft and on the last lost four, though again it should be emphasised that all the pilots escaped with minor injuries. Exhaustion was slowing their reactions.

As the German raids continued, so the pressure on the defences mounted. On 31 August, 85 Squadron was almost caught on the ground by an attack on Croydon. As the Hurricanes took off, they were shaken by the blasts of bombs bursting about them; turning in the cockpit, Squadron Leader Townsend 'saw the rest of the squadron emerging from a vast eruption of smoke and debris'. All got airborne safely but fought at a considerable disadvantage, being attacked as they climbed. They claimed two 109s and a 110 but lost two of their own number. Pilot Officer Worrall baled out unhurt, but Townsend was taken to hospital where the nose-cap of a shell was removed from his foot.

This raid had taken place at about 1300, but it was only the start of 85's day. At about 1730, the squadron was again in the air destroying three 109s for the loss of one Hurricane, which force-landed without injury to the pilot. Two hours later, 85 again saw combat. This was a classic encounter: nine Hurricanes versus nine Messerschmitt Bf 109s from JG77. It therefore deserves another detailed report from the squadron diary:

> Nine Hurricanes took off Croydon 1917 hours to patrol Hawkinge and then were ordered to intercept Raid 18c. The first indication of position of E/A was given by AA fire from Dover and then nine Me (sic) 109 were seen flying at about 15,000 feet. The squadron circled out to sea as E/A passed on left, then wheeled in and caught them by surprise when individual combats ensued.
>
> Pit Off Allard opened fire on E/A from 150 yards astern and parts of the wing appeared to break

off. E/A dived down and crashed near Folkestone either on land or just in the sea.

Fg Off Woods-Scawen carried out beam attack causing E/A to dive steeply, then gave a further burst from astern and E/A went down on fire with wing tank burning – confirmed by Plt Off Lewis.

Plt Off Lewis fired a four-seconds burst at E/A from 150 yards on the beam and from slightly below. Black smoke billowed out and E/A dived steeply. Plt Off Lewis followed it down to 5,000 feet making sure it was done for and rejoined squadron. Position then above sea near Folkestone.

Fg Off Gowers fired two bursts of five seconds and seven seconds, caused a large piece to blow out of port wing of E/A. Petrol streamed out as E/A dived vertically and when Gowers left him at 4,000 feet he was still diving straight down and by then was in flames. Confirmed by Plt Off Lewis.

Nine Hurricanes landed Croydon 2005 hours to 2022 hours.

Enemy Casualties: 4 Me 109 destroyed.

Our Losses: Nil.

All four enemy fighters claimed are confirmed in the German records. The 'inferiority' of the Hurricane to the 109 is again worthy of note.

However, as the damage to the sector stations increased, control of the squadrons declined. There were far fewer successful combats of the type just described; far more instances of squadrons fighting under disadvantageous circumstances. This was the period when Fighter Command mourned the loss of some of its most gallant officers – such as Pilot Officer Anthony Woods–Scawen of 43 Squadron, whose eyesight was deteriorating badly, 'but,' he pleaded to his flight sergeant to whom he was apologising for bringing back a shot-up

Hurricane, 'don't breathe a word or they might whip me off Ops.' He had a lucky parachute: four times it had saved his life. On 2 September, he baled out a fifth time but before his parachute could open his Hurricane exploded immediately above him.

Some twenty-four hours earlier the luck ran out for his elder brother Patrick, leading 85 Squadron in Townsend's absence – it ran out for 85 as well. Pilot Officer Allard shot down a Domier Do 17 but then landed at Lympne with engine trouble. While his aircraft was being serviced, an attack on the base destroyed it – at least the pilot was unhurt. Four more Hurricanes perished in the air. Flying Officer Woods-Scawen's body was found nearly a week later, his parachute unopened. The parachute of Sergeant Booth, described by Townsend as 'a tall quiet youth' caught fire: he lingered for six-and-a-half months before dying. Sergeant Ellis was never seen again.[14] Flying Officer Gowers baled out with wounds in hands and feet and severe burns – he found some consolation, in his amusement over his removal for treatment to Caterham Mental Hospital.

There were plenty of other pilots in hospital: Squadron Leader Krasnodebski of 303, barely conscious, under morphia, flayed by fire that had burned off his trouser legs up to the knees; Squadron Leader Gleave of 253, so burned that the skin from arms and legs hung down like hideous white paper bags, so burned that he could only just see through his swollen eyelids, so burned that his wife could not recognise him.

The absence of such leaders placed an even greater burden on the knowledgeable survivors in the squadrons. Though new recruits arrived to replace the dead and injured, they had hardly any experience of modern fighters, none of combat technique.

---

[14]It was believed that Ellis had been lost in the sea but in fact his body had been found and buried as that of an 'unknown airman'. His identity was not discovered until 1992.

The loss rate among such replacements was tragically high. On the morning of 2 September, for instance, young Pilot Officer Rose-Price joined 501. That afternoon he was killed in action. The average age of the dead was twenty-one years.

As casualties mounted, mauled, exhausted squadrons had to be pulled out of 11 Group to recover in the north or west. 32 Squadron, which had started this phase badly by losing five aircraft (but no pilots) on 24 August, was the first to go, bidding farewell to its place of honour and danger at Biggin Hill three days later. By 6 September, 56, 85, 151 and 615 had also retired together with several similarly depleted Spitfire squadrons. Within a few days, 43, 111 and 601 would follow them. The squadrons that came to relieve them often suffered heavy losses in their first combats – for example, 73 reached Debden on 5 September, eager for action. By evening, one pilot was dead, another in hospital and four Hurricanes were destroyed.

As if 11 Group did not have enough problems, Park found at this time that he had serious grounds for complaint about the support he was receiving from his fellow commanders. Though Brand in 10 Group invariably provided reinforcements as needed, Park was forced to state on 27 August that requests to Leigh-Mallory's 12 Group had not resulted in the same co-operation; when 12 Group had been asked to patrol over Park's sector stations, they had not done so – with the consequence that those bases had been heavily bombed.

Though Park and Leigh-Mallory were not particularly friendly, the real point at issue was their different opinions of the best tactics to adopt. Leigh-Mallory was a highly aggressive officer whose praiseworthy aim was to destroy the largest possible number of enemy aircraft. He enjoyed the support of the Deputy Chief of Air Staff, Air Vice-Marshal Sholto Douglas, who believed that as long as the Germans were shot down in sufficient quantities it did not matter if they were not intercepted before they reached their targets. It seems that in his eagerness to bring the enemy to action, Leigh-Mallory was sending his pilots to hunt for trouble, scorning what he regarded as the 'defensive' protection of 11 Group's airfields.

At the best means of bringing down enemy machines in great numbers, Leigh-Mallory favoured the use of a wing of three or even five squadrons, led by Squadron Leader Bader, whose views were similar to Leigh-Mallory's own. When this 'Duxford Wing' – so-called after the 12 Group sector station – made contact, which was by no means invariable since such a large formation took some time to assemble, the usual exaggerations inherent in air combat increased to such an extent as to give a completely false picture of its effectiveness. In reality, there is nothing to show that the five squadrons in the wing destroyed any more of the enemy than they would have done flying singly.

As it was, the supporters of the 'big wing' ideas criticised Park for not adopting them also. Even had he been able to do so it is by no means certain that the results would have been acceptable, for the probability is that while such large formations intercepted some enemy raids, others would have broken through quite unscathed. But, in practice, there simply was not enough available time for Park to muster such a large body of fighters.

This was proved conclusively the following year when Leigh-Mallory, then in charge of 11 Group, in a 'war game' exercise attempted to meet some actual German attacks that had taken place in September 1940 by use of the 'big wing' tactics. These were so ineffective that the 'enemy' were judged to have bombed his three main airfields while his fighters were still on the ground. No more such 'war games' were played.

This dispute further weakened the British air defences just at the moment when they seemed ready to collapse. The damage to the sector stations by 6 September, had, according to Park, resulted in 'an almost complete disorganisation of the defence system (which) made the control of our fighter squadrons very difficult'. He again bitterly attacked 12 Group for 'persistently declining to give fighter cover to my sector aerodromes'.

Also during the fortnight to 6 September, 295 Hurricanes and Spitfires had been totally destroyed while 171 more had been put out of action temporarily. Replacements, including repaired machines, were 269. Reserves were at a record low of 127. Worse still, in the same period 103 pilots were killed, 128 wounded – this out of a strength of under 1,000. Some squadrons had a deficiency of ten pilots. And new pilots, coming from training courses that had been greatly shortened, were woefully lacking in combat knowledge. Across the Channel, the German Naval Staff had requisitioned 168 transports, 1,910 barges, 1,600 escort vessels and 419 tugs, while the Stukas gathered in the Pas de Calais awaiting the order to clear the way for the invasion on the final disintegration of Fighter Command.

The Command's survival owed much to the durability of its principal fighter. It will have been noticed that a large proportion of Hurricane pilots survived because their battered aeroplanes either were still able to return to base or at least flew for long enough to enable them to escape by parachute, later to rejoin the fight. In contrast, if a German warplane came down over British soil, then, whatever the fate of its crew, they would be lost to the Luftwaffe. Already orders had had to be issued forbidding the inclusion of more than one officer in a bomber's crew.

Among instances of the Hurricane's ability to keep flying may be quoted that of Sergeant Lacey of 501 who, on 30 August, landed a Hurricane with a shattered radiator and 87 bullet holes in wings and fuselage. Next day, Pilot Officer Hodgson, a New Zealander with 85 Squadron, had just shot down a 109 when a cannon-shell set his engine on fire. With smoke pouring into the cockpit, Hodgson was halfway out of his aircraft before he realised he was over the Thameshaven oil storage tanks. Not wanting to allow a blazing fighter to go down among these, he remained at the controls to try to land in open country.

Switching off his engine, he side-slipped violently to keep the fire under control. He finally succeeded in finding a large field but this contained anti-glider obstacles, which he just managed to avoid while making a wheels-up landing. The Hurricane turned over but the unselfish airman scrambled out unhurt.

Two days earlier, Pilot Officer Constable-Maxwell of 56 had stayed with his crippled fighter, knowing that if he baled out it would fall among houses. With his engine stopped, he crash-landed in a field, narrowly missing some telegraph poles. The Hurricane was a write-off but the pilot escaped unhurt, to make his way back to base by taxi, the driver of which refused to accept a tip from the man who had just risked his life for him.

Yet another example was afforded by Squadron Leader Kellett of 303 on 6 September. This officer was slightly wounded; part of his machine's starboard aileron was shot away; great strands of fabric were torn from the tailplane; both ammunition boxes exploded, ripping 'holes a man could have leapt through' in the wing surfaces. The Hurricane continued to fly. Kellett came down at Biggin Hill at 160 mph. He was unhurt.

When the battered Hurricanes returned to base, the ground personnel fell upon them to repair them as quickly as possible, a task made easier by the simple design of the Hurricane but carried out in hard, difficult conditions, sometimes under fire. They received little official praise and few decorations, but when a squadron took off with an extra aircraft or two or even three, which the fitters and riggers had worked on all night to make battle-worthy, the pilots at least would have a proper sense of the debt owed to their brave, skilful and loyal ground crews.

When damage was too severe to be dealt with on the aerodrome or when the machine had come down in open country, an elaborate, widespread, highly efficient repair organisation went into action. The damaged aircraft would be taken to the nearest unit in the repair complex, be this an aircraft factory, general engineering works or even furnishing company. The depot would notify the extent of the injuries to Hawkers, who would supply any requisite equipment or, in complicated cases, issue a special repair scheme.

Such was the success of the network that 35 per cent of the fighters issued to squadrons during the battle were repaired aircraft. Of those Hurricanes that came down on land, at least 60 per cent were sent back into service, while many of the remainder provided spare parts for other machines. Thereafter the scheme continued to expand. At the end of the year for instance, Taylorcraft Ltd of Rearsby, having inspected two Hurricanes sent for their information, accepted work as a large-scale repair unit. This firm would in due course repair 368 Hurricanes. Another firm with a splendid record was Rollasons of Hanworth, but this suffered considerable damage in an air-raid towards the end of the battle. During the war, the Hurricane Repair Network was to make more than 4,000 machines available to the RAF.

No doubt it was partially frustration at their failure to knock out Fighter Command in the air that led the Luftwaffe to strike at its sources of aircraft production. On 4 September, the enemy engaged in a highly complicated series of feints designed to mask a raid by the Bf 110s of ZG76 on the Hawker factory at Brooklands. As the Messerschmitts neared their target they were pounced upon by the Hurricanes of 253, which shot down six of them, all confirmed on the ground, without loss.

The survivors reached Brooklands but, shaken by the Hurricanes' onslaught, they bombed the wrong factory. Although they caused appalling damage to the Vickers works making Wellington bombers, the production of which could not be resumed for four days, and inflicted some 700 casualties, nearly 100 fatal, the Hawker factory was completely untouched.

Two days later, the Luftwaffe tried again. This time it was the Hurricanes of 1 Squadron that intercepted Bf 110s of ZG26, two of which they shot down. Slight damage was caused to the Brooklands works, but this had no effect on the delivery rate of

Hurricanes, which on the contrary increased by 10 per cent in September.

If it was appropriate that Hurricanes should defend a Hawker factory successfully, it was even more symbolic that the production of Hurricanes should not be harmed. For Fighter Command was able to maintain its resistance through the period of its greatest crisis just long enough for salvation to come from a totally unexpected direction.

–

For Hermann Göring, gorgeously arrayed in a pale blue and gold uniform, wearing pinkish top-boots with spurs, bedecked with medals, sucking his giant, jewel-studded baton, the afternoon of 7 September was the greatest moment of his life. Standing on Cap Gris Nez, he watched nearly 350 of his bombers, escorted by more than 600 fighters, thunder overhead en route for London. Codenamed 'Loge' after the god who had forged Siegfried's sword, this operation was planned as the final blow that would break the British defences, leaving the country incapable of resisting the long-awaited invasion.

Seizing a radio microphone, the Reichsmarschall bellowed triumphantly: 'I personally have taken command of the attack… which for the first time (has) struck the enemy right to the heart.'

The chain of events that triggered Loge was a singularly ironic one. On the night of 24/25 August, a few German bombs had fallen, quite accidentally, in the centre of London. Next night, Bomber Command retaliated with a raid on Berlin, repeating the treatment twice more before the end of the month. In view of the distance to the target, the RAF could not carry enough bombs to hope to cause more than token damage, but the raids stunned the German public who had not believed that they could possibly happen.

Hitler, understandably enough, was enraged at this affront to his capital. He gave permission for attacks to be made on

London by way of reprisal. This did not necessarily mean that such had to be launched; if Göring had been an able officer, he would have allowed nothing to deflect him from his raids on the vital sector stations that had come so near to success, particularly since he could, quite plausibly, have claimed that the attacks on Berlin were the last despairing lunges of a beaten foe. In practice, he was entirely in favour of a change of tactics. Now, as at Dunkirk, the Luftwaffe chief gave his leader the worst possible advice.

The Reichsmarschall's motives were mixed. There is no doubt that he was infuriated by this reversal of his loudly proclaimed boast that no enemy aeroplane would ever fly over German territory. On the other hand, there were proper military reasons for attacking London. It was Britain's control-centre. It was Britain's most vital port. Perhaps bombing would break the morale of the people. Perhaps it would shatter communications. Perhaps it would compel the British Government to quit their capital.

Yet if these were valid aims, they could have been achieved without relaxing the stranglehold on the sector stations. On 3 September, Göring called a conference at the Hague with Kesselring, Sperrle and his Intelligence chief Colonel Schmid. Sperrle, the most experienced of Göring's subordinates, argued that the attacks on the airfields must be continued by day at least; London should be bombed only at night.

However, Göring would accept no such compromise. He wished for daylight attacks on London, since he believed that these would force Dowding to bring into action his final reserves of aircraft, which, the Reichsmarschall thought, were being held back deliberately. Kesselring gave enthusiastic support. Even if the airfields around London were destroyed, he claimed, Fighter Command would simply withdraw to those to the north.

In reality, Dowding could not have operated in this manner since the control of his squadrons was based on the sector stations whose loss would have paralysed communications and, without which, there could be no liaison with radar. Also, fighters from the north would not have found it easy to linger long enough in the critical combat areas in the south-east because of their short range. Schmid, whose job it was, appreciated neither these factors nor the harm already inflicted on the control system. His only contribution was a declaration that the British had only a handful of fighters left.

Thus encouraged, Göring brushed aside Sperrle's arguments. In the last resort it seems that the unspectacular hammering of the sector stations had no real appeal for him; he wanted to take personal command of a dramatic assault on the British capital.

That assault had, at first, more luck than perhaps it deserved. Park was away from his headquarters attending a conference at Stanmore. In his absence, his officers prepared to guard those airfields that appeared likely targets. As a result, most of the defenders were unable to intercept Göring's armada before bombs rained onto London's East End; about 330 tons of high explosive; more than 1,000 incendiaries. Nine vast fires were left raging. In the Quebec Yard, Surrey Docks, the biggest single fire recorded in Britain blistered the paint off the fire-boats on the far bank of the Thames. The very surface of the river, covered with liquid sugar, was on fire. Firemen literally dropped from exhaustion; 448 people died. More than 1,300 were seriously injured.

When the Hurricanes did arrive, they were at first too few in number to have much chance against fearful odds. 249 Squadron, attempting to get at the bombers, was overwhelmed by about sixty Bf 109s, which shot down five Hurricanes in about fifteen minutes – though only one pilot lost his life. Later, 249 lost a sixth aircraft, brought down by the anti-aircraft defences. Happily, the pilot baled out unhurt.

Another Hurricane squadron that met trouble was 43, now led by the brilliant, boisterous enthusiast from Southern Rhodesia who was probably Fighter Command's most popular pilot: Squadron Leader Caesar Hull. Flying as his No 2 was Flight Lieutenant 'Dick' Reynell, the Hawker test pilot who had voluntarily joined 43 to report to his company on the Hurricane's combat performance. As well as sending back valuable information, Reynell had already shot down a 109 on 2 September. 43 also tried to reach the bombers, only to be 'bounced' by 109s, a large number of which cut off Reynell from the rest of the squadron. With his Hurricane on fire, he baled out, but his parachute failed to open. Though Hull had previously used up his ammunition, he had at once dived to Reynell's aid. He too was shot down and killed.

This was perhaps the nadir of Fighter Command's fortunes. It had not only failed to prevent the attack on London but had suffered heavy losses. Thereafter, events would slowly improve. As the enemy formations retired, so avenging Hurricanes tore into their ranks. The Poles of 303 sighted a large body of aeroplanes, which they believed were Dorniers but which probably were Bf 110s. While Hurricanes of 1 Squadron engaged the escorting 109s, the Poles ripped the mass of 110s to pieces, then turned on the 109s as well. They destroyed or damaged about ten enemy aircraft. Four Hurricanes were badly mauled but two were able to return to base, while the others at least flew on until their pilots had got clear by parachute.

However, it was not the enemy casualties that marked the turn of the tide, it was the shift of target from the sector stations to London. Had the Luftwaffe persisted in the assault on the fighter bases for just a little longer, the defence, in Park's words, 'would have been in a perilous state'. As it was, not only did the Command have a chance to recover, a chance to rest exhausted pilots, even to mount training sorties for new arrivals – since it was now no longer directly under attack, it had far more time in which to bring off favourable interceptions.

In consequence, raids on London over the next few days were so hounded by fighters that many were dispersed before they could reach their objective. 253 brought down five Junkers

Ju 88s on 9 September, while only one Hurricane was damaged, force-landing without injury to the pilot. 303 destroyed or damaged six more enemy aircraft the same day. One Hurricane was shot down but the pilot baled out; a second had to force-land. Two days later, the Poles destroyed or damaged another six but this time lost two pilots, killed. Because longer warning of raids could now be given, Park had taken to operating Hurricane squadrons in pairs. On 11 September, 17 and 73 routed a formation of 110s, destroying at least five for the loss of one of 73's machines, the pilot of which escaped by parachute. Again, however, it seems questionable whether the squadrons did any more damage than they would have done flying singly.

One individual encounter of this period should also be mentioned. On the 13th, a Heinkel He 111 from KG27 bombed Buckingham Palace. As it was making off, it was spotted by Sergeant Lacey of 501, flying Hurricane P2793. He attacked, killing the rear gunner with his first burst. The bomber then dived into cloud with Lacey in pursuit. Just as he was about to renew his attack, another member of the Heinkel's crew, who had pulled the dead gunner aside, fired at him 'at a range', he later announced, 'of, literally, feet,' blowing away 'the entire radiator'. Lacey kept shooting until both the Heinkel's engines were ablaze, by which time the Hurricane was on fire also. Both machines crashed. The bomber's crew died but the Hurricane's pilot was able to bale out with minor burns.

While Fighter Command fought off enemy attacks, Bomber Command's Battles, Blenheims, Wellingtons, Whitleys and Hampdens, and Coastal Command's Blenheims, Hudsons and Albacores – the last-named on temporary attachment from the Fleet Air Arm – struck at the 'invasion coast': the harbours, the shipping, the communications. On the evening of 13 September, the Channel ports were heavily hit. Eighty barges were sunk at Ostend. The waterfront at Boulogne burned for fifteen hours. The Calais docks were also set alight. An ammunition train was blown up. Already the orders to execute

Operation Sealion had been postponed once; on 14 September, Hitler again put off his decision until the 17th.

Yet also on the 14th, the Luftwaffe's daylight raids broke past the defenders, whose interceptions were badly co-ordinated. The RAF lost fourteen fighters, half of them Hurricanes. 73 Squadron was hard hit, losing five machines (but only one pilot). Although the official figures state that fourteen enemy aircraft were also destroyed, this is highly misleading – only eight were combat losses. Hitler regained his optimism. The achievements of the Luftwaffe were, he proclaimed, 'above all praise'. Perhaps Fighter Command was at last at breaking point. He ordered all preparations for invasion to continue. Excellent weather was forecast for the 15th. This would be the chance for Göring's men to make one last massive effort, which would shatter the defences for ever.

–

The early morning of Sunday 15 September was cool and misty, but visibility quickly improved though a few scattered patches of cloud persisted. Over Bolt Head in the south of Devon, the Hurricanes of 87 Squadron shot down a lone Heinkel He 111 on an early morning weather flight. At Hendon, Squadron Leader Sample introduced the pilots of 504 to Generals Strong and Emmons, US Army Air Corps, and Rear-Admiral Ghormley, US Navy, who had arrived, complete with cine-camera, to observe the life of a British fighter squadron. At Uxbridge, Park too was receiving distinguished visitors. The weather on this day seemed suitable to the enemy. Mr Churchill wrote later; he thought it might prove interesting if Mrs Churchill and he should call at 11 Group Headquarters.

He had chosen well. Shortly before his arrival, at about 1030, the first enemy 'plots' appeared on the operations table. It was soon clear that a colossal formation of about 100 Dornier Do 17s, escorted by more than 300 fighters, was assembling over the French coast. This leisurely build-up was a great error, for it gave Park ample time to prepare his reception as well as to ask Leigh-Mallory and Brand to stand by with reinforcements.

All over south-eastern England the fighters scrambled. 504's Hurricanes got off in 4 minutes 50 seconds by an American stop-watch – but although this time is usually quoted with bated breath, it seems that some more experienced squadrons were airborne nearly two minutes quicker.

Thus, as Kesselring's men crossed the coast just after 1130, they were met by the Spitfires of 72 and 92; then by the Hurricanes of 253 and 501; then by three more Spitfire squadrons. As they flew over the Medway, the Hurricanes of 229 and 303 joined in, the Poles rushing head-on against the Dorniers. As the formation spread out over an ever-larger area, with damaged aircraft lagging behind, the Bf 109s – scattered above, on either side of, and even below the bombers – found it ever more difficult to give the necessary cover. Yet still the raid moved inexorably towards its target. Shortly before noon, London was in sight. So were its protectors.

To the Germans, brave, capable men though they were, the sight must have been appalling. No fewer than nine fighter squadrons were attacking them more or less simultaneously: four more Hurricane squadrons from 11 Group and the five-squadron-strong Duxford Wing from No 12, led by the legless, indomitable Squadron Leader Douglas Bader. As the wing's two Spitfire squadrons engaged the 109s, its three Hurricane squadrons broke past the escort to smash into the flank of the bomber formation. At the same time, the 11 Group Hurricanes charged it head-on.

The Hurricane squadrons in question might have been selected to typify Fighter Command. Bader's own 242 was a mainly Canadian outfit, representing the contribution by what were then called the Dominions. 12 Group's other Hurricane squadrons represented the 'Free' Air Forces from the occupied countries: the Poles of 302; the Czechs of 310. Of the 11 Group squadrons, 17 and 73 were veterans of the Battle of

France. 257, by contrast, had suffered severely during the early attacks on Britain but had since been revitalised by the inspiring example of Squadron Leader Stanford Tuck; it had flown nearly sixty miles from Martlesham to North Weald before turning south to join the attack. Finally, there was 504, the County of Nottingham Squadron, to symbolise the Auxiliary Air Force.

The Hurricanes' onslaught utterly broke the bomber formation, which flew into pieces in all directions like a shattered pane of glass. The Dorniers, dropping their bombs at random, fled for cloud cover with the Hurricanes snapping at their heels. Sergeant Holmes of 504, flying Hurricane P2725, having used most of his ammunition helping to shoot down two of them, now found a third Dornier in its sights. Determined on a certain kill this time, he pressed his attack so close that when the bomber exploded the Hurricane was dragged down by the blast: The Dornier crashed in the forecourt of Victoria Station while its crew parachuted down to the Oval cricket ground. Holmes also baled out. He landed on a roof in Chelsea from which he slid, to finish up, most unromantically, in a dustbin.

As the broken enemy ranks fled in disorder, the defending fighters returned to their bases where the ground crews rushed to refuel and rearm them. As the pilots hastily ate a snack lunch, they could reflect jubilantly that they had prevented the Luftwaffe from attaining any worthwhile results. Though they probably guessed there would be further fighting that afternoon, they were more than ready for it.

Morale in Kesselring's camp must have been correspondingly low, the more so as the clouds had thickened by midday, which would make detection less easy but reduce the chance of accurate bombing. However, shortly after 1400, massive waves totalling 150 bombers, escorted by over 300 fighters, again approached the Kent coast. This time not only Dorniers but Heinkel He 111s took part, while behind, throttled back to keep station, came the faster Junkers Ju 88s. The enemy advanced in three separate groups, while a diversionary raid was launched, with only minor success, on Portland.

Because the Germans on this occasion had not wasted time forming up, the afternoon's interceptions were not as good as those of the morning. However, the Hurricanes of 213 and 607 from Tangmere, flying as a pair, met one of the formations, which they attacked so fiercely that some of the bombers turned tail, jettisoning their weapons as they did so.

It seems that the morning disaster had shaken the Luftwaffe. This was confirmed by the next incident. The Northolt Station Commander, 43-year-old Group Captain Stanley Vincent, who had seen service with the Royal Flying Corps, was heading towards the enemy in the Hurricane allocated for his use. He was rather apt to do this, in order, he would explain, to see how his squadrons were getting on. But of course, if he encountered Germans, he would have to defend himself. On the 15th, he met a group of eighteen Domiers. As luck would have it, he was 'in an ideal position for a frontal attack on the bombers ... before the escort could interrupt'. The frontal attack was duly made, whereupon, to Vincent's delight, the Domiers hastily turned away to the south.

The remaining enemy aircraft reached London at about 1500, when they bombed with rather more accuracy (or luck) than in the morning. Within half an hour, eight squadrons from 11 Group, five from No 12 and two from No 10 had engaged them. Despite all the efforts of the 109s, the bombers were again scattered, making for safety in small groups as best they might. Carrying out a point-blank attack, Pilot Officer Cooper-Slipper of 605, flying Hurricane L2012, collided with a Domier, both machines falling enveloped with smoke, but not before Cooper-Slipper had baled out, his only injury three lost fingernails ripped off as he escaped from his cockpit.

There was one last raid on Southampton by Erprobungs-gruppe 210. The Germans made a superb attack in the face of heavy AA fire, then were off like lightning before Brand's fighters could intervene. It made no difference. It was the fighting over London that was decisive, and that had already been won and lost.

It was decisive because, afterwards, the Luftwaffe knew without a shadow of doubt that it had been outfought. For the bombers, defeat had been total. Apart from those that had gone blazing to their doom, there were many heading home slowly losing height, perhaps with an engine burning, perhaps with a gunner convulsing in agony in his cockpit, his blood smeared over the Perspex. For the fighters, there was the grim realisation that they had failed to give sufficient protection, nor did it appear likely that they could ever do so in the future in the face of such numbers of RAF interceptors.

Apart from two flying-boats lost in accidents, 15 September cost Göring fifty-eight fighting aeroplanes: twenty Messerschmitt Bf 109s, three Messerschmitt Bf 110s, three Junkers Ju 88s, ten Heinkel He 111s, no fewer than twenty-two Dornier Do 17s. Most of their crews were also lost – dead or prisoners. Of this total, four were wrecked in such mishaps as ditching when out of fuel. Anti-aircraft fire brought down one. Fifty-three were destroyed by British fighters.

On a day when confused combats had spread over a huge area, obviously it is impossible to give a detailed account of personal achievements. However, by seeing what successes can definitely be credited to individual pilots or to squadrons, by examining the reports of veteran pilots noted for accurate claims who would be unlikely to be in error, and by checking the types of enemy machines claimed by different formations making proper allowance for duplications, it seems likely that Spitfires made at least ten 'kills' but probably fewer than twenty, while Hurricanes brought down at least thirty of the enemy but probably fewer than forty.

It was almost inevitable that Hurricanes should achieve a good majority of the victories, since not only did nineteen squadrons see action as against twelve of Spitfires, but on average there were more machines on the strength of Hurricane squadrons at this time – though it should be stressed that once more the most successful of all units was 303 (Polish) Squadron, which, at the cost of one man dead and another in hospital, brought down eight raiders at the very least.

The other side of the coin was that the Hawker fighters suffered greater losses. Of the twenty-six RAF fighters that fell, twenty were Hurricanes. However, it might be noted that whereas four of the six Spitfire pilots were killed or died of their injuries, twelve out of the twenty Hurricane pilots survived.

Two days later, the lessons of 15 September were discussed at Hitler's War Headquarters. The German War Diary recorded the most obvious one: 'The enemy air force is still by no means defeated; on the contrary it shows increasing activity.' It added a few face-saving remarks about the weather, then came to the logical conclusion: 'The Führer therefore decides to postpone Sealion indefinitely.' Within a few days, a noticeable dispersal of the barges and transports from the invasion ports had begun.

During the Battle of Britain, Sydney Camm's Hurricane had shot down more enemy aircraft than all other fighters and all forms of ground defence all put together. 1940 indeed was primarily the Hurricane's year. The Battle was primarily the Hurricane's battle. The victory was primarily the Hurricane's victory.

Chapter Five

Night-Fighter and Intruder

The Hurricane's exploits for the two and a half months culminating on 15 September have been described in detail because this period represented the peak of its attainments. If not one Hurricane had flown after that date, it would still have ranked among the most important fighting aeroplanes of all time. In reality, of course, its services were only just beginning.

For a start, the daylight raids on Britain continued, though with what purpose it is difficult to say. Certainly, Göring still hoped that Britain's morale would collapse but 15 September had already shown that attacks by day could never bring about such a result. Unless a very large bomber force was sent, the weight of the assault would not be sufficiently heavy, but Göring just did not have enough single-engine fighters to provide proper protection to a very large bomber force. The only real value that such raids could show would be their part in damaging Britain's military potential.

Therefore, the most meaningful strikes at this time were against aircraft factories. The Supermarine plant near Southampton was hard hit on 24 and 26 September, though since the Germans believed this was a bomber factory, they never knew they had struck a heavy blow at the production of the Spitfire. On the 25th the Luftwaffe badly damaged the Bristol factory at Filton. Previously on the 21st, a lone Junkers Ju 88 had at last hit the Hawker works at Brooklands. Three of its four bombs exploded harmlessly in open ground but a delayed-action one landed in the main Hurricane assembly shop, from which it was removed to a safe distance by Lieutenant Patton of the Royal Canadian Engineers, a feat for which he was later awarded a George Cross.

Attacks against Fighter Command also continued with occasional results. 56 Squadron had a dreadful day on the 30th, losing five Hurricanes shot down with two more damaged – though not one pilot was even wounded. On 25 October, a dusk attack on 111's base at Montrose destroyed seven fighters on the ground. Yet the damage done to the Luftwaffe was usually much heavier. 238 caught the - Filton raiders as they escaped, shooting down four Heinkel He 111s, damaging a fifth, but losing only one Hurricane force-landed. On 27 September, Hurricanes of 46 Squadron downed at least five raiders; Hurricanes of 17, at least four. On 7 October, an attack on the Westland factory at Yeovil was broken up by 238 and 601, which for the loss of one aircraft, the pilot of which baled out, claimed eight of the enemy.

However, by this time such tales of destruction can perhaps be taken for granted. Suffice to say that Fighter Command slowly grew stronger, a progress emphasised by the first victory, on 8 October, of a second Czech Hurricane Squadron, No 312. The victim was a Junkers Ju 88 on a reconnaissance flight. It put up an admirable resistance, damaging no fewer than three Hurricanes, all of which nevertheless returned safely to base.

By contrast, the Luftwaffe was operating with ever-diminishing resources. It had long since been shown that the Stukas could not live in the same sky as interceptor fighters. The Bf 110s, the once-famous 'Destroyers', had ceased to appear in the fighter role. After 15 September, apart from those in a few specialised groups, they were no longer used as bombers either. The mounting losses among Domiers and Heinkels saw them also gradually disappear. Even the faster Junkers Ju 88s had suffered unacceptable casualties by the end of September.

Thus, when October brought the end of what for once had been highly unwelcome good weather, the Reichsmarschall had only one card left – the splendid Bf 109s. These were now adapted as fighter–bombers, but, fitted with a crude, improvised bomb-sight and denied any opportunity of practising the new techniques, they could not hope to perform any useful function.

As Mr Mason states in *Battle over Britain*:

> The fundamental lesson of October was that it did not *matter*. Isolated groups of Messerschmitts roaming at 28,000 feet could hold the attention of Fighter Command, could slip through the gaps in the defences, could exhaust and infuriate the pilots and could kill – a few people; but they could not destroy those defences, and they could not damage England's capacity for war. They had no realistic strategic purpose; despite Göring's grandiloquent ravings, they were no more than nuisance raiders. In four months RAF Fighter Command had so won the balance of power in the skies over England that they had reduced the proud eagles of Blitzkrieg to so many thieves in the shadows.

The German fighter leaders were not unaware of the futility of these strikes, but protests were met with snarls from Göring that a fighter arm that could neither protect the bombers nor use bombs of its own had best be disbanded. What the wretched Messerschmitt pilots felt is difficult to imagine. Their comrades who had perished in the earlier fighting had at least fallen in pursuit of a definite, attainable objective, but the deaths of the 109 pilots in October were merely to bolster the crumbling prestige of the Reichsmarschall, smarting under Hitler's justifiable accusations of failure.

In view of the height at which the raiding 109s operated, their interception was normally entrusted to Spitfires, for the Hurricanes with their slower rate of climb obviously were more vulnerable. Also, at maximum altitudes, the Hurricane Mark I responded rather sluggishly to the controls. Despite this, the Hurricanes still managed to have some memorable clashes with the 109s. On 7 October, No 605's Scottish flight commander, Flight Lieutenant McKellar, enjoyed a notable individual triumph when he was credited with the destruction of five 109s. Though a check of post-war records suggests that he achieved, in fact, not five 'kills' but five victories, namely two Messerschmitts destroyed, a 'share' in a third destroyed and two more damaged (one of which crashed on landing), few would grudge him the award of the DSO, which he received shortly afterwards.[15]

The same day, 501 Squadron also showed that Hurricanes flown by capable airmen were a match for 109s, by destroying four of these, though at the price of a pilot killed. One of the Messerschmitts fell to the guns of Pilot Officer Kenneth Mackenzie, who became separated from his colleagues during the combat. While still alone, he sighted eight more 109s. Attacking these, he damaged one that fled low over the sea towards France with the Hurricane in pursuit.

Having attained a perfect firing-position, Mackenzie, who was not a Scot but a fighting Ulsterman, found to his fury that he had exhausted his ammunition. Determined to achieve another victory, he formed the extraordinary resolve to knock his opponent down. Flying just above the 109, he banged his starboard wing across its tail. The Messerschmitt went into the sea while three feet of the wing-tip of Hurricane V6799 hurtled into the air, luckily snapping off cleanly without damaging the aileron.

Elated at his novel method of destroying enemy fighters, Mackenzie turned his damaged machine for home, only to be

---

[15]'Archie' McKellar was one of the most successful pilots of the Auxiliary Air Force, bringing down about fourteen enemy aircraft. He was killed in action on 1 November when his Hurricane dived into a country garden. A 109 crashed nearby. As no other pilot claimed this, it was presumably McKellar's last victim.

jumped by two more 109s, which chased him back over the Channel until just short of the cliffs at Folkestone, when they broke off the attack, presumably due to shortage of fuel. With smoke pouring from his engine, Mackenzie crash-landed in a field, receiving minor facial injuries – also, compensation for them in the award of a DFC.

October also saw the continuation of the struggle between the Hurricanes and Erprobungsgruppe 210. On the 5th this unit encountered foemen worthy of its steel in the Poles of 303 Squadron. The honours went to the Hurricanes, which, without loss, shot down two Bf 110s, damaged two more and killed Oberleutnant Weimann, the fourth leader of Erprobungsgruppe 210 to perish since the start of the Battle of Britain.[16]

On the 29th, the Gruppe hit back directly at the Hurricanes, when its 109 Staffel, equipped with bombs, struck at North Weald. 257 Squadron was just taking off when the raiders flung their bombs marginally wide of the Hurricanes' wing-tips. Most of these came through but the machine of Sergeant Girdwood was hurled over onto its back, exploding into a mass of flames in which the pilot died. The aircraft flown by Pilot Officer Surma was badly damaged but he managed to gain sufficient height to bale out safely The Hurricanes of No 249, which shared the base, also got into the air. They gave chase to the Messerschmitts, one of which was shot down by Flight Lieutenant Barton. The Staffel's leader, Oberleutnant Hintze, escaped by parachute to become a prisoner of war.

The fortunes of this finest of Luftwaffe Gruppen may be taken as symbolic of the course of the fighting. It had carried out all manner of difficult attacks with admirable skill,

---

[16]Hauptmann Lutz, the unit's third leader, had died on 27 September. It is usually stated that Spitfires were responsible but in view of the Luftwaffe's 'Spitfire snobbery' and often poor aircraft recognition, it is quite likely that the victors were really Hurricanes, probably those of 504 Squadron.

determination, and courage, despite the loss of four Gruppen-kommandeure and its three most experienced Staffelkapitäne. In every respect, it was a unit of which any air force in the world could justly be proud. Yet the opposition had been too much for it, and, in practice, on at least four out of every five occasions the opposition had consisted of the pilots of Hawker Hurricanes.

Yet, in general, the Luftwaffe's activities in October were so futile that accidents proved more dangerous to the Hurricanes than the Germans did. On the 16th, 302 Squadron flew into the London balloon barrage in heavy cloud. A cable struck the starboard wing of Hurricane P3935, flown by Sergeant Kosarz, with such force that it penetrated as far as the aileron hinge, ripping off over a foot of wing-tip. The aircraft spun, but the Sergeant regained control, making a faultless landing at Heston.

Only two days later, fog brought real catastrophe to 302 when four aircraft crashed into high ground in Surrey, all the pilots losing their lives. Earlier, on the 8th, the other Polish squadron, 303, had suffered a cruel blow when the wing of a Hurricane landing at Northolt hit a rise in the ground; it somersaulted, burst into flames. In the wreckage died Sergeant Joseph Frantisek, the little Czech whose seventeen confirmed victories had made him the most successful pilot in the most successful squadron in Fighter Command.

On 10 October, there was a sad loss of a different kind; Hurricane L1547, serving with the Czech Squadron No 312, caught fire for reasons unknown while on a routine patrol. The pilot, Sergeant Hanzlicek, took to his parachute but by tragic mischance drowned in the Mersey. The Hurricane also crashed into the river. L1547 had been the very first production Hurricane. It was one of a considerable number of aged Hurricanes that had been kept in service by continued modifications. Another of these veteran aeroplanes, L1808, finally retired to a training unit at the end of October, having seen distinguished service with four different Hurricane squadrons – 229, 85, 17, and, finally, 73.

The Battle of Britain officially ended on 31 October 1940 – a singularly arbitrary choice of date. The decisive defeat of Göring's bombers, which resulted in the abandonment of Sealion, had been on 15 September. The final day of September had seen the last really massive daylight raids. On the other hand, several days in November were more active than most, if not all days, in October. Even in December, the Hurricanes were able to bring down a raiding 109 or a reconnaissance machine from time to time.

8 November saw particularly widespread action, which included the return of the Junkers Ju 87s, a fact much appreciated by the Hurricane pilots who had missed their 'Stuka parties'. Flight Lieutenant Bayne, in temporary command of 17 Squadron, was lucky enough to corner a formation of Ju 87s bombing two destroyers, before their escort could intervene. His pilots broke up the attack, destroying at least four of the vulnerable Stukas.

Three days later, another flight commander led a Hurricane squadron to an ideal interception. Flight Lieutenant Blatchford of 257, detailed to meet a raid moving against shipping off Harwich, was first surprised to spot ten bombers of a type that he was unable to identify, then astounded to discover that they were escorted by about forty biplane fighters. The bombers were Fiat BR 20s; the fighters Fiat CR 42s. The Italian Air Force, the Regia Aeronautica, had entered the fray.

It is to be wondered whether the Luftwaffe liaison officers had briefed the Italians on the calibre of the resistance they were likely to encounter. If so, they were endowed with considerable courage to come at all. Certainly, they showed no lack of this quality, for when 257 appeared the bombers held their formation, firing back with great determination, while the CR 42s did their best to engage. They were hopelessly outclassed. Blatchford, flying Hurricane V6962, on running out of ammunition, rammed one of the biplanes, tearing at its top wing with his metal propeller. Three of the CR 42s were lost while the rest scattered all over the sky.

Then 257 went for the bombers, being joined by the Hurricanes of 46 Squadron climbing up from below with jettisoned bombs hurtling past them. The enemy formation disintegrated, three BR 20s being shot down. Not one Hurricane was lost. Never again did the Italians mount a major bomber-raid against Britain during daylight.

A fortnight later, Air Chief Marshal Dowding left Fighter Command, receiving scant recognition for his monumental achievements. At the end of September, he had been named a Knight Grand Commander of the Bath – typically, he wrote that the decoration should be cut up so as to be distributed to his pilots 'who are the ones who really earned it' – but this was the kind of award that any competent officer of such high rank might reasonably have expected on the eve of his retirement. He was sent on a technical mission to the United States, for which he was quite unsuited by temperament, thereafter being allowed to slide into obscurity.

Dowding's place was taken by the former Deputy Chief of the Air Staff, Air Marshal Sir William Sholto Douglas, an advocate of the 'big wing' tactics. It is perhaps not surprising, therefore, that early in December, Leigh-Mallory took over No 11 Group. Park was sent to run a flying training group. Seldom can such devoted services have been repaid by such blatant ingratitude as that accorded to Dowding and Park.

The justification of their methods could be seen most easily in the state of their beloved Command at the time of their departure. From its nadir in early September, this had gained strength slowly but surely. By the end of 1940, the Hurricane was more than ever numerically the most important RAF fighter. Though the Hurricane flight of 263 had disappeared, eight new Hurricane squadrons had been formed, including a second Canadian unit, two more of Poles and No 71, the first of the famous 'Eagle Squadrons' of American volunteers.[17] Also No 247, the last Gladiator unit in Britain, which from the small aerodrome at Roborough, Devonshire, had guarded the naval base at Plymouth during the battle, had finally converted to Hurricanes. By the end of March 1941, five further squadrons, including three Polish ones, had also converted to or been formed with the Hawker fighters.

Furthermore, the existing squadrons were adding to the Command's strength by re-equipping with new, improved machines. The Hurricane Mark II had arrived.

–

From the moment it made its first flight, the Hurricane was the subject of continuous improvements as well as of modifications, actual or projected, to meet specific requirements. Even as it fought in the skies above south-east England, its parent company was still working to make it a more effective warplane by the provision of increased armour to guard the pilot, additional protection for the engine and Linatex covers to make the fuel tanks self-sealing. However, though undeniably valuable these benefits appear almost trivial compared with the major developments envisaged by Sydney Camm on the dual themes of greater performance from the engine and greater power for the guns.

Though the Hawker Design Staff had investigated the possibility of using engines other than the Merlin, it was realised that the ideal powerplant would be an improved version thereof, since this would necessitate the minimum modifications of the airframe and so could be introduced more easily into the

[17]Contrary to widespread belief, this unit did not participate in the Battle of Britain – indeed, it did not gain its first confirmed victory until 21 July 1941.

production lines. Working as always in close liaison with Rolls-Royce, Hawkers had already flown Hurricanes with Merlin VIII, Merlin XII and Merlin 45 engines, before, on 11 June 1940, the famous test-pilot, Philip Lucas, took up an eight-gun Hurricane I, P3269, fitted with the engine ultimately adopted, a Merlin XX, which gave it a top speed of 348 mph; because later alterations slightly reduced the speed, this was in fact the fastest armed Hurricane to fly.

The Merlin XX was a two-stage supercharged engine that drove a Rotol constant-speed three-blade propeller. It developed 1,185 bhp in its early form but this was later increased to 1,280 bhp. Best of all, having been designed with the importance of easy production always in mind, it could be brought out in massive numbers.

With regard to the Hurricane's fire-power, as early as January 1940, Camm had proposed the installation of four additional Browning guns, but this was not approved by the Air Ministry until April. Shortly afterwards, the heavy losses of Hurricanes in France led to a postponement of the scheme while every effort was devoted to increasing the output of the tried, tested Mark Is. Thereafter, it was decided that it would be most convenient if the twelve-gun wing could appear on the Hurricane simultaneously with the Merlin XX engine.

Another modification that was influenced by the fighting on the Continent was the development of the Hurricane as a long-range fighter, for it will be remembered that the lack of such a weapon had been a fatal flaw in the conduct of the Norwegian campaign. Hawkers therefore began experimenting with means by which they could adapt their famous fighter to carry auxiliary fuel-tanks under the wings. At first, these were fixed 44-gallon ferry tanks, but later jettisonable 45-gallon tanks were introduced. The slight loss in performance likely to result was clearly more than compensated by the advantages to be gained, so it was decided that attachments for such long-range tanks should also be standard on the new version of the Hurricane.

As it transpired, these various alterations did not take place at the same time. An anticipated shortage of Browning machine-guns (which in practice did not occur) led to the further postponement of the twelve-gun wing. The scheme for the fuel-tank attachment points also did not reach fruition immediately. However, since the Merlin XX gave the Hurricane not only a top speed of 342 mph but also a higher service ceiling of 36,000 feet as well as a much-improved rate of climb, thereby rectifying the main disadvantages of the Mark I, it was ruled that this alone should be introduced without further delay.

As a result, the first 120 Mark IIs, known as Hurricane IIAs Series 1, carried a Merlin XX engine and incorporated certain small additional improvements such as an enlarged rear-view mirror, but lacked the wing attachments and had only the normal armament of eight guns. Early in September, the first of these were delivered to 111 Squadron, which had once been the first to receive Mark Is, but shortly afterwards 111 withdrew from the battle area following heavy losses, leaving its Mark IIs to be distributed as replacements to other squadrons as required.

They were followed a month later by Hurricane IIAs Series 2, which did have provision for the fitting of fuel tanks. They also had an extra fuselage bay and frame immediately forward of the cockpit, which lengthened the nose by about seven inches. Over the winter of 1940-41, IIAs entered service with eleven Hurricane squadrons.

Early in 1941, the Hurricane IIB began to reinforce Fighter Command. This was the version with twelve machine-guns, so it had an increase of 50 per cent in fire-power, though the extra weight in the outer wings did reduce the aircraft's rate of roll. Apart from this, its performance was almost identical to that of the IIA. By mid-summer, IIBs were serving with twenty Hurricane squadrons.

The next main Hurricane variant was the end result of a long, somewhat checkered series of events. The question of mounting four cannon in their fighter had first been considered by Hawkers in late 1935. However, at that time when only the old prototype with its wooden propeller was flying, it was believed, quite rightly, that the weight of these would reduce the speed of the Hurricane unacceptably. On the other hand, the advantage of cannon shells, which explode after impact, over machine-gun bullets, which do not, was obvious to Hawkers. In the face of official indifference, they persisted in continuing research on the use of the more advanced weapons by the Hurricane.

Fortunately, the Air Ministry was interested in the potential installation of Swiss Oerlikon and French Hispano 20 mm cannon in twin-engine fighters, such as the Westland Whirlwind, and by the end of 1938, arrangements for their manufacture under licence in Britain were well advanced. On 24 May 1939 Philip Lucas took off in a Hurricane, L1750, armed with two Oerlikons, one under each wing. Top speed of this machine was 302 mph. Although it was intended merely as a vehicle for air-testing the guns, it became, in practice, the advance guard of an immense number of aeroplanes.

Though previously it has been stated that L1750 remained safely locked up in a hangar during the Battle of Britain, an examination of squadron records shows that this was far from being the case. Early in July, Flight Lieutenant Smith of 151 collected it from Martlesham Heath and on the 14th he saw action in it for the first time, damaging a Messerschmitt Bf 109, which crash-landed at its base. Almost a month later, on Eagle Day, Smith went one better by shooting down a Domier Do 17 of Fink's KG2 in flames.

In the meantime, encouraged by the promise shown on L1750's air firing trials, Hawkers were continuing with their experiments. A number of Hurricanes that had previously suffered from battle-damage in the wings were converted to carry four of the new cannon. One such machine, V7360, fought in the Battle of Britain under the control of Flight Lieutenant Rabagliati of 46 Squadron. Though the speed of this aeroplane was reduced to about 290 mph, its cannon proved lethal on 5 September, when a three-second burst of fire blew up a Messerschmitt Bf 109.

Three days later, Rabagliati had a more frustrating experience when his cannon jammed just as he flew into the middle of a formation of Dorniers. Flinging his fighter from side to side in the hope that this would persuade his guns to work, Rabagliati failed in his attempt, but at least his violent activity avoided all the bombers' crossfire, thereby demonstrating that the cannon-Hurricanes lacked nothing in manoeuvrability.

With the development of the Merlin XX, it became apparent that the Hurricane's powerplant would now enable it to carry four cannon without too great a performance loss. So, with the fervent backing of Lord Beaverbrook, the prototype Hurricane IIC, V2461, equipped with the new engine and the four-cannon wing, first flew on 6 February 1941 in the hands of test-pilot K.G. Seth-Smith. The IIC had either Hispano or Oerlikon guns, originally with belt ammunition feed though later drum-feed gear became standard, a maximum speed of 330-336 mph and a service ceiling of 35,600 feet. There were problems in the supply of the cannon caused by priority being accorded to the twin-engine Beaufighters, to say nothing of bomb damage to the factory of the main Hispano producers, the British Manufacturing and Research Company at Grantham, but, nonetheless, the new Hurricanes began entering service in the late spring of 1941. By midsummer they were equipping seven squadrons – but that was only the start. In all, 4,711 Hurricane IICs were built.

Simultaneous with the production of new varieties of the Hurricane came a change in its role. Admittedly, Hurricane pilots had previously engaged enemy raiders operating under cover of darkness. Back on the night of 26/27 June, Squadron Leader Aitken of 601 had earned a DFC by shooting into the Channel a Heinkel He 111, which had been caught in a searchlight beam. Thereafter, when Hurricanes could be spared from more urgent tasks, they made sorties after dark – called 'cat's eye' patrols much to the pilots' disgust – occasionally earning a reward for their persistence, as when Pilot Officer Cock of 87 destroyed a Dornier Do 17 on 26 July, or when a Junkers Ju 88 was brought down by Flight Lieutenant Sanders of 615 on 25 August. It gives some idea of the difficulties faced, to remark that both Cock and Sanders mistakenly identified their victims as Heinkels.

However, with the coming of the winter of 1940/41, the whole pattern of events changed. Though the Luftwaffe now scarcely dared venture across the Channel in daylight, it struck continuously at night, the main target being London, which endured its worst ordeal in three hours of incendiary attacks after dark on 29 December, though Birmingham, Bristol, Coventry, Glasgow, Hull, Liverpool, Manchester, Portsmouth, Plymouth and Sheffield all suffered grievously. Normally, these raids were directed against targets of economic importance, but inevitably any bombs that missed their mark took their toll of civilian lives, property, or both. Anti-aircraft guns inflicted some losses, though their chief value lay in forcing the attackers to bomb from high altitude with consequent reduction of accuracy. Jamming the Germans' direction beams or lighting decoy fires also helped to diminish their efficiency; but the only weapon by which real numbers of the raiders were likely to be destroyed was the night fighter.

In the future the main task would fall on the Beaufighters, fitted as they were with AI (Airborne Interception) radar sets, but for the present these were not very effective, for the aircraft were slow coming into service, the sets were still suffering from teething troubles and the operators had not yet learned how to master their temperamental equipment. It was the Hurricane that stepped into the breach. First 73 Squadron, then 85, 87, 151 and 96 (one of the new squadrons that formed in December) converted completely to night operations, while experienced pilots from most other Hurricane units were called upon to meet specific raids after dark as the need arose.

At first, they too had little effect other than causing some of their less resolute opponents to turn back early or jettison their bombs at random. When the Luftwaffe devastated Coventry on the night of 14/15 November, Squadron Leader Douglas Bader led three Hurricanes of No 242 over the stricken city but saw only 'the sea of flame below'. No 85 flew more hours on night-fighting duty than any other squadron in the three months from November 1940 to January 1941, but lost several aircraft in accidents due to bad weather with nothing to show for its efforts. It was not until 25 February that Squadron Leader Townsend, who had resumed command after recovering from his injuries, gained 85's first night victory when he shot down a Dornier Do 17.

Contrary to the complaints of understandably frustrated pilots, the Hurricane was, in fact, the most suitable single-seat night-fighter. Its main disadvantage was its short range, which naturally reduced the length of its patrols. However, it was blessed with several qualities, chief of which were its fine forward view, its steadiness, its viceless response to control, and, a cynic might add, that wide, sturdy undercarriage, which would stand up to the most forceful landing without ill-effects.

The real problems lay elsewhere. The inaccuracies of the early warning radar system, which might be remedied in daylight by able formation leaders who could see the situation that had developed, were vastly increased after dark when no such correction was possible. The ground controllers appear not to have appreciated at first that much greater precision was needed for an interception at night, when, for instance, a head-on encounter was quite useless as well as highly dangerous. The searchlights tended to focus on the Hurricanes, thereby blinding the pilots in addition to warning enemy raiders of their presence. The anti-aircraft guns also often failed to distinguish friend from foe; the flashing of identification signals from lamps fitted to the fighters worked only in ideal conditions.

These difficulties were immense, but they were overcome. Recognition of the problems brought about an increasing co-operation between the airmen and the ground defences, but the essential progress came from the patient, painstaking resolution of the pilots to learn from their mistakes. They discovered how to make fullest use of such light as was available.[18] They trained their eyes, some by wearing 'dimmer' glasses – goggles with dark lenses – others by keeping away from all strong lights at all times. They learned how searchlights or bursts of AA fire could act as guides to the progress of enemy bombers. They became practised in picking out raiders silhouetted in the glare of fires below.

As the pilots improved, so did the Hurricanes. It was found that the exhaust stubs from the engine, six of them on each side of the nose, spurted flames in front of the cockpit as the fighter closed in on its target at full throttle, thereby eliminating forward visibility just when it was most needed. This defect was corrected first by fitting shrouds to damp down the flames, then by adding a small, metal anti–glare shield between the exhausts and the windscreen.

In consequence, the fighters gained steadily in effectiveness. In January 1941, they shot down only three enemy aircraft at night, in February only four, in both cases considerably fewer than those brought down by the AA guns. However, by March, the airmen had overtaken the gunners with a dramatic increase to twenty-two night raiders destroyed. In April, the figure rose again to forty-eight.

Many of these successes were gained by the AI Beaufighters, but the Hurricanes also were scoring freely now. Their night patrols ranged from Lowestoft, over which Squadron Leader Stanford Tuck of 257 Squadron destroyed a Junkers Ju 88 on 9 April, to Liverpool where, on 12 March, Sergeant MacNair of

---

[18]Contrary to general belief, it is physiologically impossible for men (or even cats!) to see in utter, complete darkness, but in practice there is always a small amount of light on even the blackest of nights.

96 downed a Heinkel He III, after which he had to manoeuvre his way clear of the balloon barrage. Over Belfast, another Heinkel fell on 8 April to Squadron Leader Simpson, the CO of No 245, this being the first night victory over Northern Ireland.

In May, the Luftwaffe made its final full-scale assaults. Thereafter, the weight of the attacks greatly decreased as the bulk of the bombers turned their attention eastward. It was well for them that they did. Despite a falling-off in the intensity of the raids during the second half of the month, the Luftwaffe lost through various causes during the nights of May, 138 aircraft. Ninety-six of these fell to the night–fighters, a total divided about equally between the RAF's single-engine and twin-engine interceptors.

Several of the Hurricane squadrons gained especial fame in May. For example, No 43 flying Hurricane IIBs in defence of Glasgow, claimed to have shot down five enemy aircraft at night during the month plus two more in daylight. Squadron Leader Dalton-Morgan was personally credited with three of the night victories. By the end of the year, he had brought down three further victims after dark to become one of the most successful of the night aces.

However, the finest display was put on by the unquenchable pilots of 1 Squadron now flying a mixture of Hurricane IIAs and IIBs. The night of 10/11 May saw the last really massive raid on London. The Luftwaffe chiefs had chosen this date for a major effort because good weather and a full moon made for unlimited visibility – but the same factors gave No 1 its great opportunity. In the course of the attack, it lost one pilot, killed, hit apparently by AA fire, but it was credited with the destruction of no fewer than seven Heinkel He IIIs and one Junkers Ju 88. The greatest achievement was that of the Czech, Sergeant Dygryn, who, in three sorties, destroyed two Heinkels and the Ju 88.

Both 1 and 43 also had a part to play in connection with another device used by the defenders at night. This was the

'Turbinlite', an airborne searchlight that was carried in the nose of an AI equipped twin-engine aircraft to illuminate enemy raiders. Since the Turbinlite so weighed down the aeroplane that carried it that it could have no capacity for guns, destruction of the intruder was to be left to an accompanying single-engine fighter, known as a 'satellite'.

This scheme promised much on paper and from May 1941 onwards Turbinlite Flights were formed with American Douglas Havocs, a conversion of the Boston light bomber, which had a performance slightly inferior to that of the Beaufighter. The fighter chosen as the 'satellite' was the Hurricane, several of the most famous Hurricane squadrons, including not only 1 and 43 but 3, 32, 87, 151, 247, 253 and 257, co-operating with the Havocs in their complicated task.

However, results were meagre. The difficulties of operating two very different types of aeroplane in unison at night proved formidable – so much so that it frequently transpired that the fighter was out of position when the target was illuminated. Even if this did not happen, the unmasking of the searchlight robbed the attacker of the vital element of surprise. A bomber thus exposed would take violent evasive action and, once it had escaped the beam, the Hurricane pilot was helpless.

In consequence, it was not until almost a year later, on the night of 30 April/1 May 1942, that Flight Lieutenant Winn, at the controls of a Turbinlite Havoc of 1459 Flight, illuminated a Heinkel He 111, which was then destroyed by a Hurricane of 253 flown by Flight Lieutenant Yapp. 253 later added scores of a 'probable' and a 'damaged' – which may not sound very impressive, but was more than any other Turbinlite formation could boast.

It was felt that one important reason for the scant results achieved by these Flights was that, being dependent on regular day-fighter squadrons, they had missed many opportunities because the Hurricanes had been required for duties elsewhere. To remedy this, in September 1942 the Flights were expanded into no fewer than ten full squadrons, equipped with their own Hurricanes, usually Mark IICs, as well as with Havocs. All ten squadrons disbanded early in 1943. They had gained no successes whatsoever.

The failure of the Turbinlite scheme emphasises that the Hurricane night-fighter triumphs were individual ones. It seems appropriate therefore to conclude them with the story of the supreme lone wolf, Pilot Officer Richard Stevens of 151, whose exploits took place throughout 1941. He was a very different figure from the gay, light-hearted young man enshrined by legend as the typical defender of the British Isles. He was older for a start, having joined the RAF after the outbreak of war at the age of thirty-two, which was the maximum for pilot training. As a pre-war commercial pilot on the Croydon-Paris air-route, he had flown 400 hours at night. This would stand him in good stead, but it was a grimmer experience that made him such a dangerous foe: his wife and children had been killed in one of the early night attacks on Manchester.

Thereafter, Stevens' hatred of night raiders was implacable. It is said that he would scream with rage when he sighted one but since, presumably, he did not do so over the radio, it may be questioned how this could possibly be known. There is no question that he pressed home his attacks with total ferocity. In one case, after blowing up a bomber at point-blank range, he returned with his Hurricane's wings stained by German blood. He refused to allow the grisly traces to be removed.

Stevens exacted the first instalment of his vengeance on the night of 15/16 January 1941, when he sighted a Dornier Do 17. His first burst of fire hit his target, for oil spurted back onto his windscreen, but the bomber was not badly hurt. Hoping to shake off its attacker in the dark, it spiralled away in a twisting dive, but Stevens hung on to it relentlessly. The Dornier then pulled out, climbing steeply. Determined to remain in touch, the British pilot wrenched Hurricane V6934 out of its dive so violently that the underside of the fuselage was cracked, while fluid formed in one of Stevens' ear drums. Catching the Dornier at the top of its climb, Stevens sent it hurtling to earth in flames.

Having landed to refuel and rearm, Stevens took off on a second patrol. This time he found a Heinkel He 111. Despite heavy return fire, he continued firing at it until he ran out of ammunition, by which time the raider was diving earthwards with smoke pouring from both engines. Observer posts confirmed that it had crashed.

Stevens was awarded a DFC for his feat, but the fluid in his ear so reduced his sense of balance that it grounded him for some time. On his return to the fight, he destroyed two more Heinkels in one night over Birmingham on 8/9 April. Two nights after that, he gained a third double success, bringing down yet another Heinkel and a Junkers Ju 88. By the end of May he had shot down ten enemy night raiders, including another pair of Heinkels in one night on the 7th. By the end of October, he had achieved a further four victories, which put him well ahead of any of the radar-assisted pilots. In November, having by now reached the rank of flight lieutenant, he transferred to another Hurricane squadron, No 253, which at that time was also hunting at night over the enemy's own aerodromes.

Although Stevens' attainments had been amazing, they had been due mainly to a complete disregard for his own life. He flew in the most difficult weather conditions. He moved into areas where anti-aircraft bursts were thickest, believing these provided the surest pointers to the location of his prey. It was generally felt that his luck could not last. On 12 December the award of his DSO was announced. Three nights later, he set off, alone as always, on a sortie over enemy territory, from which he did not return.

–

By day the Hurricanes were already ranging far afield to seek out their foes. In part, this was an extension of their former interceptor role, for such was now the dominance of Fighter Command that it had become responsible for the protection of all shipping within forty miles of the coast — a task in which it worked in happy co-ordination with the Admiralty. No 87 Squadron detached a Flight, led by the CO, Squadron Leader Gleed, to the Scilly Islands, from which it added a steady sequence of Dornier Do 18 flying-boats, to the list of Hurricane victims. Most convoy patrols were unexciting, however. The Czech Squadron, No 312, in five months on such duties encountered only a solitary Junkers Ju 88, which was promptly shot down.

In any case such activities appear minor when compared with the offensive tasks that the Hurricanes now commenced. Though originally they were not well suited for intruder-work because of their short range — neither were the Spitfires for the same reason — this defect was remedied by the introduction of the two jettisonable 45-gallon long-range fuel tanks on all versions from the later Mark IIAs onwards. Thus sustained, the Hurricanes were ready for the execution of Air Marshal Sholto Douglas' policy of 'leaning forward into France'. 1 Squadron, true to its motto, began to 'lean' on New Year's Day 1941 when three Hurricanes, led by Flight Lieutenant Clowes, strafed enemy positions between Calais and Boulogne for twenty minutes, unhindered by hostile fighters.

Such operations by small numbers of intruders were mysteriously code-named 'Rhubarbs'. For these the Hurricanes, according to Air Vice-Marshal Wykeham, were 'invaluable', especially after the IIBs or IICs with their heavier armament began reaching the squadrons. Sometimes they did not even need to cross the enemy coast before finding trouble. On 12 January, Squadron Leader Douglas Bader and Flight Lieutenant Stanley Turner of 242 Squadron sighted two E-boats, the German equivalents of Motor Torpedo Boats. These they took completely by surprise. After two strafing runs each, they headed for base though not before they had observed crew-members jumping overboard.

Other Hurricanes shot up barges or merchant vessels or port installations. On occasions, they even attacked flak ships, although since these, as their name suggests, were designed solely to bring down RAF aircraft with their formidable battery of guns, they were hardly recommended targets.

Having passed over the coast, the Hurricanes found before them even more varied opportunities, mainly on the ground, though they had the odd encounter with 109s or engaged Junkers Ju 52s on transport duty or Heinkel He 59 seaplanes on air-sea rescue work. On 7 April, five of No 1's Hurricanes caught three 109s preparing to take off from Berck airfield, near Le Touquet. They destroyed all three as well as riddling the luckless ground crews with bullets. Earth-bound targets attacked included motor vehicles, petrol stores, wireless stations and columns of troops.

Larger operations in which British bombers took part were codenamed 'Circuses'. The first such was arranged on 10 January. Six Blenheims from 114 Squadron were sent to bomb gun emplacements in the Forêt de Guînes. The Hurricanes of 56 flew with them as close escort, while three squadrons of Spitfires provided top cover. The Hurricanes of 242 and 249 were ordered to sweep ahead of the bombers, creating a diversion by attacking the aerodrome at St Inglevert.

At about 1300 hours, the Blenheims dropped their bombs, which started fires. 242 and 249 also attacked at low level. The former could not find a target but 249 strafed a line of reconnaissance machines parked under trees. The Hurricanes were then attacked by 109s, two of which they shot down, though they lost one of their own number, whose pilot was forced to bale out. All the other aircraft involved in the raid returned safely.

Circus 2, which was launched on 5 February, was less satis-factory. Though the Hurricanes of 302 and 601 and the Spitfires of 610 kept guard over the twelve Blenheims detailed to attack St Omer airfield, six other squadrons missed the rendezvous for various reasons. The bombers hit their target but a strong force of 109s now appeared, attacking the Spitfire squadron, which, hampered by the need to stick close to the Blenheims, lost four pilots killed. Only one Messerschmitt was brought down in return.

Later Circuses, learning from the mistakes made, were more profitable. The Hurricanes escorted Blenheims, Hampdens, or, from July, four-engine Stirlings, in raids against airfields, factories or shipping. These resulted in a number of clashes with 109s, 317, one of the Polish squadrons, destroying two of these on 10 July in defence of a Blenheim formation, while on 21 June, 1 Squadron also had a successful encounter when guarding Blenheims – for the loss of two Hurricanes, one of whose pilots was later picked up by a rescue launch, it claimed to have shot down four 109s, of which two were the new 109Fs, an improvement on the 109E, which had been the Luftwaffe's main weapon in 1940.

Not all combats were as favourable. 242 Squadron, for example, had had only six pilots killed from June to December 1940. In the first four months of 1941, on convoy patrols, Rhubarbs and Circuses, it lost nine including Squadron Leader Treacy who succeeded Bader as CO. In practice, during the first half of 1941, losses of aircraft on each side were about equal, but Fighter Command now suffered the disadvantage that even if a pilot baled out, he would become a prisoner of war.[19]

---

[19]There were a few exceptions to this rule. For instance, Squadron Leader Whitney Straight who took over 242 after Treacy's death, was brought down by a flak ship on 31 July. He crash-landed his Hurricane near the coast, evaded capture and, after numerous adventures throughout the length of France, finally reached the Pyrenees only to be arrested by the Vichy authorities. He escaped from them after several attempts, ultimately reaching Gibraltar safely.

It might therefore be argued that such actions, like those of the 109s the previous October, had little meaning. It is certain that they had no decisive results, though the Circuses in particular did more damage on the ground than the Messerschmitts had ever managed. They also kept some extra aircraft, airmen and AA guns in the west, which otherwise would have been employed in the Mediterranean or Russia, but their achievements in this respect were rather disappointing. They provided invaluable experience of offensive operations, but whereas later raids would prepare the way for the invasion of Europe, the sorties in 1941 at best prepared for the preparations.

Yet some intangible but important gains were made. The RAF pilots became accustomed to exercising the initiative, while the Luftwaffe had to adopt an increasingly defensive posture. For the inhabitants of occupied countries, the raids had an especial meaning. Some Hurricanes dropped propaganda leaflets through a chute on the port side of the fuselage originally designed for the use of parachute flares at night – but the finest propaganda was the Hurricanes' own presence, clearly proving that whatever the Germans might claim, the RAF had gained command of the air. 257 over Holland saw people in the village streets doff their hats to the fighters. 601, over the Belgian town of Ghent, could not fire at German troops because of the delighted civilians who poured onto the pavements to wave to them. It was not coincidence that at this time came the first flickers of organised resistance in north-western Europe. The Hurricanes brought hope.

They gave equal encouragement to their own pilots. Its power-plant might change, its armament might alter, its role might vary, but the reliability of the Hurricane remained.

On 21 June, Squadron Leader Robert Stanford Tuck, CO of 257, was flying a Hurricane IIC on a routine patrol over the sea about 100 miles from Southend, when he was 'bounced' by three Messerschmitt Bf 109s. Shells ripped through the Hurricane, shaking it like a leaf – but it stood up to the hammering. As the leading 109 hurtled in front of him, Tuck was on its tail, one burst from the Hurricane's 20mm cannon sending the German machine plunging into the sea. Another 109 was now firing at Tuck but the Hurricane out-turned it. This enemy also Tuck blasted out of the sky.

He then 'pulled up quickly and was immediately hit from the left'. The throttle lever was wrenched from his hand. His forehead was cut by a flying fragment of the reflector plate. The cockpit hood was blown off. The third Messerschmitt now raced past but, with no throttle and his engine missing badly, Tuck could not pounce on this one. The 109 turned to charge head-on. It received a dose of cannon fire, which caused it to make for base, spilling out glycol.

The Hurricane still 'stumbled on'. For all the battering it had received it stayed airborne until the coast came into sight, at which point the starboard aileron fell off, while flames burst through the floor of the cockpit. Tuck baled out. Landing in calm water, he inflated his rubber dinghy. Two hours later he was picked up by a coal-barge from Gravesend, to be greeted with a mug of rum, which did much to console him.

Tuck's comment on this episode could have been echoed by the hundreds of pilots who owed their lives to the staunch-ness of their aircraft. 'My God,' he exclaimed, 'the punishment these Hurricanes could take. The air-frame could stand almost anything!'

Chapter Six

The Mediterranean

In the Mediterranean theatre, the Hurricane received a different but equally sincere compliment from a more senior officer. On hearing that the King had acknowledged his services by creating him a Knight Grand Cross of the Bath, Admiral Cunningham remarked: 'I would sooner have had three squadrons of Hurricanes.' This statement adequately sums up the value of the Hurricane. Its presence was vital, its roles were varied, but its numbers were never sufficient.

Britain's original enemy in this theatre was Italy, which, when Mussolini declared war on 10 June, had possessions in three main areas. First there was the Italian peninsula thrusting into the centre of the Mediterranean, flanked by Sicily and Sardinia, and, across the Adriatic, by Albania, which the Duce had annexed in April 1939. Next, in North Africa threatening the British bases in Egypt, was Libya, divided into the provinces of Tripolitania in the west and Cyrenaica in the east. Finally, in East Africa were Eritrea and Italian Somaliland, between which was Abyssinia, a picturesque but backward country swallowed up by Mussolini in 1936.

The Italian forces were vastly superior in numbers. The Viceroy of Abyssinia, the Duke of Aosta, an honourable, humane man who had done much to reconcile the local population to their conquerors, had well over 250,000 troops and 150 aircraft, compared with 20,000 from the British Empire and Commonwealth backed by a remarkable collection of out-dated aeroplanes, the majority biplanes, in the Sudan, Kenya and British Somaliland. From his central position, he could put pressure on all these colonies, capturing Kassala, Karora, Kurmak and Gallabat on the Sudanese border, and the frontier post at Moyale in Kenya. British Somaliland, surrounded by hostile territory, was totally indefensible. On 16 August, after a delaying action that inflicted heavy casualties on the invaders, Major-General Godwin-Austen, much to the annoyance of Churchill, withdrew his forces by sea, ultimately to Kenya.

On the frontier between Libya and Egypt the Italians made no progress whatsoever, although they numbered almost 250,000 men supported by 282 aircraft, whereas General Sir Archibald Wavell, General Officer Commanding in Chief in the Middle East, had only 36, 000 troops and Air Chief Marshal Sir Arthur Longmore, the Air Officer Commanding in Chief, only about 150 aircraft of comparatively poor quality. The main Italian bomber, the Savoia Marchetti SM79, though not quite as fast as the Blenheim had a greater range and carried a heavier bomb-load, while the Fiat CR42 biplane fighter was definitely superior to the Gladiator.

Yet, with astonishing audacity, Wavell ordered Major-General Richard O'Connor, the commander of his Western Desert Force; to take the initiative. Beyond all expectation, British patrols captured Forts Maddalena and Capuzzo from an enemy who greatly overestimated their strength. Incredibly, for the first three months of fighting, it was the Italians who were on the defensive.

In his attacks, O'Connor had the strong support of Air Commodore Raymond Collishaw, a Canadian who had been one of the greatest World War I fighter pilots but who now brought a similarly aggressive spirit to his command of No 202 Group, the Royal Air Force units in the Western Desert, sending his Blenheims to bomb the enemy's main air base at El Adem within a few hours of the outbreak of hostilities. However, such were the RAF's maintenance difficulties that Longmore was compelled to restrain his eager subordinate. Collishaw perforce resorted to bluff in order to keep his opponents in check. As an example of this, he flew a single Hurricane from landing–ground to landing–ground in the hope of convincing the Italians that large numbers of modern fighters were reaching North Africa. Certainly, the Regia Aeronautica seemed unwilling to venture over the British lines, much to the delight of the RAF personnel who bestowed on the Hurricane in question the nickname of 'Collie's Battleship'.

There are two candidates for this title, both of which have an honoured place in the Hurricane's exploits in any event. L1669 was the first Hurricane in the Middle East, having been sent to Khartoum via Egypt at the end of 1939, fitted with a tropical filter over the carburettor air intake to protect the Merlin engine from harmful dust or sand particles. As a result of trials with this aircraft, a standard filter, the Vokes Multi-Vee, was fitted to Hurricanes detailed for use in the Desert. Hanging down under the nose, by creating additional drag this reduced the Hurricane's speed to 312 mph as well as restricting its rate of climb. Certain pilots who flew Hurricanes both in Britain and in the Desert have said that the 'tropical' versions were like carthorses compared to thoroughbreds. However, since the alternative to the filter was the rapid destruction of the engine, there is no doubt that its adoption was a vital measure.

Having completed its tests in July 1940, L1669 was sent to the Western Desert. It is recorded on the strength of a fighter squadron, No 80, but since by that time its guns had been rendered inoperative by the hostile environment, it cannot have seen combat. It therefore appears almost certain that Collishaw used this machine to bluff the Italians in view of its inability to play a more positive role. This supposition is confirmed by the fact that L1669's arrival coincided with Longmore's orders for every possible economy in the use of fighting aeroplanes.

P2638, on the other hand, though almost certainly not the 'Battleship' as some accounts have claimed, had the double distinction of being the only Hurricane in Egypt at the time that hostilities with Italy commenced, having been shipped out a short time previously, and the first to be flown in action in this theatre. It too was allocated to 80 Squadron but when it saw combat on 19 June, its pilot Flying Officer Wykeham-Barnes[20] was accompanying the Gladiators of No 33. A force of Fiat CR42s was encountered, two of which became the first Italian victims to fall to a Hurricane's guns.

Although the demands for Hurricanes elsewhere were great, it was considered vital that they be sent to the Middle East. Small numbers were therefore ferried through the chaos of France, thence via Tunisia and Malta to Egypt. These flights, made to the limits of the Hurricane's range, were expensive – of about fifty fighters thus dispatched, only a handful reached North Africa, and four of these were sent back to provide protection to Malta.

The remaining machines joined 80 Squadron, which now moved to Amriya near Alexandria where it guarded Britain's most precious possession in the Middle East, Admiral Cunningham's Mediterranean Fleet. It was while covering the fleet as it returned from a bombardment of Bardia on 17 August, that Flying Officer Wykeham-Barnes and Flying Officer Dowding shot down a scouting Cant Z501 flying-boat; but it presumably radioed news of its sighting first, for not long afterwards Flying Officer Lapsley spotted a group of Savoia Marchetti SM 79s about to attack. He shot down three in quick succession, all of them being confirmed in enemy records.

Two days later, No 80 gave up its Hurricanes to form the basis of a new squadron, No 274 – its CO, Squadron Leader Dunn, assuming command of this unit. Several pilots with experience on Hurricanes accompanied him, including Lapsley, who, on 10 September, gained 274's first success by

---

[20]This officer enjoyed a distinguished career in the desert, which included the command of 73 Squadron during the fighting around Tobruk. He later dropped the name 'Barnes' and as Air Vice-Marshal Wykeham he has already been mentioned as the author of *Fighter Command*.

bringing down two more SM 79s despite damage to his own aircraft.

In the meantime, on 28 June, the Italian commander Marshal Balbo had been killed by his own AA gunners when his aeroplane had arrived at Tobruk on the heels of a British air-raid. His replacement, Marshal Graziani, after much pressure from Mussolini, lumbered forward into Egypt on 13 September, the small British forces withdrawing before him. Four days later, exhausted by having captured the few white, mud-brick buildings, mosque and landing-ground elevated in their communiques to the status of the 'town' of Sidi Barrani, the Italians halted, building a chain of forts stretching south-westward into the desert. These were stocked with every available luxury but were neither properly defended nor mutually supporting.

Yet it was clear that reinforcements would have to reach Egypt if Graziani was not to resume his advance sooner or later. The collapse of France having ended the former supply route, a handful of Hurricanes were shipped out in crates through the Mediterranean. Some were virtually useless since they lacked air filters, but those that had been suitably modified equipped 33 Squadron. However, as this sea crossing became more perilous, it seemed that any fighters sent in future could only travel by the time-consuming journey round the Cape of Good Hope.

There was one alternative. On 14 July, Group Captain Thorold had led a working party to the port of Takoradi in what was then the Gold Coast. It was proposed that short-range aircraft should be shipped here, thereafter flying across Africa by way of a series of intermediate landing-grounds.

By 24 August, Thorold had created the equivalent of an RAF maintenance unit in the little port by the provision of such essentials as roads, hangars, runways, workshops, store houses and living accommodation. Hurricanes would be delivered here, usually in a dismantled condition – they would be reassembled, without guns in order to save weight but with two fixed long-range tanks, 44-gallon ones at this stage – though the Mark IIs, which used the route later, carried 90-gallon tanks. Then, guided by a twin-engined aircraft, they would move off to a sequence of airstrips that Thorold had somehow managed to convert into efficient staging posts.

The perils of the flight are graphically described in the *RAF Official History*.

> The first stage, 378 miles of humid heat diversified by sudden squalls, followed the palm-fringed coast to Lagos, with a possible halt at Accra. Next came 525 miles over hill and jungle to an airfield of red dust outside Kano, after which 325 miles of scrub, broken by occasional groups of mud houses, would bring the aircraft to Maiduguri. A stretch of hostile French territory[21] some 650 miles wide, consisting largely of sand, marsh, scrub and rocks, would then beguile the pilot's interest until he reached El Geneina, in the Anglo-Egyptian Sudan. Here, refreshed with the knowledge that he had covered nearly half of his journey, he would contemplate with more equanimity the 200 miles of mountain and burning sky which lay between him and El Fasher. A brief refuelling halt, with giant cacti providing a pleasing variety in the vegetation, and in another 560 miles the wearied airman might brave the disapproving glances of immaculate figures in khaki and luxuriate for a few hours in the comforts of Khartoum. Thence, with a halt at Wadi Halfa, where orange trees and green gardens contrast strangely with the desert, and a

---

[21]Thus, the Official History but in practice before the flights started, French Equatorial Africa was no longer hostile having accepted the leadership of General de Gaulle.

house built by Gordon and used by Kitchener shelters the passing traveller, he had only to fly down the Nile a thousand miles to Abu Sueir. When he got there his airmanship would doubtless be all the better for the flight. Not so, however, his aircraft.

This last qualification was a very necessary one. On 5 September, the first six Hurricanes arrived at Takoradi in crates. A fortnight later, they had been assembled. A week after that, five had reached Abu Sueir but since they lacked air filters their worn-out engines urgently needed replacement. It was not until December that the first one, V7295, was ready for action. It was now fitted with a forward-facing camera that entitled it to be known as a Tac R (Tactical Reconnaissance) Mark I. As such, it was handed over to 208 Squadron, which later received other similarly modified Hurricanes to replace its Lysanders on 'recce' sorties.

The day after the first Hurricanes reached Takoradi, the little aircraft carrier *Argus* arrived carrying thirty more. These were fitted with air filters but lacked long-range tanks, the installation of which required considerable work on the Mark I whose wings, unlike those of the Mark II, were not specially adapted for their attachment. However, by mid-November enough had reached Egypt to bring 33 and 274 up to full strength.

Also by mid-November, preparations were well under way for the dispatch to North Africa of 73 Squadron, whose Hurricanes had won an enviable reputation in France and the Battle of Britain. The ground crews were taken through the Mediterranean in the cruiser *Manchester*, reaching Alexandria on the 30th after an exhilarating clash with the Italian fleet off Cape Spartivento. On the 27th at Takoradi, thirty-four Hurricanes, intact apart from long-range tanks, were flown off the carrier *Furious*. One crashed into the sea, though the pilot was rescued unhurt, but the others made it safely.

They were soon moving over the supply route, but on 1 December the Blenheim leading the first six lost its way on the flight to El Fasher, forcing all seven machines to crash-land in the desert on the approach of darkness. One pilot was killed and all the Hurricanes were badly damaged, two of them being written-off permanently.

Thus, only twenty-seven Hurricanes got to Egypt where they had to be checked after their gruelling trip, stripped of the tanks fitted at Takoradi and issued with new guns. By mid-December twelve were ready for action, being attached temporarily to 274; but not until the end of the month was 73 able to take its place in the fighting as a squadron.

Having delivered 73 Squadron, *Furious* returned to Takoradi on 10 January 1941, with a further forty Hurricanes.[22] Thereafter, despite the effects of the hostile environment on men as well as on aircraft, fighters continued to reach Egypt, if scarcely in a flood, at least in a steady stream. They were soon to be desperately needed.

During the pause in the ground fighting following Graziani's initial advance, the Hurricanes flew as escorts for Blenheims or for Lysanders on photographic sorties, but normally in their traditional role of interceptors. No 33's first major action with Hurricanes came on 31 October, against a raid on British forward positions by Savoia Marchetti SM 79s, escorted by CR 42s. It destroyed or damaged five bombers and one fighter. One Hurricane was lost, its pilot tragically being killed when his parachute failed to open. A second crash-landed but the pilot of this was unhurt.

By December, having received reinforcements, especially of tanks that were rushed out to him in fast merchantmen, Wavell, though still outnumbered, instructed O'Connor to advance. The latter's plan of attack was based on an extraordinary fact:

---

[22]She had had an eventful voyage also, since the convoy with which she had been sailing had been attacked at dawn on Christmas Day by the German heavy cruiser *Admiral Hipper*. Fortunately, the raider was driven off after causing only minor damage.

that south of the fort of Nibeiwa, the Italians had left a gap in their defences fifteen miles wide, neither defended nor even patrolled, but through which their camps could be assaulted from the rear in succession.

8 December, therefore, found O'Connor's men concentrated opposite this gap, sheltering under such cover as they could find, while above them the Hurricanes of 33 and 274 prowled in search of enemy aircraft. Undetected from the air, the troops pushed forward that night, the rattle of their Infantry or I tanks – the famous Matildas – drowned by the noise of British bombers overhead.

In the morning of 9 December, they took Nibeiwa completely by surprise, killing its commander, General Maletti, as he tried to rally his men. Thereafter, O'Connor, who had superior numbers of tanks though of nothing else, struck the other forts in sequence, the Italian artillerymen being demoralised when they discovered their shells literally bounced off the Matildas' armour. By the 11th, all the Italian positions from Sidi Barrani southwards were in British hands.

Code-named Operation 'Compass', this, the first major Allied offensive in North Africa, had resulted in the capture of 73 tanks, 237 guns, more than 1,000 vehicles and more than 38,000 men including four generals – all at a cost of 624 casualties. According to General Sir William Jackson in *The North African Campaign 1940-43*, it was 'one of those rare battles that did go according to plan. It was fought with professional standards that were never again achieved by the British in the Western Desert until Montgomery won El Alamein.'

Nor were the deeds of the RAF unworthy of the efforts of the Army. Just after midday on the 9th, 274's Hurricanes broke up a raid of SM 79s, claiming at least four of these, whose destruction was confirmed by the ground forces. Later that afternoon, it was again in action, this time claiming five CR 42s. In neither combat was any Hurricane lost.

While 274 protected the advance, 33 was operating farther afield. During the day it ranged over enemy positions near

Bardia, strafing motor transport, providing vital information of the Italians' movements and even finding sufficient time to down three CR 42s.

Yet the most amazing exploit of any Hurricane pilot came on the 11th. Flying Officer Charles Dyson of 33 Squadron, on a lone patrol, emerged from cloud to find that he was immediately behind six Fiat CR 42s flying in two Vic formations of three machines each, all acting as escorts to a single SM 79. Dyson fired a long burst at each formation in turn. He claimed that he had hit all six enemy fighters.

However, Dyson had little time to assess the damage he had done, for he was now attacked by another group of CR 42s. After a brief combat, he was forced to crash-land in the desert. He did not rejoin his squadron until six days later, when he found that his achievement had taken place in full view of army units who had reported not only the destruction of all the Fiats but that one of them, in falling, had smashed into the Savoia, bringing that down also. If this report was correct, then Dyson in one sortie had accounted for seven enemy aircraft confirmed – a feat that no other RAF pilot would ever equal.[23]

The next stages of the campaign were less spectacular – but still pleasantly satisfactory. By 17 December, the capture of Sollum had cleared Egypt of the enemy. Bardia, bombarded by the Royal Navy and the Royal Air Force, was taken on 5 January 1941. Tobruk, after similar treatment, was stormed on the 22nd. Dema fell on the 30th.

These advances were made under the protecting wings of the Hurricanes. 33 Squadron during December claimed thirty-six enemy aircraft destroyed, with ten probables and eleven

---

[23]It has been asserted frequently and strongly that the report was not correct. Certainly, these losses cannot be confirmed in Italian records but there are occasions when they have been shown to be incomplete. The report could have been mistaken but, in that case, it is surprising that it should have referred to the destruction of the Savoia, which Dyson had never claimed or apparently known about.

damaged – figures that seem to have been exaggerated only slightly, since most combats were individual ones where error was less likely to occur. The squadron's finest efforts were on the 13th, when it claimed two SM 79s and three CR 42s without loss, and on the 19th, when the totals were three SM 79s and four CR 42s – again without loss. 274 did even better on the 14th, claiming six SM 79s on one patrol and five CR 42s on another. It then achieved further victories in January, notably on the 5th, when it claimed six more enemy machines.

By this time, reinforcements had arrived. On 14 December, the Canadian Flight Lieutenant Smith shot down a Savoia Marchetti SM 79 – the first success for a pilot of 73 Squadron, though at the time he was temporarily attached to 274. 73's most expert pilot during this early period was Sergeant Marshall, who was soon established as a specialist in bringing down SM 79s: two on 16 December, three more on 3 January 1941, and a sixth the next day.

The Regia Aeronautica was also looking for aid. As it had become obvious that the Fiat biplanes could not match the Hurricanes, Fiat G50 monoplanes were ordered to Libya; but since these had a top speed of less than 300 mph, were not particularly manoeuvrable and were armed only with two 12·7 mm machine-guns, the Hurricanes outclassed them also – they suffered their first losses on 6 January, when Warrant Officer Goodchild of 33 shot down a couple of them.

The Hurricanes' dominance was now such that they had to seek far behind enemy lines for hostile aircraft. 274 was especially noted in this work, due largely to an alliance between the volatile Liverpudlian Pilot Officer Ernest Mason and the quiet South African 2nd Lieutenant Robert Talbot. On 8, January these officers went off on patrol over Martuba, each of them destroying a CR 42. Next day, they made two flights to Derna, shooting down a SM 79 apiece on the first one, while Mason claimed another SM 79 and Talbot a G50 on the second. Much to the regret of 274, Talbot was posted home on the 25th, though he was later to return to the Western Desert with I SAAF Squadron, again flying Hurricanes, but Mason, though bereft of his partner, destroyed three CR 42s single-handed next day.

Having achieved the first essential, the security of their own troops, the Hurricanes concentrated against their enemy on the ground. 274 guarded Fleet Air Arm Swordfish in attacks on Bardia. 73 protected RAF Blenheims raiding Tobruk. Flying almost at ground level, Hurricanes shot up troops, staff-cars, transports. They brought back Intelligence of enemy move-ments. Most of all, perhaps, they strafed hostile machines on their own landing-grounds. 274 destroyed eight SM 79s on the ground and damaged five more on 8 January. 73 in similar raids destroyed eight aircraft on 3 January, four more (plus several lorries) on I February, another eight on 5 February.

Under this pressure, the Italian Air Force for all practical purposes disintegrated, whereupon, deprived of its protection, the Italian Army on 3 February began pulling out of Cyrenaica. The retreat was revealed by air reconnaissance with the result, according to Alan Moorehead in *The Desert War*, that 'the Italians were never given a moment's rest. Through every daylight hour Hurricanes were swooping on them at three hundred miles an hour, or the Blenheims were bombing'.

On the 4th, after consultation with Wavell, O'Connor unleashed his forces in two different directions. While 6th Australian Division pushed along the coast road, through a fearful jumble of lorries wrecked by the strafing Hurricanes, to Benghazi, which was occupied on the 6th, 7th Armoured Division raced straight across the Cyrenaican 'bulge', cutting the Italian line of retreat at Beda Fomm.

The 6 February therefore saw the most savage conflict of the entire campaign as the Italians strove to break out of the trap. They had a heavy numerical superiority in tanks but instead of concentrating these, attacked piecemeal, dashing themselves vainly on British tanks firing from protected positions and backed by artillery. On the morning of the 7th, their last effort was broken by the anti-tank guns.

Twenty thousand prisoners were taken, as were 120 tanks and about 200 guns. A hundred more tanks were found wrecked on the battlefield The Italian leader, General Tellera, had been killed in action. Next day Marshal Graziani resigned, while the capture of the frontier post at El Agheila completed the conquest of Cyrenaica.

In almost exactly two months, the British and Commonwealth force, though never exceeding two divisions, had utterly routed an army of nine divisions, capturing 130,000 men, 1,300 guns and 400 tanks. Its own losses were 500 killed, 1,373 wounded, fifty-five missing.

There were many reasons for this wonderful result. It owed much to the strategic ability of General Wavell and the tactical brilliance of General O'Connor; much to the fighting skills of the men they commanded. It owed perhaps even more to the fact that the Italian army was neither trained nor equipped to oppose a modern mechanised force.[24]

Yet most of all, success came through total control of the air. The Italian moves were invariably reported, almost invariably attacked. By contrast, those of the British were rarely detected, let alone disrupted. This ascendancy was due mainly to the Hurricanes, which were so completely superior to anything the Italians could put against them that their pilots could do as they liked. No army in the desert would ever again achieve so unqualified a triumph – but then, none would ever again enjoy the measure of dominance in the air that was won by the Hurricanes during the first conquest of Cyrenaica.

---

[24]As will be seen later, O'Connor's career in the Desert was only too brief, for he was captured on 7 April 1941. He was released from a prisoner of war camp in Italy on the surrender of that country in 1943, rendering further distinguished services in Normandy, where, however, he learned how different it was to fight German soldiers using German weapons.

However, a similar command of the skies was gained by the Hurricanes in East Africa – a still more remarkable achievement this, since there were even fewer of them, but one that again had gratifying consequences for the ground troops.

The first move in what became a series of breathtaking advances was a singularly modest one on 6 November 1940, envisaging only the re-capture of the Sudanese frontier fort of Gallabat followed by the seizure of the corresponding Italian post in Abyssinia, Metemma. Even allowing for the advantages enjoyed by defenders, it must have seemed that a British and Commonwealth force of about equal numbers, obtaining complete surprise, backed by tanks that the Italians did not possess, and led by an officer of the calibre of Brigadier Slim, later the commander of Fourteenth Army in Burma, could hardly avoid achieving these limited objectives.

It was not to be. Gallabat was taken under cover of attacks by British aircraft, but before the troops could mount an assault on Metemma, it was discovered that the most important factor of all had been overlooked: the Regia Aeronautica's command of the air. First, superior numbers of Fiat CR 42s drove away the protective cover of Gladiators, shooting down five of them without loss. Next, Italian bombers pounded the British forces, whose advance broke down amid circumstances of disgraceful panic. Then, all the following day, they turned against Gallabat such an accurate bombardment that Slim abandoned this position also.

Thus, entirely due to the enemy's air superiority, the first British offensive against the despised Italians had been an ignominious failure. If Slim had any doubts remaining as to the value of fighter protection, they were removed a couple of months later when a strafing CR 42 put him out of the campaign with a bullet in the buttocks. No wonder that when he commanded in Burma, he regarded air power as all important.

In December, the entire situation changed. Their defeats in Cyrenaica undermined the morale of the Italians in East Africa also. At the same time, Hurricanes began to reach No 1 Squadron South African Air Force. These ultimately came to the massive total of ten, but to the local commanders, Lieutenant-General Platt and Air Commodore Slatter, each one was very precious. They began by attaining command of the Sudanese air-space. On the 16th Captain Driver shot down a Savoia Marchetti SM 79. Next day, he damaged two others.

By mid-January, 1941 the Duke of Aosta had become highly pessimistic. Correctly foreseeing a British offensive to re-capture Kassala, he ordered a withdrawal to Agordat and Barentu, medium-sized towns set in mountainous country, which threatened to prove excellent defensive positions, particularly if the Italians could oppose the advancing troops with a weight of air-attack similar to that brought against Slim's men.

Fortunately, the co-operation, born of mutual confidence, between Platt and Slatter, exceeded even that between O'Connor and Collishaw. On 19 January, Platt occupied the airfield at Sabdaret, Eritrea. To it came the Hurricanes of 1 SAAF. On the 26th and 27th, they made dawn strikes against the main Italian aerodromes at Asmara and Gura, which lay just within their range. As in Cyrenaica, the Hawker fighters proved highly effective against aircraft on the ground, destroying about fifty. This in turn so discouraged the Italian Air Force Commander, General Pinna, that he withdrew his remaining machines from both bases, to which (apart from rare individual sorties) they never returned.

These attacks snatched air superiority away from the Italians – they thereby ensured the capture of Agordat and Barentu, which fell by 2 February. At the latter was an airfield, to which No 1 SAAF now moved, and from which it took off to strike Makelle, the only remaining enemy air base in Eritrea, with results that were equally satisfactory – for the Hurricanes.

General Wavell now ordered Platt to thrust eastward to Asmara, the Eritrean capital, and Massawa, its port, which dominated the Red Sea. But halfway to the capital was the stronghold of Keren where the only way through a wall of mountains, rising more than 2,000 feet, was a single deep, narrow gorge, which had been blocked by the Italians. Furthermore, Aosta felt that the loss of Asmara and Massawa – particularly the latter, which contained his main naval base and army installations – would end all chance of effective resistance. He therefore committed to the defence of Keren the bulk of his central reserve, including three battalions from the Savoia Division, the finest troops under his command.

They did not fail him. All Platt's efforts to force a passage, commencing early on 3 February, proved in vain. By the 12th, mounting casualties brought about a pause during which supplies were slowly built up for the next round of assaults.

This lull gave the Regia Aeronautica a splendid opportunity of attacking Platt's lines of communication, maintained over country that was rough to say the least. It failed miserably in its attempt because it was quite unable to break the grip established by the Hurricanes – at least eight Italian fighters fell to 1 SAAF during February, while several more were damaged. On the contrary, it was the RAF that struck at the enemy army, strafing positions, reporting movements and dropping leaflets encouraging desertions – though this last-named task proved largely a waste of time.

On 15 March, under cover of RAF raids, the attack on Keren reopened. Despite frantic appeals from army leaders, General Pinna's airmen were powerless to intervene. On 21 March, CR 42s tried to strafe British positions only to be pounced on by the Hurricanes, Lieutenant Pare shooting down two of them. On the 27th, a squadron of 'I' tanks finally forced the roadblock, whereat the Italians retired in good order. Platt had lost more than 500 dead and more than 3,000 wounded, but the conquest of Eritrea was assured. Asmara surrendered on 1 April, by which time the Hurricanes were already machine-gunning lorries fleeing down the road to Massawa. The port fell a week later, enabling President Roosevelt to declare the Red Sea no longer a combat zone, thus allowing US ships to lift some of the supply burdens from the British mercantile marine by sailing direct to Suez.

Platt had not only over-run Eritrea, he had helped to prepare the way for a still more spectacular advance from Kenya. Here the British and Commonwealth forces were under Lieutenant-General Sir Alan Cunningham, the brother of the naval commander. His target was Italian Somaliland but he had first to cross the all-but-waterless Northern Frontier district of Kenya. Yet instead of waiting for the spring rains as originally intended, he in fact began to advance, on 24 January. While Churchill's claim that his 'prodding' prompted this decision is accepted by Liddell Hart among others, it does less than justice to Wavell and Cunningham, who took far more note of the discovery of limited supplies of water across the frontier in the line of the advance and the decline in Italian morale caused by Platt's early successes. It seems at least probable that Cunningham was also encouraged by the spirit being displayed by a SAAF contingent under the command of his colleague Air Commodore Sowrey.

This included No 2 Squadron SAAF, which flew an extraordinary mixture of Hurricanes, Gladiators, Gauntlets and Furys, and No 3 SAAF, equipped with Hurricanes alone. They were soon in action, though No 2 shortly transferred its Hurricanes to the care of No 3.

The pilots of this squadron did fine work in the role of interceptors, particularly Captain Frost who on 3 February won a DFC for destroying three Caproni Ca 133s and a Fiat CR 42, all of them confirmed on the ground. They also made effective raids on enemy aerodromes. On 10 February, that at Afmadu was the target of an effective attack, which was followed by an equally useful strike at Gobwen.

Once more, these actions gave the Hurricanes air superiority, which enormously simplified Cunningham's advance.

Crossing the Juba River, with Sowrey's squadrons strafing before him, he pushed on to Mogadishu, the capital of Italian Somaliland. It fell without a struggle on the 25th, leaving in the victors' hands vast quantities of stores, including 350,000 gallons of petrol and 80,000 gallons of aviation fuel, which enabled the advance to be followed-up without delay. Also, twenty-one wrecked aircraft were left on the aerodrome.

Cunningham then turned inland, striking into Abyssinia from the south. Since Aosta's finest troops were now engaged at Keren, resistance on the ground was negligible, but the Regia Aeronautica operating from Diredawa showed spirit until the Hurricanes paid this base a visit on 13 March, shooting down two CR 42s and badly damaging one SM 79, which force-landed, but losing Lieutenant Dudley, killed in action. On the 15th, 3 SAAF returned, this time catching the Italian aircraft on the ground. Thirteen of these were destroyed or damaged but a single machine-gun post shot down two Hurricanes. Captain Harvey was killed but Captain Frost crash-landed without injury. On seeing this, Lieutenant Kershaw landed beside Frost and took off again safely with his colleague sitting on his knees – a feat for which he was awarded a DSO.

Cunningham's advance had by now convinced the Italians of the need to abandon British Somaliland also. A force from Aden landed at Berbera on the 16th, quickly recapturing the whole colony, thus enabling Cunningham to re-route his supplies through this much nearer port. Pressing on with the utmost speed, he brushed aside feeble stands at Marda and Babile Passes, potentially as formidable defensive positions as Keren, to enter Harar on the 25th.

There now came the final blow to the enemy's nerve. At Diredawa, fifty miles north-west of Harar, the native troops turned on the Italian civilians committing all manner of atrocities especially against women. In response to a plea from the authorities, a South African force rushed to the town, where it took a day and night of street fighting to quell these thugs. This episode convinced Aosta that an attempt to hold Addis Ababa would only result in similar horrors on a greater scale. At the same time, Cunningham felt able to advance on the capital since the capture of Diredawa aerodrome had given his Hurricanes an advanced base.

As the final dash to Addis Ababa began, the Hurricanes again played their part. They provided fighter cover, shot up Italian lorries, escorted ex-South African Airways Junkers Ju 86s in bombing raids, which broke a planned stand on the Awash River, and especially strafed airfields. On 4 and 5 April, attacks on that at the capital destroyed twelve enemy machines, and these were only the last of a series of raids – when the base was taken, there were thirty-two wrecks encumbering it. Next day Addis Ababa fell – to the relief rather than otherwise of the Italian population.

The capture of the capital did not stop the Hurricanes' pressure against the remaining hostile forces. On the 6th, they raided Dessie airfield at which ten Italian aircraft were destroyed on the ground, while three CR 42s were shot down or crash-landed. On the 10th, they followed this with attacks on the airfields at Shashamanna and Jimma, destroying two more enemy machines in the air and eight more on the ground.

These strikes virtually wiped out the Italian Air Force in East Africa. On 30 April, Captain Frost completed the story when he shot down the last SM 79 that was still airworthy. The Hurricanes of 1 SAAF in Eritrea now moved to the Middle East, leaving those of 3 SAAF to represent the type at the elimination of the remaining Italian garrisons.

Of these the chief was at Amba Alagi, whither Aosta had retired. Here he was penned in from all sides, including by local 'patriots' whose ill-treatment of prisoners unnerved the Italians. On 19 May, the Viceroy surrendered to a force of South Africans from whom he could rely on proper treatment for his men. Other scattered enemy groups were mopped up during the summer so that by early July there remained only the mountainous district surrounding the fortress of Gondar, the conquest of which was postponed by the coming of the rainy season. In August, 3 SAAF converted to Curtiss Mohawks, handing over its few remaining Hurricanes to another South African squadron, No 41, which flew them, together with aged Hawker biplanes, on the occasional close-support mission. Ground operations against Gondar resumed in late September – they brought about the surrender of this last outpost by 28 November.

The pilots of 41 SAAF headed for the Western Desert where there was a more formidable foe – the Luftwaffe.

–

The Luftwaffe, in the Mediterranean area, had first appeared in the skies around Malta. The German airmen must have been highly annoyed to find that the Hurricanes had preceded them there.

On the outbreak of hostilities with Italy, the Air Officer Commanding, Air Commodore Maynard, had a headquarters, a radar station, airfields at Luqa, Hal Far and Takali; but only a small Station Fighter Flight under Squadron Leader Martin to take advantage of these facilities. It is widely 'known' that the air defence of Malta was entrusted to three Sea Gladiators called 'Faith, Hope and Charity' by 'the grateful people of Malta'.

It is still a delightful story although tiresome historians have shown that there were about half-a-dozen Gladiators on hand (though probably never more than three airborne at any one time) and that the names were the invention of enterprising pressmen in Britain, were not used by the Maltese during the period in question and became known to the AOC only after he had left the island the following May. More to the point, from very early on, the Gladiators were supplemented by Hurricanes.

It has already been mentioned that before the fall of France, Hurricanes were passing through Malta en route to the Middle East. Pleas that they be allowed to remain were refused at first, since priority had to be given to the protection of the Fleet Base at Alexandria. However, by 28 June, the higher authorities had relented, for on that day, four Hurricanes under the command of a South African, Flight Lieutenant Barber, returned to Malta from North Africa. They carried six Browning machine-guns only since it was thought that the loss of weight would be of value in the reduced take-offs necessary on Malta's short runways.

No further reinforcements arrived for another month but though the Italians, who could muster about 200 machines in Sicily, a mere sixty miles distant, made daily air raids, only one Gladiator and one Hurricane were destroyed in combat. The Hurricane was lost on 16 July, together with Flight Lieutenant Keeble, but the same day saw the Fighter Flight's greatest achievement, ten Italian aircraft being claimed as destroyed or damaged.

However, it was clear that if Malta was to be held, let alone built up as a base from which naval or air striking or even reconnaissance units could operate, then an adequate fighter defence must be provided. The fall of France having severed the old supply route, it was decided that the fighters would be conveyed in an aircraft carrier to within striking distance of the beleaguered island, whence they would fly the rest of the way.

Accordingly, 418 Flight was formed at Abbotsinch with 12 Hurricanes, which were embarked on the carrier *Argus* on 20 July. On the 30th, she reached Gibraltar from which the ground staff and stores were dispatched to Malta by submarines. Next day, *Argus* sailed with her precious cargo to a take-off point south of Sardinia, about 200 miles from Malta. The importance of her task may be judged from the fact that she was guarded by the carrier *Ark Royal*, battleships *Valiant* and *Resolution*, cruisers *Enterprise* and *Arethusa* and ten destroyers. The mission, with nice double-meaning, was given the code-name Operation 'Hurry'.

The launching position reached on 2 August, two Fleet Air Arm Skuas took off to act as guides. Then it was the Hurricanes' turn. They went off the deck like birds. All reached Malta safely, though one crashed on landing, happily without serious injury to the pilot. They amalgamated with the old Fighter Flight to form 261 Squadron under Squadron Leader Balden.

In the face of this opposition, the Italians were forced to mount ever larger escorts for their bombers. They also tried their hands at dive-bombing the airfields in September, using Junkers Ju 87s purchased from the Germans, but after the Hurricanes had shot down two, such raids died away. By the end of November, the Malta fighters were credited with having destroyed or damaged a total of thirty-seven of their enemies, prompting the Regia Aeronautica to appear only after dark. Even this did not give complete immunity, for on 18 December, Sergeant Robertson shot down a Savoia Marchetti SM 79 at night.

The Hurricanes' effectiveness was the more remarkable in that, although inevitably they suffered occasional losses, there was no further attempt to provide reinforcements until 17 November. Codenamed Operation 'White', this launch of twelve Hurricanes and two Skuas from *Argus* seemed a repetition of the August mission, but it was attended with very different results.

It appears that on this occasion the staff work had been badly neglected. The aircraft on Operation Hurry had been equipped with variable-pitch propellers but those on the November mission carried constant-speed propellers with which the pilots were quite unfamiliar. They were never briefed on the very different adjustments of pitch and of throttle needed to ensure economy of fuel consumption – vital for a flight of almost 450 miles. Also, the fighters' best altitude for such a trip was 10,000 feet but their pilots were briefed to fly at 2,000 feet, where in the 'heavier' lower air, their range was much diminished. Also, the extent to which the wide tropical radiators would reduce speed and range was not fully appreciated.

Yet all might still have been well if a following wind at the time of take-off had not veered to blow almost exactly against the aircraft, bringing with it patches of thick sea-mist. When the first section of one Skua leading six Hurricanes, made a rendezvous with a Sunderland flying-boat from Malta, it was already 25 minutes behind schedule. On the final stage of the journey, two Hurricanes ran out of fuel to plummet into the sea. The Sunderland was able to pick up one pilot but the other was lost. The four surviving Hurricanes reached Luqa with only enough fuel left for another five minutes' flight.

The observer in the Skua guiding the second section was on his first flight out of training school. In the increasingly difficult conditions, he never met up with his Sunderland. Nor did he contact Malta, presumably because of a fault in his radio. Since there was no time to be lost, the flight tried to find the island on its own. The Skua finally made a landfall in Sicily, where it was shot down by anti-aircraft guns. The Hurricanes were never seen again.

At least important lessons were learned. Later missions were conducted with far greater efficiency. Never again were such risks taken. Never again did such a tragedy recur.

Yet even without reinforcements, the Hurricanes' dominance was a fact. Mention has been made of the presence on Malta of Sunderlands. Wellington bombers were also now based there; so were Swordfish; so also Glenn Martin Marylands for long-range reconnaissance. In November, these performed their most valuable task – which earned a letter of thanks from Admiral Cunningham – by getting detailed information of the dispositions of the Italian fleet in Taranto harbour. On the night of 11 November, the carrier *Illustrious* flew off twenty-one Swordfish. At the cost of two aircraft, the crew of one of which survived as prisoners of war, these crippled three Italian battleships and dealt a shattering blow to Italian morale.

In one respect, the attack was too successful. Hitler, dissatisfied with his ally's conduct of the fighting, determined to provide a stiffening, which the Italians were very willing to accept. His first step was to send Fliegerkorps X to Sicily.

Commanded by the aggressive General Geissler, this numbered some 350 aircraft specially trained and equipped for attacking ships. By 8 January 1941, the first waves of this force, including the crack StG 1 and StG2, had arrived.

During the 10th and 11th of the month, they bombed warships escorting a convoy to Malta – carrying among other things twelve crated Hurricanes – sinking cruiser *Southampton* on the 11th.

The day before, the Stukas had taken revenge for Taranto. In a brilliant attack they hit *Illustrious* six times, causing damage that must surely have sunk any carrier not possessing an armoured flight deck. As it was, though hit once more in a later raid, she managed to limp to Malta. It was clear that Fliegerkorps X would try to finish her off before she could effect the temporary repairs needed to take her to safety.

The storm broke on 16 January. A massive raid of about eighty Stukas and Ju 88s was met by a determined, extremely well-directed anti-aircraft barrage, above which the Hurricanes, aided by three Fulmars from *Illustrious*, savaged the attackers. Though there were now fifteen Hurricanes in Malta, such were the difficulties of servicing them – which frequently necessitated their using parts from various other types of damaged aircraft – that there were never more than half-a-dozen of them in the air at any one time. However, despite the odds against them, they claimed to have destroyed or damaged five enemy machines. It seems that they also helped to put the Stukas off their aim, as only one bomb hit the carrier, causing minor damage – but those that missed laid waste the neighbouring dockyards and residential areas, killing nearly 100 civilians.

Certainly, the disruption caused by the fighters was not appreciated by General Geissler. Two days later, his forces returned, to strike not the carrier but the airfields at Hal Far and Luqa – Takali was already inoperative after heavy rain. This raid put Luqa out of action for a time and wrecked six aircraft on the ground, but those that were airborne claimed to have destroyed or damaged seven of the raiders.

On the 19th Fliegerkorps X resumed its assault on *Illustrious*. Two near-misses caused underwater damage but this time there were few civilian casualties. Once more the defending fighters performed their primary task of breaking up the co-ordination of the attack. They also claimed to have destroyed or damaged eleven of the attackers.

For all their numbers it was the German pilots who were now exhausted. Apart from many damaged bombers temporarily out of action, 16 enemy machines were total losses, about five falling to the AA guns, the rest to the Hurricanes. No further attack was made before, after dark on the 23rd, *Illustrious* slipped out of Grand Harbour to reach Alexandria, battered but intact, two days later. Only one Hurricane and its pilot and three Fulmars and one Fulmar pilot had been lost in combat.

To help Maynard make good these losses, Longmore took the opportunity, on 30 January, of sending him six Hurricanes under the command of Flight Lieutenant Whittingham from newly captured bases in Cyrenaica. However, as that great admirer of the Hurricane, Admiral Cunningham remarked: 'That is no good. He ought to have two full squadrons and at once.'

In February, Maynard was promoted to Air Vice-Marshal but he did not get his two squadrons. Yet these were more than ever required, for having recovered from their losses in what the Maltese called the '*Illustrious* blitz', and no doubt spurred on by the establishment at the island of submarines to prey on the Axis supply lines, Fliegerkorps X resumed its attempt to neutralise Malta – though fortunately the German High Command did not appreciate the far greater desirability of eliminating it altogether. The raids concentrated mainly on dockyards and airfields. On 26 February, Luqa was put out of action for forty-eight hours, six Wellingtons being destroyed on the ground. The Hurricanes claimed to have shot down or damaged seven of the attackers. This was followed at the beginning of March by a strike on Hal Far, which wrecked three Swordfish and rendered the aerodrome temporarily unserviceable. Again, 261 Squadron met the raiders, being credited with destroying or damaging another seven for the loss of one Hurricane together with its pilot.

The Luftwaffe was now also carrying out sweeps by Bf 109s, which, being less severely handicapped by tropical equipment than the Hurricane Is, could outfight them, particularly at higher altitudes. By catching the Hurricanes as they climbed, the Messerschmitts inflicted heavy casualties on them – though 261 still downed the occasional bomber, among its pilots being Sergeant Robertson, who by the time he was posted to the Middle East early in April, had been credited with twelve 'kills'. However, under the constant pressure the Sunderlands and Wellingtons had to be withdrawn to Egypt. At the same time, the islanders were suffering real hardships as essential stocks fell steadily lower.

Yet the Hurricanes hung on grimly until matters improved. On 23 March, a convoy from Alexandria arrived. Thirty Junkers Ju 87s, escorted by twenty 109s, made a determined attempt to smash the ships in harbour, but although they damaged two, they were thwarted by 261, which shot down or damaged some seven of the Stukas.

Then, on 2 April, twelve more Hurricanes flew off the carrier *Ark Royal*. These were Mark II As, which, with their improved climb, could meet the enemy more effectively. This is not to say that 261 found the fighting easy. On 11 April, for instance, a savage combat saw the loss of two Hurricanes with their pilots, while four more were damaged, one so badly that it was wrecked on landing, although the pilot escaped with slight injuries. During April, the defenders lost, in all, seven aircraft and three pilots.

However, it was also on the 11th that a flotilla of destroyers was based at Malta. By the middle of the month, the Sunderlands and Wellingtons were back and the latter were striking at Tripoli, a fact that suggests that the Mark IIs had reduced the enemy's ability to bomb accurately. On 27 April, *Ark Royal* flew in a further twenty-three Hurricanes and, shortly thereafter, Fliegerkorps X began an unobtrusive withdrawal to the Balkans. Malta had weathered the storm – the first storm at any rate.

Chapter Seven

The Balkans and The Middle East

Other storms had already broken over the unhappy countries of the Balkans. Hitler, a superb professional politician, believed these could be forced to submit by peaceful means. By October 1940, Hungary and Rumania were firmly under German domination, while effective pressure was being brought upon Bulgaria and Yugoslavia. Mussolini, however, obsessed with thoughts of personal greatness, was determined on a cheap military victory. On 28 October, Hitler met the Italian dictator in Florence. As he stepped off the train, Mussolini greeted him with the words: 'Führer we are on the march!'

Before dawn, the Italian army in Albania had swarmed into Greece. Cheered by propaganda promising an easy conquest, its regiments, which boasted such blood-curling titles as the Wolves of Tuscany, the Hercules of Ferrara, the Demigods of Julia and the Red Devils of Piedmont pressed forward confidently, but within a week the Greeks, fighting savagely on their wooded mountain slopes, had halted every advance. They then counter-attacked, driving out the last Italians by 23 November, after which they in turn invaded Albania, where they made slow but steady progress.

In these struggles, the Greeks enjoyed the support of the British Air Forces in Greece (BAFG) under Air Vice-Marshal D'Albiac, which by the end of November consisted of three squadrons of Blenheims and the Gladiators of No 80, which soon built up an impressive record against heavy odds. For the moment the Greeks, who believed rightly that Hitler was not pleased with the Italian action, showed no wish for help from the British Army, though British detachments did take over the defence of Crete. However, since it seemed probable that Germany ultimately would feel compelled to aid her miserable ally, it was decided that strong British reinforcements would go at once to Greece's aid if Bulgaria or Yugoslavia was occupied.

This possibility prompted a realisation that the Gladiators would be inadequate to deal with German aircraft, while the appearance in February 1941 of Fiat G50 monoplanes further stressed the need for more modern fighters. On 7 February, therefore, 'B' Flight of 80 Squadron, stationed in the mountain valley at Paramythia, thirty miles south of the Albanian frontier, received six Hawker Hurricanes.

After a period of intensive training on the new types, the flight was ready for action on the 20th. That day, it escorted Blenheims in a raid on Berat, which hit supply dumps and a major road bridge. As the bombers completed their attack, a formation of G50s appeared but was driven off by the Hurricanes. Flight Lieutenant Woods shot down an enemy fighter; Sergeant Casbolt shot down two. Yet the very first Hurricane victim in the Balkans was the Fiat leader, which exploded after receiving a vicious burst of fire from the Hurricanes' commander, Flight Lieutenant Pattle.

No man could have been more worthy of this distinction. 'Pat' Pattle, as he was called throughout the RAF – he was wise enough to guess that Christian names of Marmaduke Thomas St John would bring on a certain amount of leg-pulling – was a South African from Butterworth, Cape Province. Although a small, quiet, calm man, he was recognised as an outstanding leader, a brilliant tactician and a superlative pilot; in one instance he landed a Gladiator that had lost its port wheel so beautifully that it suffered only minute damage and indeed was airborne again later that day, with Pattle, incidentally, again at the controls.

When to these skills are added exceptional eyesight and phenomenal shooting ability, there can be little wonder that Pattle was deadly in combat. He had already gained fame flying Gladiators in the Western Desert as well as in Greece, but after he converted to Hurricanes his exploits became unbelievable. He is now officially recognised[25] as the top-scoring RAF ace with forty confirmed victories, but even that does not give full weight to his achievements, for Pattle was no boaster and there is small doubt that more than once he claimed only a 'probable', which in fact crashed – indeed, on at least two occasions a count of enemy wrecks on the ground appears to have confirmed this. In addition, he 'shared' in the destruction of four other hostile machines. Finally, it should be added that the records of the Greek campaign are by no means complete; many of Pattle's colleagues believe his final total may well have been nearer sixty.

Pattle and his companions soon received reinforcements – the Hurricanes of 33 Squadron, like No 80, transferred from the Western Desert. The fighters strafed ships in the main Albanian harbour of Valona, shot up targets on land such as lorries or troops concentrations or headquarters buildings, and escorted the Blenheims. Their work was soon to reach an exceptional peak.

Aware of the likelihood of Germany entering the struggle, the Greeks attempted to break Italian resistance by a push towards Valona, with the BAFG providing close support. On 27 February, Hurricanes from both 33 and 80 covered Blenheims in a raid on the port, beating off an attack by CR 42s in the process. The biplanes stood no real chance. It is reported that seven were shot down though there may well be some duplicated claims in this figure. Two more collided. The Hurricanes had no losses.

Next day provided the greatest moment of all. Sixteen Hurricanes from 33 and 80, plus a dozen Gladiators from 80 and

---

[25]Even in that ultimate arbiter of argument *The Guinness Book of Records*.

112 – which last-named squadron was another recent reinforcement – providing cover to the Greek Army, now approaching Tepelenë, engaged large numbers of Italian machines in continuous air battles from the mountain ranges in eastern Albania to the coast in the west. In the course of them, the pilots claimed that they had brought down twenty-seven Italian aircraft; but although the advancing Greeks later reported that every one had been confirmed on the ground, enemy records indicate that they again made many duplicated claims. No Hurricane was lost. One Gladiator crashed but the pilot baled out unhurt.

During the conflict there were two remarkable individual efforts. Pattle destroyed two Fiat BR 20 bombers, but then had to return to Paramythia since oil from his victims had splattered over his windscreen reducing visibility almost to nil. Re-armed and re-fuelled, he was soon back in the fight, this time downing at least two, probably three, CR 42s near Valona. Flying Officer Cullen also claimed five enemy aircraft: two Savoia Marchetti SM 79s, two CR 42s, one BR 20.[26] Signals of congratulation, from Longmore, from D'Albiac, poured into Paramythia.

With the aid of the RAF, the Greeks reached Tepelenë but were unable to take Valona in the face of mounting opposition. In any case, their attention was now distracted to their north-eastern frontier, for, on 1 March, Bulgaria, which had now joined the Axis Pact, was occupied by Germany. In response, a British force of one armoured brigade and three infantry divisions under General Wilson was shipped to Greece.

Thus, March passed tensely. The Hurricanes of 80 Squadron again had many clashes with the Regia Aeronautica, Pattle destroying three G50s on the 4th. 33 Squadron, based now at

---

[26]In contrast to Pattle, 'Ape' Cullen, an ex-racing motor-cyclist, was a completely reckless flier who scorned tactics. On 3 March, he destroyed four Cant Z1007s to bring his total of victories to sixteen, but next day he did not return from a patrol and was never seen again.

Larissa in the east, had a less easy time. Experiencing problems in keeping its aircraft serviceable, it claimed only seven enemy aircraft during the month. However, on the 23rd, it strafed the airfield at Fieri, destroying or damaging a number of G50s on the ground.

At the end of March, events suddenly moved with dizzy rapidity. On the 25th Yugoslavia joined the Axis Pact. Two days later, a dramatic military coup in Belgrade forced the Prime Minister, Mr Cvetkovic, to resign. The Regent Prince Paul was exiled. The young King Peter, who had escaped from Regency control by sliding down a rain-pipe, was installed in the exercise of full regal powers. The streets of the capital were filled with delirious patriots shouting: 'Better war than the Pact; better a grave than a slave.'

Such a choice soon proved more than academic. The defection of Yugoslavia brought the possibility of an attack on the wretched Italians in Albania from the rear. Then on the night of the 28th/29th at Cape Matapan, Admiral Cunningham dealt a smashing blow to the Italian Fleet, which ensured British mastery of the reinforcement route to the Balkans. On learning of these events, Hitler exploded into one of his volcanic rages. Ordering his planned attack on Russia to be postponed for 'up to four weeks', he announced that Yugoslavia must be crushed 'with pitiless harshness'.

On 6 April, the onslaught began. The German Twelfth Army under Field-Marshal von List, supported by Fliegerkorps VIII, struck into the Vardar valley from Bulgaria to drive a wedge between Yugoslavia and Greece. Three days later, the German Second Army, under General von Weichs, crossed Yugoslavia's northern frontier with Hungary, aided by Luftflotte 4, which had previously been engaged in Operation 'Punishment', a two-day bombardment that had reduced Belgrade to a mass of flaming rubble in which lay the corpses of more than 17,000 civilians.

At this time, the Royal Yugoslav Air Force possessed thirty-eight Hurricanes equipping three fighter squadrons. Some of these engaged the Belgrade raiders – though in the confusion many unhappily attacked Yugoslav Messerschmitts also trying to protect the capital – but they could do little in the face of impossible odds. For a week they battled with almost a thousand enemy aircraft or strafed the advancing German troops. It was a hopeless gesture. Belgrade fell on the 12th. King Peter left the country two days later. The Yugoslav army capitulated five days after that.

By then, the Yugoslav Hurricanes were already no more. Most had had to be destroyed on the ground as their bases were over-run. A few escaped to Greece, reaching Paramythia on the 15th together with remnants of other squadrons, but hardly had they landed when a vicious German strike wiped out the whole lot on the airfield.

For on the same day that he attacked Yugoslavia, List had also invaded Greece, his task being made easier by the fact that the defenders were widely scattered in three main areas; fifteen Greek divisions on the Albanian front, three others plus General Wilson's forces on the Yugoslav frontier and three more Greek divisions in the east opposite the southern frontier of Bulgaria. Taking full advantage of this, one German column thrust down the Vardar valley to cut off the eastern divisions, which surrendered by 9 April, while on the 12th, after a three-day resistance by British, Australian and New Zealand troops, another broke through the Monastir gap, thereby separating the other two Allied forces. General Wilson's command began a rapid retreat to Thermopylae with its old tanks breaking down by dozens – fifty-two were lost, only one by enemy action – but the Germans trapped the Greek Army in Albania, which was forced to capitulate by the 20th.

As the German soldiers had taken over from the Italians, so the Luftwaffe had taken over from the Regia Aeronautica. The Hurricanes still had some encounters with the latter, as on the 6th, when the Canadian Flying Officer Woodward of 33, although having taken off with only four of his guns loaded, shot down three Cant Z1007s; but in the main their enemies now were Germans. On the night of the 6th, a Luftwaffe attack on Piraeus harbour hit an ammunition ship, causing an explosion that wrecked ten more vessels, virtually put the port out of action and broke windows in Athens, seven miles away.

Nonetheless, it is interesting that this attack was made after dark. During the day, an offensive sweep by 33 over Bulgaria had brought another notable combat with 109s. Of the twelve Hurricanes, none was lost. Of the twenty Messerschmitts, five were shot down, two by Flight Sergeant Cottingham, one by Flying Officer Wickham; but the first German aircraft – the first two in fact – to be destroyed by the Hurricane in the Balkans fell, like the first Italian machine, to the extraordinary Pattle, now a squadron leader and 33's CO.

However, as the land forces retreated, the story followed the tragic pattern of earlier Blitzkriegs. The Hurricanes strafed enemy columns, shot up 109s on their own airfields, protected the retiring armies. On the 14th Flight Lieutenant Dean and Flying Officer Woodward of 33 spotted six Junkers Ju 87s about to attack Allied troops. When the Hurricanes appeared, the escorting 109s unaccountably disappeared into some clouds, whereupon Dean and Woodward, amid the enthusiastic acclamation of the soldiers, shot down three Stukas, all confirmed on the ground, and damaged the other three.

Yet such successes could do little more than bring down on the RAF the full fury of the Luftwaffe. Without radar cover, without even an early-warning observer system, which had either broken down or been over-run, the British bases were dangerously vulnerable to surprise attack.[27]

---

[27]Also, once more the odds against the Hurricanes were colossal. On the 13th, for instance, Flying Officer Dyson of 33 had met a formation eighty-six strong. Feeling that such opposition was a little too formidable, he 'withdrew strategically into the sun'.

On 15 April, fifteen Messerschmitt Bf 109s pounced upon Larissa just as three of 33's Hurricanes were taking off. Pilot Officer Cheetham was brought down and killed almost at once. Flight Lieutenant Mackie fastened onto the 109 responsible, shooting it down as well. The pilot baled out but was riddled with bullets by Greek machine-gunners protecting the aerodrome – an action that, to their credit, infuriated the squadron's pilots far more than the raid, which they regarded as a legitimate act of war. Mackie was also killed immediately afterwards, his machine bursting into flames as it hit the ground, but Sergeant Genders was able to keep up a running fight with the attackers, one of which he destroyed. He landed without a bullet-hole in his Hurricane.

D'Albiac, who arrived at Larissa just after this strike, now determined to withdraw the BAFG to the area around Athens. All had retired by the 17th apart from 208 Squadron, which was now based at Paramythia, having been sent to Greece with its reconnaissance Hurricanes after the German invasion had begun. It fell back two days later in the very nick of time, for the Hurricanes had hardly got airborne when a devastating raid hit the airfield, wiping out the Greek Gladiator squadron that had shared it with them. Even further south the Hurricanes were not safe. At dawn on the 19th, a raid on Eleusis by Junkers Ju 88s demolished seven damaged machines awaiting repair in a hangar.

From their new airfields, the fighters kept up their hopeless struggle. Officially they now numbered two squadrons, 33 and 80, but the aircraft of 208 were also pressed into service. No 80 gained high honours on the 17th, when it caught a force of Ju 88s attacking a factory, claiming to have downed four without loss. On the 19th it claimed four Stukas and a 109, again without loss. The same day 33, having butchered a wretched Henschel Hs 126 observation aircraft, met a formation of 109s of about equal strength. In the resulting maul, three of the German fighters were destroyed. Three Hurricanes were badly damaged but were able to crash-land. Two of the airmen were unhurt but the Hurricane of Flying Officer Holman, an ex-policeman from Rhodesia, turned a somersault in a marshy field. The pilot, who had loosened his safety harness, broke his neck. When the adjutant arrived to take charge, he found the local people had already laid out the body, which they had covered with flowers.

That evening, the squadrons joined to disperse a formation of Ju 88s, destroying two without loss – but it was clear that attrition would shortly reach its inevitable end. On the morning of the 20th, two Hurricanes of 80 Squadron were lost as a result of combat. One was shot down, causing the death of Pilot Officer Still. The other carried Sergeant Bennett safely back to base but was so badly damaged that it was never repaired. That brought the total number serviceable down to fifteen.

All fifteen Hurricanes took off that afternoon, the squadrons now banded together under the inspirational leadership of the indomitable Squadron Leader Pattle who although suffering from a high temperature, had refused to demoralise his men by retiring to a sickbay. Their duty was to intercept a raid of Ju 88s, escorted by Bf 109s and Bf 110s, on the shipping in Piraeus harbour. They found about 100 enemy aircraft filling the sky over the port. Directing three of his pilots to tackle the Junkers, Pattle led the remainder against the escort.

A tremendous dog-fight broke out in which it is claimed that the Hurricanes shot down fifteen of the enemy. Five Hurricanes were also lost. Sergeant Cottingham of 33 helped to shoot down three 110s but was then hit by the rear-gunner of a fourth and baled out, wounded. Flight Lieutenant Kettlewell and the Frenchman Flight Sergeant Wintersdorf, both of 80 Squadron, also escaped by parachute after each had claimed a 110.

Two of the Hurricane pilots were killed, both from 33 Squadron and both as a result of complete unselfishness. The South African Flight Lieutenant Starrett, after taking part in the destruction of two Ju 88s, had his Hurricane set on fire.

Since it was not burning badly, knowing that every fighter was precious, he tried to save it by a crash-landing at Eleusis. The Hurricane had come almost to a standstill when it exploded into a mass of flames. His clothing on fire, the pilot got clear but could not rid himself of his burning parachute. He died of his injuries two days later.

The other pilot who was lost was Pattle. Having been seen to destroy a 110 for certain and probably a 109 as well, Pattle spotted a Hurricane in trouble with a 110 on its tail. Although there were other enemy machines overhead, which he must have known would pounce on him, he dived to his colleague's aid, shooting down the 110 in flames; but two more 110s closed in from above, sending the Hurricane hurtling into the sea with a dead man slumped over the controls.

Thus, sacrificing his life to help a friend, there perished the top-scoring RAF fighter ace and the greatest Hurricane pilot of them all.

–

There now followed a series of disasters for the Hurricanes, as for British fortunes in general. Flight Lieutenant Woods of 80 Squadron, the pilot whose life Pattle had saved at the cost of his own, was killed in action later the same day. By the 21st, General Wilson had decided that nothing but immediate evacuation could save his command. D'Albiac concurred. Within two days all his aircraft had left for Crete or Egypt except the Hurricanes, which fell back to Argos, a training airfield in the Morea from which they could cover the army's retirement.

In the meantime, the efforts of the ground crews together with new replacements from Crete had brought the total of Hurricanes to twenty-three – though in the circumstances it is difficult to see how a greater number on hand meant more than a greater number at risk. By the afternoon of the 23rd, by which time only twenty were still serviceable, their base was raided by a force of about forty Bf 110s. Four of the Hurricanes managed to get airborne but were unable to record any victories – the remainder were blasted on the ground. Thirteen were wrecked. At dawn the next day, the last seven Hurricanes in the Balkans took off for Crete.

Though the German army, as in France, had out-run its supply lines, so could not corner the retreating Allies, the withdrawal from Greece in the absence of fighter cover had perforce to be carried out at night. Early on the 27th the Dutch transport *Slamat*, which had lingered too long to rescue troops, was sunk by the Luftwaffe. Her escorting destroyers, *Diamond* and *Wryneck*, picked up survivors only to be sunk as well. From the three ships, forty-two sailors and eight soldiers were saved. The evacuation was completed by the 28th, although as at Dunkirk, all heavy equipment had to be abandoned.

Army, Navy and Air Force all rallied in Crete. It was to prove a deceptive refuge, for after initial reluctance, Hitler had yielded to General Student, the commander of Fliegerkorps XI's 500 transport aircraft and seventy-two gliders, who was urging him to make an airborne attack on the island. The support role was allotted to General von Richthofen's Fliegerkorps VIII, which at the end of April, from airfields in Greece or the Dodecanese Islands, then Italian possessions, began to hurl continuous raids against the defences of Crete or the supply ships arriving there.

To meet these attacks, the only modern fighters available at first were the seven Hurricanes that had survived from Greece. Two of them were quickly lost in combat; but seven reinforcements soon flew in from Egypt. Thereafter, further Hurricanes arrived to replace losses, but the numbers present at no time rose above sixteen.

Since all 208's machines had been lost in Greece, its remaining pilots were withdrawn but the 'Hurricane Unit, Crete' was manned by members of both 33 and 80 Squadrons, the most notable of whom was perhaps the Australian Flying Officer Vale of No 80, who followed a distinguished career in Greece by claiming seven enemy aircraft in Crete. In addition, the pilots of 112, whose three remaining Gladiators had been grounded as quite outclassed, also flew sorties in Hurricanes, as did the Fleet Air Arm pilots of 805 after their Fulmars had been lost early on – though the inexperience of the naval airmen proved costly on 16 May, when in shooting down two enemy machines they lost two Hurricanes with both their pilots.

The spirit with which the RAF met its challenge can be seen from an incident on 14 May. Squadron Leader Howell, who had taken command of 33, was being shown the layout of a Hurricane's cockpit – he had never previously flown the type – when he heard 'the roar of engines starting up'. Looking round he saw two other Hurricanes making a rapid take-off. A force of 109s was about to strafe Maleme airfield.

Howell made hasty preparations to follow. He recounted later:

> I opened the throttle and saw a string of five Messerschmitts coming in over the hill firing at me. It seemed an age before my wheels came off the strip. I went straight into a turn towards the approaching 109s, my wing tip within inches of the ground. The faithful old 'Hurribus' took it without a murmur, the enemy flashed past and I went over instinctively into a steep turn the other way.

Despite such handicaps as having to locate the positions of unfamiliar controls, an unadjusted rear-view mirror that prevented him from seeing behind, and a large borrowed helmet that persisted in falling over his eyes, Howell out-turned his assailants, of which he destroyed one and damaged another. He then retired to Maleme to find that both his companions had been shot down, though each had previously been seen to destroy two 109s. Sergeant Ripsher was killed but Sergeant Reynish baled out of a blazing Hurricane into the sea, where he was rescued by a fishing boat.

However, the handful of Hurricanes could scarcely be expected to continue in action indefinitely against a German force of more than 600 aircraft, particularly in circumstances where there were neither radar stations nor communications with the ground controllers. Also, there was a shortage of ammunition, which meant that the fighters often went into action with only six, sometimes only four, of their machine-guns loaded.

So, for all their bravery, by 19 May, only four Hurricanes plus the three grounded Gladiators remained fit for action. The senior RAF officer, Group Captain Beamish, with the full agreement of the defence's commander, General Freyberg, therefore ordered the survivors back to Egypt to prevent a further useless sacrifice.

Unhappily, perhaps because the island's garrison was only too well aware of what the absence of the Hurricanes would mean, it was hoped that they might somehow come back at a later date. This led to the disastrous decision not to obstruct the three main landing grounds at Maleme, Retimo and Heraklion so that they would be available for the RAF on its return.

Yet it was precisely on those airfields that the Germans had planned their airborne landings. The day after the Hurricanes left, gliders or paratroopers rained down on all three, backed by the full weight of Fliegerkorps VIII. Yet even with 100 per cent command of the air, the Germans found the task immensely difficult. The attacks at Retimo and Heraklion were contained. Only at Maleme did the enemy obtain a perilous ascendancy. Even there, prompt counter-attacks could have restored the situation, but these were delayed because Freyberg, learning from Intelligence reports of an impending seaborne invasion, kept back his reserves to meet this.

Ironically, on the night of 21/22 May, Admiral Cunningham's fleet destroyed or dispersed the German convoys, though the casualties inflicted were vastly less than was thought at the time. The response from the Luftwaffe on the next three days was so savage that, having had cruisers *Gloucester* and *Fiji* and four destroyers sunk, with numerous other ships damaged, Cunningham was forced to the 'melancholy conclusion' that 'losses are too great to justify us in trying to prevent seaborne attacks on Crete'.

In any case, the decisive assault would come from the air. A flood of Junkers Ju 52 transports rolled into Maleme utterly regardless of casualties. By the 27th the German airborne troops, supported by relentless bombing or strafing, had worn down the defenders' resistance. Once again, evacuation was ordered but only half the forces on the island were brought out, and by the time operations ceased on 1 June, Cunningham had lost AA cruiser *Calcutta* and two more destroyers, while. in the course of the campaign, the aircraft carrier *Formidable*, battleships *Warspite, Valiant* and *Barham*, five other cruisers and eight other destroyers were damaged, many beyond local repair.

Amid these perils, demands were made for Hurricanes in the vain hope that they might somehow turn the tide. On the morning of 23 May, 73 Squadron was ordered to send six Hurricanes to Heraklion, still covered with wreckage from the thwarted German attack. En route, these were fired on by naval vessels, which so scattered them that all except one had to return to North Africa, having lost contact with their escorting Blenheims. The survivor, flown by Flight Sergeant Laing, got to Heraklion safely but was promptly destroyed by an air raid, though its pilot was unhurt.

In the afternoon, the other five pilots plus a replacement, this time guided by Marylands, again headed for Crete. They reached Heraklion in time to drive off a raid by Stukas, two of which they damaged. After landing on the atrocious surface, they kept up patrols over the airfield for the rest of the day, but their position was clearly untenable. Next morning, with Flying Officer Goodman sitting on Laing's lap, they took off for the last time, regardless in two cases of damaged tail-wheels. After emptying their guns into enemy troops near the landing-ground, they made their way back to North Africa. Here they found a sandstorm blowing. Only two machines, one being that carrying Goodman and Laing, reached their base. Two more force-landed in the desert, of which one was written off. The other two were never seen again.

On the 27th, an even more costly effort was made to supply Hurricanes to Heraklion. The Royal Navy again broke up the flight, shooting down two of the fighters. The remainder dispersed, only four reaching their destination, where they were destroyed by air attacks before they could even refuel.

Meanwhile Hurricanes with fixed 44-gallon fuel tanks were flying from Egyptian bases to participate in the fighting. A few such sorties were made by 1 Squadron SAAF, newly returned from East Africa, which downed two enemy aircraft on the 31st, but the majority were entrusted to the airmen of No 274.

Although the Hurricanes' endurance allowed them only a brief time over the island, they enjoyed a surprising degree of success. The leading pilot of this period was Flight Lieutenant Honor who was credited with having destroyed or damaged at least six enemy aircraft either on the ground or in combat before he was shot down on the 26th. Baling out into the sea, he swam ashore, to be evacuated later in a Sunderland flying-boat.

Also on the 26th, a New Zealand pilot of 274, Flying Officer Tracey, claimed a 109 without firing a shot. He had just brought down a Junkers Ju 52 transport when the Messerschmitt engaged him. Diving steeply with his attacker in pursuit, he pulled out just above water-level. The German fighter did not and it crashed into the sea.

Such exploits indicate that the fighting for Crete was not all disaster. Fliegerkorps XI had all but ceased to exist. The German losses of more than 6,000 airborne troops meant that they would never again risk a similar major operation, for example against Malta, the capture of which would have been of far greater value to them. In combat the Hurricanes, for all their problems, had again proved their worth by destroying or damaging about seventy-five enemy machines.

However, it may be questioned whether the Hurricanes' own losses were not equally harmful. The fighting in Crete cost the RAF twenty-eight of them, together with sixteen of their pilots. To this must be added another dozen fighters destroyed in later attempts to reinforce the island – such as the misfortunes of 73 already described. When Hurricanes were in such short supply, these casualties could not easily be borne. Furthermore, the diversion of aircraft as well as troops to the Balkans and Crete had already had disastrous repercussions in North Africa, for here also Hitler had come to the aid of. the Italians. On 6 February 1941, he personally instructed General Erwin Rommel to proceed to Libya forthwith. By the 12th that officer was in Tripoli, to be followed two days later by the first elements of his new command. They included fifty Stukas and twenty Bf 110s plus some reconnaissance aircraft, all under General Frölich, who was entitled the 'Fliegerführer Afrika'.

The RAF Commander in Cyrenaica, Group Captain Brown, had little with which to provide opposition, let alone give powerful support to the Army. In addition to the squadrons sent to Greece, 274 had returned to Egypt for a well-deserved rest leaving Cyrenaica in the care of only two Hurricane squadrons, 73 and 3 RAAF, which latter had received its first such fighters only on 29 January. There was also a Free French Hurricane flight, Groupe de Chasse I, formerly attached to 274 but now flying with 73, while, on 1 March, No 6, an Army Co-operation squadron, received a few Hurricanes to supplement its Lysanders.

The fighter squadrons were soon in action with the Luftwaffe. Flying Officer Saunders of 3 RAAF made the first interception on 15 February, shooting down a reconnaissance Junkers Ju 88, while on the 17th, his squadron joined 73 to bring down eight Stukas from a formation trying to bomb British positions. This combat was not marred by the loss of a single Hurricane. Indeed, the Luftwaffe's first successes did not come until the 19th, when 3 RAAF lost two aircraft and one pilot, though it destroyed a 110 and a Stuka in the same encounter.

However, it was impossible for the fighters to ward off all German attacks especially those on the port of Benghazi, which was bombed so persistently that it had to be abandoned as a supply base in favour of Tobruk, 200 miles to the east. This was only part of the dire situation in which the British and Commonwealth forces were placed. As with the RAF, it seemed that every experienced unit that had not gone to Greece was re-equipping in Egypt. In particular, 7th Armoured Division had been replaced by 2nd Armoured Division, new to the Desert manning worn-out tanks and short of transport. Even O'Connor had returned to Egypt, leaving Lieutenant-General Neame in command. Yet Wavell was confident from Intelligence reports that the enemy would not attack before May. He had not appreciated either the driving energy of Rommel or the admirable speed with which the Panzer units, the famous Afrika Korps, would adapt to the needs of the desert fighting.

After a preliminary capture of El Agheila on 24 March, Rommel on the 31st began a relentless, audacious advance, whereupon, on 3 April, the main Allied petrol dump at Msus was set on fire by its guards on the rumoured approach of a German armoured column. This wrecked any chance of resistance, for those tanks of 2nd Armoured that did not run out of fuel were forced to turn towards the coast in search of it, uncovering the desert flank they were supposed to be protecting.[28]

Thereafter, Rommel raged on unchecked. Benghazi fell on the 4th. On the 7th Neame and O'Connor, who had joined him as an adviser, were captured. By the following day, 2nd Armoured, for all practical purposes, had ceased to exist. Within

---

[28]Rommel's admirers make great play with his perennial fuel difficulties. They lay much less stress on the point that in his first offensive, lack of petrol was a crippling handicap faced by his opponents.

a fortnight Rommel was over the Egyptian frontier and had re-taken almost all the gains made by the first British offensive – with one vital exception.

As the Afiika Korps advanced, the German airmen were meeting with difficulties. On 3 April, 3 RAAF pounced on a hostile formation, destroying three Stukas along with five of their escorting 110s, no fewer than four of these falling to Pilot Officer Turnbull. On the 5th the Australians teamed up with 73 for an even more memorable day, during which the squadrons shot down fourteen Stukas between them. Two Hurricanes were lost, as were their pilots.

Then the paths of the squadrons parted as 3 RAAF retired to Egypt. From here it made further sorties against the enemy – notably that of the CO, Squadron Leader Jeffrey, on the 15th. Sighting four Junkers Ju 52 transports about to land near Fort Capuzzo, he shot down one, then attacked the other three on the ground, destroying them also. However, five days later, 3 RAAF withdrew to Palestine where it re-equipped with Tomahawks. It handed over its Hurricanes to 274, which had already moved up to rejoin the fighting on the 16th.

No 274 was quickly in action again, attacking Axis machines on their landing grounds as well as in the air. This was especially the case with Flying Officer Weller, who having wrecked six CR 42s and one SM 79 on the 18th, destroyed or damaged seven Ju 52s embarking troops on the 27th, for which he was awarded a DFC. The enforced decline in enemy activity in the air helped stabilise the fighting just west of Sidi Barrani. Yet the main reason for Rommel's inability to push further into Egypt was that his rapid conquests had not included the crucial port of Tobruk, which he needed to receive supplies and from which his vital lines of communication could be disrupted.

Tobruk was guarded by 9th Australian Division under Major-General Morshead, who quickly began to receive reinforcements shipped out to. him by Wavell. He also enjoyed the support of 73 Squadron and the Hurricane flight of No 6, operating from landing strips within the defended perimeter in the fighter and army co-operation role respectively. The fortress was completely cut off by the 10th but, even before this, 73 was striking at enemy columns. It kept up strafing attacks against the first Axis probes during the next two days, though by now it had lost six aircraft shot down or forced to crash-land by flak, one pilot being killed and one taken prisoner.

On 14 April, Rommel made a major assault under cover of raids by Stukas, guarded by Bf 110s and Italian G50s. This failed with heavy losses, due in no small measure to 73, which, lacking proper ground control or early warning system, still destroyed six Ju 87s and two Fiats during the day, though at the cost of two Hurricanes with both their pilots.

The odds against the Hurricanes now increased with the arrival in Africa of the Messerschmitt Bf 109s of JG 27. On the 19th these attacked three Hurricanes of 274, one of which they shot down, though the pilot escaped by parachute, while another force-landed with its pilot wounded. That afternoon, Pilot Officer Spence made some amends by so damaging a 109 that it crash-landed. He was a member of 274 but at the time was attached to 73, against which the enemy fighters now turned their full attention.

The next few days were marked by continual German strikes against Tobruk, the dive-bombers now being escorted by the 109s. A brief account of the fighting cannot hope to indicate adequately the strain imposed on the Hurricane pilots, but it may at least give an impression of the efforts that they made.

Very heavy raids started on the 21st, during which 73 shot down four Stukas and a 109. Pilot Officer Spence, the 274 pilot attached to the squadron, pressed home his attack so close that he collided with a second 109, which crashed. He managed to glide his damaged machine back to the British lines. A second Hurricane force-landed but again without injury to the pilot. On the following day, the squadron took off six times, destroying two Stukas, three 109s and three G50s. The odds reached a peak on the 23rd, when seven Hurricanes met a raid of twenty Ju 87s, escorted by thirty 109s and ten 110s. They downed four Stukas and two 109s but lost three of their own number, one pilot being killed. That evening Sous-Lieutenant Denis, one of the French pilots flying with 73, had a final fling, destroying one 109, damaging a second.

Even had 73 been at full strength, its airmen would now have been too exhausted for further combat. On the 25th, its four surviving Hurricanes withdrew to Egypt. However, its ordeal had not been in vain, for if the attacks on Tobruk had been severe, they were at least not nearly so effective as would have been the case without the Hurricanes' interceptions. Also the defenders had had a chance to stabilise their positions, which would enable them to hold out against the worst that Rommel could do thereafter. 6 Squadron's Hurricanes remained in Tobruk on their army co-operation duties until 8 May when, also down to four in number, they too flew out. Three days later, though the Hurricanes' armament had been reduced to increase their performance, Pilot Officer Griffiths, jumped by three 109s, not only escaped but, in a splendid gesture of defiance, destroyed one of his attackers.

It was now the turn of the British to take the offensive. Their first attempt, on 15 May, was prophetically code-named 'Brevity'. It was as unsuccessful as it was short, but it did enable Captain Quirk of 1 Squadron SAAF, just arrived from East Africa, to win a DSO on the 16th. When a companion, Lieutenant Burger, crash-landed behind enemy lines as a result of ground-fire, Quirk put his Hurricane down alongside. Then, sitting on Burger's knees, he took off again, reaching his base safely.[29]

---

[29]This was the first incident in which a Hurricane carried two passengers in the Mediterranean theatre, preceding by a few days the rescue of Flight Sergeant Laing of 73 from Crete, previously recounted. It will be

In the meantime, on the insistence of Churchill, a large convoy known as 'Tiger' was rushed through the Mediterranean carrying 300 tanks and fifty Hurricanes. One ship was lost to a mine, but 238 tanks and forty-three Hurricanes reached Egypt on the 12th. Unhappily, having risked a great deal to deliver these, the Prime Minister badgered Wavell, against the latter's inclinations, to launch an offensive before proper time could be given either to put the new tanks into fighting condition or to train the crews to man them.

Codenamed 'Battleaxe', this operation was preceded by maximum efforts on the part of the RAF, now much stronger due to the arrival of the Hurricanes in the convoy, plus others that had flown from carrier *Furious* via Malta. By the time of the attack, three more Battle of Britain Hurricane squadrons, 213, 229 and 238, had contingents in the Middle East although, since it took some time to build them up to full strength, their pilots were temporarily attached to 73 or 274. The Hurricanes achieved a measure of local air superiority, largely as a consequence of strafing attacks on enemy airfields: 274 destroying eight Axis aircraft at Derna on 4 June, while, on the same day, 73 destroyed six at Gazala – though on its return it was bounced by 109s, losing two Hurricanes and one pilot.

These raids reached a peak on the 14th, the day before Battleaxe was launched. 73 again spent the morning strafing but lost three machines to flak, only one pilot surviving as a prisoner. In the evening, 1 SAAF shot down two Ju 87s but was then attacked by 109s, losing two aircraft together with their pilots. During the coming campaigns, the Messerschmitts would frequently allow Hurricanes to become engaged in combat before striking from above with maximum effect – but although this enabled the German fighter pilots to increase

---

remembered, however, that Lieutenant Kershaw of 3 SAAF had won a DSO for a similar feat in Abyssinia some two months earlier.

their scores more easily, they would have done their job far better had they intercepted the Hurricanes before these could tackle the valuable, vulnerable German bombers.

The Axis army, particularly the Afrika Korps, needed no lessons on how to perform its task. In three days, Battleaxe had ended in dismal failure, mainly due to the deadliness of the German anti-tank guns especially their 88 mms, which ended forever the reputation of the Matildas as the invulnerable rulers of the Desert. The British lost almost a hundred tanks, nearly twice as many as the Germans, who, since they held the battle-field, were able in any case to repair most of their knocked-out vehicles. To make matters worse, the British believed that their casualties had been caused not by guns but by enemy tanks, to which they henceforth accorded a much-exaggerated respect.

Nor was good use made of the RAF. The Hurricanes constantly seemed to be hit from above when answering desperate pleas to strafe enemy columns. 274 was caught in this way on the first day, losing three aircraft (from which only one pilot survived) while a fourth crash-landed. Next day, 1 SAAF did slightly better destroying or damaging four enemy machines before it was bounced – three aircraft and one pilot being lost.

17 June was nearly as bad for the Hurricanes as it was for the ground forces. They shot down or damaged ten of the enemy but could ill-afford the destruction of eight of their own number in combat, plus two more by flak. Such casualties moreover were largely wasted for, by nightfall, the British had been driven back to their original positions.

There then followed a 'lull' as both sides settled down to build up their forces – but in the air especially there was no cessation of activities. There were the enemy fighters that strafed Hurricanes on the ground on 7 September, destroying three, damaging twelve. There was the loss of five Hurricanes from No 73 to AA fire on 4 July. There was the success of 73 and 229 on 15 July, when they downed seven enemy aircraft at a cost of two Hurricanes. There were the exploits of new Hurricane units at night, 94 Squadron destroying four enemy machines during this period, while No 30, on 7 August, claimed one raider shot down and others only damaged, but German records show that four Junkers Ju 88s in fact fell on that night. There was the loss of three Hurricanes of 1 SAAF to 109s on 24 September, which was made less serious by the fact that Captain Van Vliet and Lieutenant Dold were able to make their way back to friendly territory, while 2nd Lieutenant MacRobert was even more lucky, for Lieutenant Liebenberg landed and picked him up – an increasingly frequent trick in the Hurricanes' repertoire.

In the main, though, this was a period of adjustment. Wavell was succeeded by General Auchinleck. Longmore had already been recalled on 3 May, his post being taken by Air Marshal Arthur Tedder. In July, Air Vice-Marshal Arthur Coningham took over Collishaw's command of the squadrons in the Western Desert. Also during this time, the old Mark I Hurricanes were gradually replaced by Marks IIA, IIB or even IIC, a factor much appreciated by the pilots who, as mentioned earlier, had found that the adaption to tropical standards imposed a considerable performance handicap on the original version.

Saying this emphasises the difficulties under which the Hurricane pilots had operated. There could be no wonder that they had had heavy losses during the spring and summer of 1941. Yet their sacrifices had not been in vain. For all their admirable tactics, the Germans had won only a limited victory. The way to Suez was still barred.

–

There were perhaps more indirect ways. On 1 April, a coup in Iraq caused the flight of the pro-British regent, Abdulla Illah, and brought to power Rashid Ali, a politician whose natural sympathies for Germany had been strengthened by generous donations. It was fortunate that Hitler's present commitments in the Balkans, to say nothing of his future plans for Russia, prevented immediate aid to Iraq, but the likelihood of this in the near future presented a grim threat to British commanders in the Middle East.

Therefore, on 16 April, Rashid Ali was informed that Britain intended, in accordance with existing treaty rights, to send troops through Iraq to Palestine. The first contingent reached Basra two days later without incident, but when reinforcements arrived on the. 29th, Rashid Ali, appreciating that the British would soon become too strong for him to oppose, decided to act without waiting for German aid. At dawn on the 30th, the Iraqi army appeared on the high ground dominating the British airfield at Habbaniyah. At the same time the British-controlled Iraqi Petroleum Company at Kirkuk was taken over by force, the flow of oil to Haifa being cut and redirected to Syria.

If the Iraqis had believed that Habbaniyah would be an easy target, they were swiftly disillusioned when on 2 May its commander, Air Vice-Marshal Smart, attacked them with every machine that No 4 Flying Training School could coax into the air – a wonderful collection of elderly aircraft, adapted with every imaginable ingenuity to carry bombs. So successful were their raids that on the night of the 5th, they forced the enemy to abandon the plateau, thereby lifting the siege. Nonetheless, it was obvious that military assistance was still needed. Since an approach to Habbaniyah from Basra was blocked by flooding, a small British contingent known as 'Habforce' set off from Transjordan, reaching the base on the 18th.

Three days earlier, the Germans had sent their first armed forces to Iraq when fourteen Messerschmitt Bf 110s and ten Heinkel He 111s painted with Iraqi insignia reached Mosul airfield via Vichy French Syria. On the 16th, three of the Heinkels attacked Habbaniyah, doing more damage than that achieved by the Iraqis during the whole of the previous fighting.

Clearly Hurricanes were needed. On the 17th, 94 Squadron with four Hurricanes and nine Gladiators reached Habbaniyah. The Hawker fighters were quickly in action, strafing Iraqi motor-transport.

However, since the German bases were outside their range, two further Hurricanes, this time fitted with fixed 44-gallon long range tanks, were sent to join 94 from Egypt.

On the 21st, Flight Lieutenant Sir Roderic MacRobert and Flying Officer Sandison took the new arrivals to Mosul and Erbil, where they destroyed a number of hostile aircraft on the ground as well as exploding two petrol lorries. Sad to relate, however, MacRobert was then shot down and killed.[30]

Four more fighter Hurricanes plus a tactical reconnaissance version joined 94 in the next few days. With ten modern aircraft on hand, the squadron covered the advance of the ground forces to Baghdad, though no air combats transpired. By the 31st, Rashid Ali had fled to Syria, an armistice had been signed and the Regent was on the way back to his capital, which he re-entered next day. Mosul and Kirkuk were occupied soon afterwards.

During the fighting in Iraq, weapons, ammunition and aviation fuel had been shipped to the Iraqis from Syria, through which also the Heinkels and 110s had been staged. Further, the Vichy French High Commissioner, General Dentz, on 12 May, frankly notified the British that he would obey any orders he received to allow a German occupation of Syria. Two days later, the RAF was authorised to attack German machines on Syrian airfields.

Though after the fall of Rashid Ali, Dentz obtained the withdrawal of the remaining Luftwaffe crews, the Allied commanders dared not take the risk of their returning at a more propitious time. On 8 June, British, Commonwealth and Free French forces invaded Syria. It was hoped that no resistance

---

[30]His elder brother, Sir Alasdair, had previously lost his life in a flying accident. His younger brother, Sir Iain, died later serving with the RAF. In their memory, their mother, Lady MacRobert, later donated to No 94 three Hurricane IICs – HL844, HL851 and HL735 – bearing the family crest, each one named after one of her sons.

would be offered but it quickly became clear that Dentz's well-trained, well-equipped troops would not yield easily, if only for the sake of their professional pride.

Supporting the advance were the fighter Hurricanes of 80 Squadron, back from its travail in the Balkans, plus a flight of Army Co-operation Hurricanes from 208 – which, however, did not operate only in the reconnaissance role but also mounted strafing raids, particularly on airfields. Shortly afterwards, the Hurricanes' responsibilities were further widened. The 15th Cruiser Squadron under Vice-Admiral King was guarding the coastal flank but the two squadrons of Fulmars providing him with fighter cover were outclassed by the French Morane and Dewoitine fighters. Some of the Hurricanes therefore were diverted to guard the cruisers. They achieved several victories, but while they were so pre-occupied, there was naturally a considerable reduction in the aid they could give to the advancing ground forces.

In consequence, it was only after much fierce fighting, with regrettably high losses, that Dentz evacuated Damascus on 21 June. Later matters got even worse, for, with the enemy showing ever greater determination to resist, the Allied advance slowed almost completely to a halt.

Fortunately, the failure of Operation Battleaxe in the Western Desert had at least enabled more aeroplanes to be moved to the Syrian front. They included the Hurricanes of 260 Squadron, recently sent out from Britain, to which was attached a contingent from a new Australian Hurricane squadron, No 450. Pilots from 213 also arrived, flying with No 80. In addition, two Fleet Air Arm squadrons, 803 and 806, were transferred to land bases. Exchanging their Fulmars for Hurricanes, they also flew as flights of RAF squadrons.

Some of the Hurricanes were engaged in escorting Blenheims but the others persisted in strikes against the Vichy warplanes on their airfields. According to General Jeannequin, the enemy air commander, these raids destroyed or damaged beyond repair fifty-five of his machines. A further thirty hostile aircraft were brought down in combat as against ten RAF fighters. These losses forced Jeannequin to withdraw his remaining machines to Aleppo in the far north, whence they were unable to provide adequate support for their troops.

As a result, the Allies were able to resume their progress towards Beirut. At the same time, two other columns pushed into Syria from Iraq, one, covered by Tomahawks, moving on Palmyra, the other heading into north-eastern Syria under the protection afforded by a new squadron, No 127, which originally contained four Hurricanes and four Gladiators. These saw a good deal of action, gaining several victories, though losing two Hurricanes on 3 July.

The final contribution by the Hurricanes to the campaign came when a number with fixed 44-gallon auxiliary fuel tanks reached 260 Squadron. Their increased range enabled them to hit the airfield at Aleppo, which hitherto had been out of reach. It seems this was the last straw for General Dentz, who now asked for terms. Hostilities ceased on 11 July, the armistice being signed three days later. The Hurricane units returned to the Western Desert, except for 127, which disbanded, its aircraft forming the nucleus of a new 261 Squadron.[31]

In order to complete the campaigns, in the Middle East during 1941, it is convenient to mention here that, following Hitler's invasion of Russia on 22 June, a joint British-Russian advance into Iran (or Persia for those who, like Churchill, preferred to use the older name) took place on 24 August, after demands for the expulsion of German nationals had been refused. Neither the legality nor the morality of this action will bear close examination, but the immensity of the struggles taking place elsewhere may at least offer some excuse.

---

[31]It will be recalled that the original 261 Squadron had been stationed in Malta, but, as will be described later, this had been disbanded in May.

It is some consolation that Iran offered little more than token resistance – also, that in contrast to the bombing carried out by Soviet aircraft, the RAF contingent saw little action before hostilities ended on the 28th, British and Russian troops entering Teheran peacefully on 17 September. The Hurricanes of 261 Squadron took part in the campaign but, although they strafed Iranian airfields, there was only one recorded combat, when Squadron Leader Mason shot down an aged Audax on 26 August.

Although limited in duration, as well as in terms of the men and resources involved, the fighting in Iraq, Syria and Iran had highly beneficial results. It prevented any attack on Egypt from the north, saved the Middle Eastern oil for the Allied cause and secured a vital supply route to the new ally, Russia, over which would eventually pass a total of five million tons of arms, aircraft and ammunition.

Chapter Eight

The Atlantic and the Arctic

The Hurricane was already helping to guard even more vital supply routes. When the failure of the Luftwaffe thwarted Hitler's direct assault on Britain, his best course would probably have been to hurl every available man into the conquest of the Middle East, but though he dispatched Rommel's forces to the aid of the Italians, he did not wish to employ the bulk of his army, which he kept in reserve for dealing with Russia, Therefore, he entrusted to the German Navy and Air Force, in particular to Admiral Dönitz's U-boat fleet, the task of defeating Britain by cutting off all supplies of food, fuel and raw materials. The Battle of the Atlantic, longest, hardest and grimmest of the war, began; the attempted destruction of Britain's merchant fleet by the submarine, the surface-raider, the bomber and the mine.

In a sense, Hurricanes had helped to provide protection against such foes from the first day of hostilities, flying patrols over coastal convoys, engaging bombers, reconnaissance machines and aerial minelayers. By the summer of 1941, moreover, they were extending such duties well away from the shores of Britain.

On the collapse of Norway, British troops had been sent to Iceland to forestall any possible German landings, but in July 1941 it was announced that their place would be taken by United States forces. To cover the withdrawal of the British garrison, during July and early August, a total of nine Hurricane

Mark IIAs were shipped to Iceland, where they formed 1423 Flight, continuing their patrols until December, when they returned to Britain. During this period the Flight sighted an occasional reconnaissance Heinkel He 111, but each time the intruder was able to make good its escape.

In October 1941, a new Hurricane unit, 128 Squadron, was formed at Hastings, Sierra Leone, to defend the ports of West Africa, a detachment being sent to Gambia about a year later. Equipped with a mixture of Mark Is and IIBs, 128 continued its protective duties until March 1943, when it was disbanded. Its potential enemies were the Vichy French not the Germans, but it had very few combats with them, Squadron Leader Drake shooting down a Martin 167 on 13 December 1941 and Sergeant Todd another one on 22 August 1942.

Such encounters in any event were trivial compared with the Hurricane's clashes with its arch-enemy in the Atlantic, the Focke-Wulf Fw 200C. Known appropriately as the 'Condor', this huge four-engined bomber had a range of 2,210 miles at an economic cruising speed of 180 mph, with a top speed of approximately 235 mph. It carried a crew of five, needing a large number if only to man its guns, which consisted of a 20 mm cannon and from three to five 7.9 mm machine-guns.

Though mercifully few in number, being produced only by one small factory at Bremen with an output of but one aircraft per week, the Condors of KG40, based at Bordeaux, soon achieved considerable results. Their vast range enabled them to fly well to the west of the British Isles, before landing in Norway, striking en route at Allied vessels far beyond the reach of shore-based fighters. KG40 began its depredations in August 1940. In two-and-a-half months it had sunk shipping totalling nearly 90,000 tons. By the end of February 1941, it had destroyed a total of eighty-five vessels including five in one day on 9 February.

From early March the giant Focke-Wulfs were also used as long-range scouts for the U-boats – for although not in direct contact with these, they reported all sightings to the Flag Officer, Submarines, at Brest, who in turn would notify the 'wolf-packs'. Poor navigation resulted in their being somewhat disappointing in this role – on occasions it was information from the U-boats that directed the Focke-Wulfs to the convoy rather than the other way about – yet once a Condor had located Allied vessels, then its ability to remain in the air for fourteen hours, if necessary, enabled it to stay in touch, sending out a stream of information on the convoy's composition, course, speed and progress, all the while circling safely out of range of the escorts' inadequate AA guns. Indeed, there is an amusing, if improbable, story of a convoy commodore signalling a request to the Focke-Wulf to fly the other way round as he was getting giddy watching it.

The only way of dealing with this menace was for the convoys to be given fighter protection. Since aircraft carriers were few in number, it was proposed that certain merchantmen should be converted to carry an interceptor of their own. The catapults normally used to launch aeroplanes from ships were too heavy, complicated and expensive to be installed on such vessels, so it was decided that a rocket catapult should be used, the necessary three-inch rocket motors being then in plentiful supply.

A quick survey of types that the Air Ministry would make available showed that the only machine with the necessary performance, reliable handling characteristics and sheer robustness was the Hurricane. In October 1940, enquiries were made of Hawkers whether its fighter could be adapted for the job. The firm replied favourably, adding that a prototype could be ready in five weeks, though strangely enough it was not until 19 January 1941 that a decision was reached and the first conversions commenced.

These aeroplanes, the first of which was ready in March 1941, were known as 'Catafighters' or 'Hurricats' or more officially as Sea Hurricane Mark IAs. They were basically old Mark Is

– which now gained a new lease of useful life – altered so as to incorporate catapult spools and attachments by which they could be lifted on to the launching-ramp and lashed thereon, if necessary, in bad weather. The fuselage was strengthened, while among other adjustments may be mentioned a heavily-padded head-rest against which the pilot would lean so as to absorb the shock of being hurled down a 70-foot-long steel ramp by thirteen rockets, which produced a speed of 75 mph at the moment the Hurricane became airborne. Unfortunately, these various modifications did have the effect of reducing the speed of the Hurricats to only about 245 mph at 3,000 feet.

Yet no amount of forethought could enable this new version of the Hurricane to land back on the ship that had launched it. Nor, in view of the areas in which the Condors operated, would it often be possible for the Catafighter to reach land, even after 45-gallon drop–tanks – which, in any case, reduced the Hurricane's manoeuvrability as well as necessitating an increase in the power of the catapult – became standard fittings.

Thus, whatever the result of his combat, the pilot would be faced with a choice of ditching in the sea or taking to his parachute. Despite the introduction of an improved method of jettisoning the sliding hood and the provision of a new one-man dinghy, it was not pleasant to land a Hurricane in the sea, since its large radiator tended to drag it under in a matter of seconds.[32] Baling out into the icy waters of the North Atlantic was not to be recommended either, particularly as it was then even more difficult to spot the pilot. And, in either case, the airman relied on rescue by the convoy's escorts, which if U-boats were in the vicinity would hardly take the risk of stopping to pick him up.

---

[32]Pilots were advised that when preparing to ditch they should adopt a nose-high attitude and allow one wing-tip to drop as the tail-wheel touched the sea in order to swing the aircraft to a stop, thereby reducing the amount of water that would be scooped up by the radiator. However, in practice, they were not always able to follow these suggestions.

It is therefore less surprising that Fighter Command felt compelled to ask for volunteers than that so many intrepid characters were prompted to face the obvious hazards – whether by a wish for excitement, a desire for change, or a feeling that the task was one of such vital importance as to justify the risks. They were formed into the Merchant Ship Fighter Unit, which was led by Wing Commander Moulton-Barrett until January 1942, when he was succeeded by Wing Commander Pinkerton. Training was carried out at the airport of Speke, near Liverpool, where a rocket-catapult was erected. This had only a six-foot clearance off the ground, which, since the Hurricane was inclined to sink slightly when launched, did not afford much margin for error.

The officer in charge of catapult development and training was Squadron-Leader Louis Strange, a veteran of World War I in which his most extraordinary exploit was falling out of a Martinsyde scout, saving his life by hanging on to the drum of his Lewis gun and somehow managing to pull himself back into his cockpit. After such an experience, presumably a catapult launching two months short of his fiftieth birthday was comparatively minor. At all events, Strange made the first such at Speke, remarking to his pilots: 'If an old boy like me can do it, it won't mean a thing to lads like you.' His demonstration went off magnificently, though the watchers experienced considerable alarm at the colossal uproar from the rockets, accompanied by vast sheets of flame, which severely damaged the blast screen.

While the Hurricanes were being adapted, so were the ships. In order not to withdraw vital vessels from convoy service, the catapults together with elementary radar and radio were fitted to freighters under construction. Originally it was intended that 200 such should sail but sober reflection showed that it would be difficult to keep so many equipped in view of the probability of corrosion from the salt water damaging the fighter's airframe or engine or both. In consequence, the number was reduced, first to fifty, then to thirty-five. Carrying normal cargo as well as the Hurricat, they were known as CAM-ships; – the initials standing for Catapult Aircraft Merchantmen.

The first RAF officer to make a trial launch from a CAM-ship was Pilot Officer Davidson who, on 31 May 1941, was flown off *Empire Rainbow*. This event very nearly ended in disaster. Two of the rockets failed to fire and the Hurricane's throttle slipped back to only half open when the pilot raised his hand to signal his readiness for take-off. Perhaps rattled by staff officers standing by with stop-watches, Davidson also failed to guard against the Hurricane's tendency to drop. All these factors caused the Catafighter to swing badly, the port wing touching the water. It says much for the Hurricane's sturdiness that the pilot was able to regain control and land safely. It says also a good deal for Davidson's own strength of character that when *Empire Rainbow* left to join a convoy on 8 June, he was still on board ready to man her fighter.

However, even before Davidson's launch, the Senior Service had maintained its proud position by being first to fly off a Hurricat at sea. Not only did the Royal Navy provide all the CAM-ships except two with their Fighter Direction Officers who guided the airmen to their targets, and their radar operators, they also found the first pilots, using men who had gained some experience of such launchings in fleet spotting aircraft.

In contrast to the RAF pilots, those of the Fleet Air Arm were not volunteers but were simply transferred to 804, the squadron allocated for 'catafighting' duties, at Sydenham, Belfast, on normal appointments. It can only be assumed that the authorities were confident that the perils inherent in the work would be unlikely to daunt any naval airman.

One member of 804 was Sub-Lieutenant Birrell, who had previously flown Hurricanes when attached to 79 Squadron during the Battle of Britain. He had already made a launch from a ramp on the ground at Gosport when he was posted to join the CAM-ship *Michael E*, from which his Hurricat was catapulted in Belfast Lough on 18 May, nearly a fortnight before

Davidson's exploit. Like Davidson, he met with trouble, for only half the rockets fired, but, he reported, he 'hauled the aircraft out of the sea' to land safely. Ten days later, *Michael E* joined a convoy, the first CAM-ship to do so, but, ironically, she passed uneventfully through 'Condor territory', only to be sunk by a U-boat – though Birrell survived to make further trips as a Catafighter pilot.[33]

As well as supplying pilots, the Navy adapted five vessels being fitted out as auxiliaries, for use in catapult work. Known as Fighter Catapult Ships, these flew the white ensign, did not carry freight and were crewed by naval personnel, while 804 Squadron manned the aircraft. In practice, four of them either made no combat launches or were equipped with Fulmars, which lacked the necessary rate of climb and acceleration. The only one from which Hurricanes were sent into action was HMS *Maplin*, formerly the fast banana-boat *Erin*, which had two of them, one on the catapult, the other stowed on deck between mast and bridge. If the first machine was launched, the second had to be swung round the mast on the aft derrick, transferred to the forward derrick and lowered onto the catapult – an immensely difficult manoeuvre in anything other than flat calm, not normally met with in the Atlantic. Later, in February 1942, *Maplin* was adapted to carry a third Sea Hurricane, continuing in this role until June 1942 – but her finest exploits were her first ones and her finest pilot was the senior officer of her early trips, Lieutenant Robert Everett.

There can have been few more interesting characters in any service than Everett, who in 1941 was forty years old. Born in Australia, he had seen two years service as a midshipman in World War I, had farmed for a time in South Africa, and had

---

[33]In all, eleven of the CAM-ships would be lost in action, one more being lost by accident, but three of these had ceased to operate as such by that time.

then returned to Britain where he learned to fly, spending his summers as a charter pilot. He passed his winters riding as a professional National Hunt jockey, gaining his greatest fame by winning the 1929 Grand National on Gregalach.[34] He stated without hesitation, however, that everything in his previous career paled into insignificance compared with the excitements of catafighting.

His first such experience came on 18 July 1941, when he was fired off to engage a Condor that had just damaged a merchantman; but as he closed for a head-on attack, the Focke-Wulf, hit by anti-aircraft fire, dived into the sea amid audible cheers from the convoy but somewhat to the disappointment of the Sea Hurricane pilot. At least the lack of combat meant that he had sufficient fuel to enable him to fly the 300 miles to Scotland, where he landed safely after being airborne for just under two hours.

On 31 July Everett rejoined *Maplin*, which was now guarding a body of ships on the route to Gibraltar, from which they would proceed to ports in West Africa or round the Cape of Good Hope to the Middle East. The voyage south was uneventful, but on 2 August, *Maplin*, together with three destroyers, was detached to escort a homeward-bound convoy.

Next afternoon, a Condor was observed but quickly departed, short of fuel. Presumably, however, it had reported its sighting for soon another Fw 200 was seen trailing the convoy, clearly scouting for U-boats. With *Maplin* 450 miles

---

[34]Admirers of steeplechasers will agree that Gregalach deserves at least a footnote of his own. Although a son of that splendid stallion, My Prince, he was not noted for looks, being a big, ungainly chestnut with a large, blotchy star on his forehead; but he was generously endowed with the virtues of strength, steadfastness, determination and durability. He survived the rigours of his sport to enjoy retirement in 1934 – which is more than can be said for many more famous horses. His greatest moment was his Grand National win, which he achieved out of a record field of 66.

from Britain, it was clear that this time the Catafighter would not be able to land anywhere except in the sea, yet it was decided, quite rightly, that the safety of the merchantmen demanded that the shadower be eliminated or at least driven away. At about 1515 Everett was flung into the air.

The big four-engine Focke-Wulf was a redoubtable antagonist, which lacked little in speed compared with the Catafighter. Everett was committed to a chase of about thirty-five miles during which he came under resolute fire from his foe's admirable defensive armament. Finally, he reached an attacking position, shooting into the Condor's side in the face of a barrage from the enemy gunners. He continued to fire until his guns were empty. At the same time, thick oil splashed over his windscreen. Though Everett believed his Hurricane was hit, this probably came from the Condor, which, as he watched, caught fire inside the fuselage – dropping a wing, it plunged into the sea.

If the Hurricat had incurred damage, this was not sufficient to prevent it from returning to *Maplin*. Everett ditched near his ship but as soon as his machine touched the water, it rolled over to sink inverted, pulling the pilot thirty feet below the waves before he struggled clear of the cockpit. Somehow, he reached the surface to be picked up by destroyer *Wanderer*. For his success in destroying the first Focke-Wulf to fall to a catapult fighter, he was awarded a DSO, which may be said to have been thoroughly well earned.[35]

The first operational launch of a Hurricat by a CAM-ship came on 1 November, when Flying Officer Varley from *Empire Foam* intercepted a Condor with its bomb-doors already open some 550 miles out in the Atlantic from the Irish coast. It

---

[35]Sad to report, the sea proved unforgiving. On 26 January 1942, Everett was flying a Hurricane from Belfast to Abingdon on a routine mission when it suffered engine failure, crashing near the Isle of Anglesey. The pilot was drowned, his body being washed ashore later.

seems from an intercepted signal that the enemy pilot had not heard about the Sea Hurricane IA. His horror at Varley's appearance can be imagined. Abandoning his bombing run forthwith, he fled into cloud where he made good his escape. Varley remained over the convoy for more than two hours, during which he twice investigated radar plots, seeing nothing, but, by his presence, forcing shadowers to remain at a distance where they could not observe the merchant ships. Then with fuel exhausted, he baled out, being picked up safely by destroyer *Broke* after only four minutes in the water.

It might be claimed that this was not a profitable mission, yet as well as thwarting one attack, Varley, by keeping away two reconnaissance machines, had deprived the U-boats of crucial information. Indeed, the main value of the Sea Hurricane IA quickly proved to be as a deterrent. After its appearance, the Condors all but abandoned their bombing raids, which had previously resulted in such destruction. Instead, they concentrated on co-operation with their submarines, but even in this role they were greatly hampered by the CAM-ships. There was no longer any question of their following a convoy – on the contrary, the mere sight of a CAM-ship would cause them to depart swiftly from the danger-zone, greatly reducing the preciseness of any information they might give. How many lives were thus saved by the protective presence of the Catafighters can never be calculated.

By early 1942, the Focke-Wulfs had largely ceased to operate against the shipping crossing the North Atlantic, limiting their efforts to the 'Gibraltar run', which was conveniently close to KG40's base at Bordeaux. In July, CAM-ships were taken off the North Atlantic route, many were re-converted to normal freighters and by autumn only eight such remained in service.

All of these were used to mount guard over vessels going to or returning from Gibraltar, in which a spare pool of aircraft was set up. It was on a homeward-bound convoy, 250 miles off the coast of Spain, that Flying Officer Taylor from *Empire Heath* won a DFC on 1 November 1942, by shooting down a Condor despite having his Hurricane's port wing riddled with bullets. He had chased his victim well clear of the convoy before making his kill, but, on his return, he was welcomed by the sirens of every single ship. He then baled out, only to find that his dinghy would not inflate. This could only too easily have proved fatal for Taylor since, extraordinary to relate, he was a non-swimmer, but mercifully his life-jacket kept him afloat until an escorting corvette picked him up, exhausted but unhurt.

Shortly thereafter, the bulk of KG40's Condors were transferred to Russia, or Italy on transport duties. They did not return until the spring of 1943, when, with the aid of a new improved bomb-sight, they were sent out in groups to attack specific targets. By this time, in any case, the increasing numbers of small, mass-produced escort carriers entering service meant that the need for CAM-ships was passing. On 15 July 1943, the Merchant Ship Fighter Unit formally disbanded – an occasion marked by a generous signal from the Admiralty expressing 'great appreciation of the services rendered by the RAF in providing this valuable defence for our convoys'.

Yet 15 July did not mark the end of the activities of the Hurricats. There were still CAM-ships at sea. On the 28th, the last two in service were about 800 miles west of Bordeaux with a convoy from Gibraltar when this was subjected to a series of attacks by Fw 200s. It has been surmised that enemy Intelligence had reported the termination of the CAM-ship scheme, so that, in the absence of any carrier, the Condors had believed they would be opposed only by anti-aircraft guns.

If so, they were to experience grim disillusionment. Pilot Officer Stewart from *Empire Darwin* shot down one, then drove away a second, which dropped its bombs at random. Pilot Officer Flynn from *Empire Tide* attacked a third. His Hurricane was badly damaged by return fire but he left the Condor losing height with smoke pouring from one engine. It appears from German records that it failed to get back to base. Both

Hurricane pilots were rescued safely after taking to their parachutes.

Thus, the final total of Condors destroyed by Hurricats was four. It does not seem a great number, even when it is remembered that each carried a highly trained crew with it to its doom. Yet these victories were merely the highlights of the Hurricanes' contribution to the struggle for control of the Atlantic, in which their main importance lay in their value as a deterrent. Churchill, who from the beginning had been a strong supporter of the catapult fighter, perhaps best summed up their achievement: 'The Focke-Wulf, being challenged itself in the air, was no longer able to give the same assistance to the U-boats and gradually became the hunted rather than the hunter.'

–

Long before the CAM-ships escorting the Gibraltar convoys had gained fame by destroying Condors, the Sea Hurricane IAs had participated in the protection of another, more grim supply line. On Hitler's invasion of Russia on 22 June 1941, Churchill at once promised Britain's new ally everything that could be spared in the way of material aid. By the end of the war, the route through Iran had become by far the most important, but because of the distances involved and the time taken to improve the Iranian transport system, it was not until late 1943 that a greater monthly volume could be delivered by way of Iran than could be carried by convoys around the North Cape of Norway.

This was indeed the shortest, most direct route, but it was one fraught with dangers. Apart from the risk of attack from German bases in Norway, the vessels had to pass through the icy waters of the Arctic Ocean, in which they were battered by great waves; in which a man would freeze to death within minutes; in which mighty winds blinded the crews by dashing into their faces rain, snow, hail or sleet.

Yet during the war, forty convoys made the Arctic run, to the ice-free Russian port of Murmansk, 200 miles east of the

North Cape at the head of the Kola Inlet, or to Archangel, a further 400 miles to the south-east, where the unloading facilities were vastly better but where even the strenuous attempts of the Russian ice-breakers were unable to keep the port open during the winter months. These convoys carried four million tons of cargo, 300,000 tons of which was lost en route.

By the end of May 1942, the PQ convoys, as they were known from their code-numbers – returning convoys naturally enough were coded QP – had already delivered 3,000 aircraft, 4,000 tanks, 30,000 other vehicles, 42,000 tons of petrol, 66,000 tons of fuel oil and 800,000 tons of miscellaneous supplies, including food, machine-tools, rubber, aluminium, tin, ammunition, blood-transfusion sets, emergency operation equipment, surgical needles, antiseptics and other medical supplies. And this of course was only the beginning.

In view of the vast scale of the conflict on the eastern front, the delivery of weapons such as tanks or anti-tank guns was probably of minor significance, but considering the titanic losses of Soviet aircraft during the first months of hostilities, every aeroplane that could be provided must surely have been of immense value, especially before the Russians could complete the evacuation of their aircraft factories beyond the Ural Mountains. The lorries, jeeps and other motor transport were of outstanding importance to judge by the extent to which they formed the equipment of the Red Army when the Allies later met it in the ruined heart of Germany. Perhaps most vital of all were the raw materials, without which the Russians could not have manufactured hardened steel or certain special alloys.

Certainly, the Germans were only too well aware that the passage of the Arctic convoys might make the difference between victory and defeat. Thus, in *Sea Warfare 1939-1945: A German Viewpoint*, Vice-Admiral Friedrich Ruge, speaking of those from August 1944 to April 1945, says:

> The weapons and equipment (carried) including
> nearly half-a-million vehicles, allowed the

Russians to equip a further sixty motorized divisions, which gave them not only numerical but material superiority at focal points of the battles. Thus Anglo-American sea power also exerted a decisive influence on the land operations in Eastern Europe.

During the period he mentions, moreover, the supplies coming through Iran far exceeded those delivered by sea. How much more important then were the convoys before August 1944, when the Iranian route had not been developed fully, but in which over twice as many ships had reached Murmansk or Archangel and at a time when Russia's need was infinitely greater.

Fortunately, the Germans did not reach this conclusion at once. The first eleven PQ convoys lost only destroyer, *Matabele*, and a merchantman. This was partly because Göring refused to co-operate with the German Navy by providing air reconnaissance, until his obstinacy was ended by a direct order from Hitler, but it was probably due in the main to the enemy's belief that Russia would in any case be knocked out within six months. When by December 1941 'General Winter' had again aided the Russians with shattering effect; when perhaps only Hitler's stubborn determination not to yield an inch of ground without a fight had prevented a more catastrophic retreat than that of Napoleon; when it was clear that a war of attrition had replaced the planned war of movement; then the Luftwaffe, the U-boats and the German surface fleet all turned on the Arctic convoys with a determination that reveals their importance. One example must suffice: the great battle-cruiser *Scharnhorst* went to her doom under the guns of the British battleship *Duke of York* in response to an order from Dönitz, then the German Navy's Commander-in-Chief, to 'strike a blow for the gallant troops on the eastern front' by destroying a convoy.

In contrast, the Russian leaders, for political reasons, tended to decry the value of the aid received from the Western Allies

– though their angry protests when deliveries were postponed for any reason are sufficient to demonstrate their falsehood. It is therefore pleasant to be able to report that in 1943, Mr Maisky, the Soviet Ambassador in London, made honourable amends when he called the Russian convoys 'a Northern Saga of heroism, bravery and endurance'.

In that saga, the Catafighter pilots played their part. The first Hurricat launch was made on a returning convoy, QP12, which on the morning of 25 May 1942 was being trailed by a number of airborne scouts, including Junkers Ju 88s that were awaiting a chance to supplement their reconnaissance role with bombing attacks. The CAM-ship *Empire Morn* decided the right moment had come. Shortly before 0900, Flying Officer Kendall was blasted off to the attack.

Climbing rapidly, Kendall first sighted a Blohm & Voss BV 138 flying-boat, which fled at his approach; then chased a Ju 88 along the entire length of the convoy before shooting it into the sea. The remaining enemy machines quickly left the dangerous vicinity of QP 12, which they never again located.

Tragedy followed. After about an hour aloft, Kendall baled out but, for some unexplained reason, his parachute did not open sufficiently. Destroyer *Badsworth*, dashing to the scene, pulled him out of the water, but he died almost at once. He was buried at sea – the only Hurricat pilot to lose his life after a catapult launch.

Later on 25 May, an even more splendid defence was put up by the Hurricane guarding PQ16. This convoy had already sighted reconnaissance machines but, in the prevailing good weather with the Arctic sun shining through the whole twenty-four hours of the summer day, it was considered pointless to use the Catafighter, since it could only make one trip, while the enemy could easily replace any 'recce' aircraft lost – better to keep it as a defence against the inevitable assault.

For Flying Officer Hay, yet another of that grand group of South African pilots who did so much for the RAF, sitting at instant readiness in the cockpit of his Hurricat on board *Empire Lawrence*, the afternoon dragged on in an atmosphere of tense expectancy. Shortly after 2030, the raiders arrived: six Junkers Ju 88s making shallow diving attacks, timed to co-ordinate with a low-level strike by five Heinkel He 111s, each carrying two torpedoes. It was against the latter that Hay was directed. With a roar and a sheet of flame, the Hurricane was launched.

Ripping into the Heinkel formation, Hay shot down one, badly damaged another and so broke up the attack that not one torpedo found its mark. The convoy's gunners, thus left free to concentrate on the Junkers, brought down two of them, though a near miss caused a damaged merchantman to be sent back to Iceland under tow. The torpedo-bombers then concentrated their fury on the Hurricane, which was badly damaged by a cross-fire, Hay being wounded in the thigh. Yet, even as he turned away, he found another enemy in his sights into which he put his remaining ammunition, though without visible result.

Hay then flew back over the convoy only to be hit again by gunners who had come to treat all aircraft as hostile. Happily, his parachute did open. His dinghy had been punctured by a bullet but prompt aid came from destroyer *Volunteer*, which picked him up after six minutes in the water.

Perhaps Hay's sortie had made the enemy airmen wary. Certainly, on the 26th their attacks were desultory, the only vessel lost being claimed by a U-boat. The following day was a different story. *Empire Lawrence*, which went down within minutes after a concentrated strike by Ju 88s, was only one of six ships sunk by the Luftwaffe. Nonetheless, another twenty-seven, carrying some 90,000 tons of cargo, including 124 aircraft, 321 tanks and more than 2,500 transport vehicles, reached Murmansk or Archangel. How many of these would have been lost but for Hay's intervention can only be conjectured. How many more would have got through if additional CAM-ships had been allocated to the convoy is perhaps a matter for sorrowful reflection.

There was to be one final launch on a Russian convoy. Since PQ18 was protected by the escort carrier *Avenger* – whose achievements will be detailed later – it might seem that the presence of the veteran CAM-ship *Empire Morn* was redundant. However, it was planned that on 16 September, a large proportion of the escort, including *Avenger*, would transfer their aid to the homeward-bound QP14. It was thought unlikely that PQ18 would be troubled at the limits of the Luftwaffe's range next day, but thereafter, as it steamed the last few hundred miles to Archangel, it would again be open to air-attacks – against which the solitary Hurricat would be the only defence.

As expected, raids duly developed on the 18th, the early ones sinking a merchant ship. That no further losses were suffered may be attributed to the fact that when, at 1150, the second attack was made by fifteen torpedo-carrying He 111s, Flying Officer Burr catapulted off *Empire* Morn dodging balloon cables as well as a hail of shells from AA gunners – whose aim, luckily, was as bad as their aircraft recognition – in order to engage them head-on. He utterly disrupted the Heinkels' formation, then swung behind one of them, firing until his ammunition was exhausted, by which time smoke was pouring from both the bomber's engines. It was seen to crash by observers in the convoy. With their combined strike thwarted, the Heinkels dropped their torpedoes individually. Not one hit was made.

Although Burr was now defenceless, he had no intention of relinquishing his duties as convoy-defender, being prepared to make mock-attacks on any other formation that showed up. However, when none had done so by about 1230, he determined to save his Hurricane by making for the aerodrome at Keg Ostrov near Archangel, though this necessitated a flight of almost 240 miles through thick banks of fog. He landed there safely at 1415 with just five gallons of petrol in his reserve tank.

–

This was neither the first nor the last time that a Hurricane would land in Russia. Churchill had promised to send 445 of them as soon as he heard of the German attack. This might not seem a vast amount when compared with the titanic losses the Soviet Air Force had suffered, but when it is recalled how often the RAF had been forced to fight vital campaigns with nothing like that number, the generosity of this gesture will be apparent. It could also be remarked that more than once even a small body of Hurricanes had had an effect out of all proportion to its size.

Indeed, the Russians would shortly be given a demonstration of this at close quarters, for it was decided that if they were to receive Hurricanes, it would be best if they were shown how to fly and maintain their new acquisitions. Also, since Finland had sought revenge for the Russian attack in 1939 by allying with Germany, the Luftwaffe had gained bases within close range of Murmansk and Archangel, which necessitated fighter cover for the Arctic convoys unloading in those ports.

Accordingly, 151 Wing came into being on 12 August 1941, under Wing Commander Ramsbottom-Isherwood, who was a New Zealander but whose family, surely, had originally hailed from the North Country. This contained two new Hurricane squadrons, 81, led by Squadron Leader Rook, which was formed by expanding a flight detached from 504, and 134, under Squadron Leader Miller, which similarly started as an offshoot of No 17. Equipped with Hurricane IIBs plus a handful of IIAs Series 2, this unit was detailed to perform the dual role of protecting the ports and training the Russians.

Thus, when after a rendezvous in Iceland, the very first convoy – prior even to the famous 'PQs' – set out for Russia at the end of August, it contained twenty-four Hurricanes in flying trim on the little carrier *Argus*, no stranger to such missions, while fifteen more went in crates on board the seven accompanying merchantmen, whose main cargo consisted of munitions. The wing's personnel, apart from the pilots in *Argus*, travelled in the *Llanstephan Castle*. A strong escort was provided by carrier *Victorious*, cruisers *Shropshire, Devonshire* and *Suffolk* and six destroyers. During the voyage, according to Air Vice-Marshal Wykeham, the airmen 'were regaled with lectures on Russia and the Soviet people... In spite of this their spirits were high'.

On 7 September, the pilots took off from *Argus* for Vaenga airfield, some seventeen miles from Murmansk. In the process, the Hurricanes of Flight Lieutenant Berg and Sergeant Campbell, both of 134, damaged their undercarriages, though happily they crash-landed at Vaenga without serious damage to either pilots or aeroplanes. However, enemy air activity prevented the merchant ships from unloading at Murmansk as planned. Instead, the remainder of the wing was diverted to Archangel, where the crated machines were dumped on the aerodrome at Keg Ostrov.

The majority of the ground crews promptly set off to rejoin the pilots in Vaenga, leaving only a small party at Archangel to erect the Hurricanes, which was not easy in the absence of such essential equipment as spanners for the airscrews. Fortunately, the Russians had mustered a group of exceptionally able technicians to receive instruction on Hurricane –maintenance, with a view to their passing on their knowledge to others later. These men now arrived at Keg Ostrov, where they carried out a remarkable improvisation of the missing tools. With their backing, the British detachment assembled all fifteen Hurricanes within the space of nine days.

They were duly flown to the squadrons at Vaenga. This, as the personnel there were already aware, had a surface of rolled sand, which, in the frequent bad weather experienced, provided only too many opportunities for the Hurricanes to prove their ruggedness. Also, the living conditions, particularly the sanitation, were primitive to a degree. On the other hand, the hospitality of the Russian pilots was overpowering – in some cases utterly so, as at the welcoming banquet that resulted in the wing spending most of the next day in bed – while the co-operation of the local commander, General Kuznetsov, was exemplary.

There were problems in the air as well. Although Churchill had promised to send to Russia 'eight- and twelve-gun Hurricanes, which we have found very deadly in action', those of 81 and 134 had at first been fitted with at most eight, sometimes only six, guns to reduce weight for their flight from *Argus*. Also, the inferior quality of Russian petrol – being only of 95 not 100 octane, as was essential for the best performance of the Merlin – caused many difficulties in the bitter temperatures encountered. Thus, when, on 11 September, four aircraft from 134 made the first patrol over enemy-occupied territory, only some thirty miles from Vaenga, Flight Lieutenant Ross and Pilot Officer Cameron both had their engines cut out, the former on three separate occasions. Not until they were within a few feet of the ground, did their feverish use of the priming pump bring the Merlin back to life.

Next day saw the first combat. Five Hurricanes of 81 Squadron, armed only with six machine-guns each, sighted a Henschel Hs 126. This they damaged, but were then attacked by its escort of five Messerschmitt Bf 109s. They destroyed three of these but Sergeant Smith was shot down and killed – the only pilot, as it transpired, who lost his life during the wing's stay in Russia.

On the 17th, 81 was detailed to cover Russian bombers returning from a raid. The squadron reached its charges just in time to prevent their interception by 109s. These engaged the Hurricanes – which were now armed with twelve guns – instead, only to lose three of their number. 81 had no casualties even though the Messerschmitts had no sooner departed than some Russian fighters, thinking the Hurricanes were hostile, also attacked them. Happily, the RAF pilots were able to take successful evasive action.

This encounter foreshadowed an increasing use of the Hurricanes as escorts to the Russian bombers on their raids.

These were certainly needed, for Murmansk was now being threatened by the heroes of Narvik, General Dietl's crack ski-troops. However, attacks on German army positions, AA batteries and aerodromes had the desired effect – by the end of September, the German advances had halted on the Litsa river.

The Soviet bomber crews are reported to have had such faith in the protection provided by the Hurricanes that they went straight for their targets without even bothering to look overhead. Although it has been suggested that such statements were put out for reasons of propaganda, it is a fact that although a Russian bomber fell to AA fire, not a single one was lost to enemy fighters in the course of the strikes, about thirty-five in all, escorted by 151 Wing. An especially satisfying clash came during a raid on 26 September, when 81 Squadron was attacked from above by a force of the improved Messerschmitt Bf 109Fs. Despite their disadvantageous position, the Hurricanes out-manoeuvred their faster opponents, shooting down three without loss.

The Hurricanes also operated in a defensive role against German raiders, though with scant rewards because the Russian control system was so abysmal. It was on one such fruitless sortie, on 27 September, that the wing suffered its only other fatalities. The bumpy surface of the airfield made it necessary for two men to hold down the tail of each machine as it taxied into position. In this case the Hurricane's steadiness proved a liability, for Flight Lieutenant Berg of 134 tried to take off without realising, as he would surely have done with most other types of aircraft, that the men were still on his tail. The Hurricane crashed from about fifty feet, killing both the ground crew and severely injuring the-pilot.

Indeed, the only really satisfactory interception took place on 6 October, when the Hurricanes could scarcely fail to spot the enemy since the raid was directed against Vaenga. Even then the warning given was so minimal that several of the fighters were still taking off as bombs from the attacking Junkers Ju 88s landed on the airfield. However, 134 was able to get among the bombers, which it dispersed, claiming, without loss, two destroyed, one 'probable' – which it appears also came down since the Russians located the wrecks of three Ju 88s on the ground – and two more damaged.

These were 134's only victories during its stay in Russia. Twelve more were gained by 81, its most successful pilot being Flight Sergeant Haw, who was credited with the destruction of three enemy aircraft, for which he received the DFM as well as the Order of Lenin, this latter award also going to Ramsbottom-Isherwood and Squadron Leaders Rook and Miller. Each victory, incidentally, brought with it a bonus of 100 roubles, then worth about £20, but the airmen's jests that such would infringe their 'amateur status' merely masked a distaste for such 'blood money' – which ultimately was sent to the RAF Benevolent Fund.

Apart from sheer bad luck, the main cause of 134's lesser number of 'kills' was that it had been mainly responsible for training the Russian pilots to fly Hurricanes. General Kuznetsov was the first to take up the Hawker fighter, followed quickly by Captains Safanov and Kuharenko. That two Hurricanes were crashed by the Russians – though both were repairable – is scarcely surprising when the enthusiastic outlook of the pupils, who would demand training in the most appalling conditions, is taken into account. One of them is reported to have flown a Hurricane for the first time in a snowstorm. That he landed safely is a tribute to both man and machine.

On 19 October, by which time an almost total absence of daylight at Vaenga, which was 170 miles north of the Arctic Circle, had all but brought activities to a standstill, both RAF squadrons handed over their machines to the 72nd Regiment of the Red Naval Air Fleet. This in turn provided the nucleus of the 78th Air Regiment, the first Russian Hurricane Wing, which was formed on 25 October under the command of Safanov, newly promoted to lieutenant-colonel.

This Russian wing was able to contain three Hurricane squadrons, whereas the RAF one had consisted only of two, because other Hurricanes were now arriving in large numbers. Indeed, in mid-November they began to equip other Air Regiments as well. By 29 November, when the RAF wing, apart from a small staff of signals personnel sailed for home, Russian Hurricanes had already destroyed a Bf 109, a Bf 110 and a wretched Junkers Ju 52 transport, which had strayed over Soviet territory after losing its way.

Thereafter, massive quantities of Hurricanes were dispatched to Russia via the Arctic convoys or from the Middle East or even from stocks in north-west India. In view of the losses suffered by the convoys, no doubt many perished en route, but the vast majority will have reached their destination – though one IIB was later captured by the Finns who used it as a trainer until the end of May 1944. When the deliveries of Hurricanes to Russia ceased in 1944, 2,952 of them had been sent there. Among these – full details are not known – were 210 Mark IIAs, 1,557 Mark IIBs or Mark XIIs, which were similar machines manufactured in Canada, 1,009 Mark IICs, sixty Mark IIDs and thirty Mark IVs.[36]

Since one-fifth of all Hurricanes built were sent to Russia, there can be little doubt that their achievements must have been considerable, but how considerable is difficult to establish. At the time, apart from a few interesting articles that somehow slipped past the censor, the Soviet authorities preferred to suppress any references to the Hurricanes, being content to grumble that too few were handed over. Reports on them were certainly made by western observers on the Arctic convoys, at the British Air Section set up in Moscow, or in the Free French

---

[36]The IID was an anti-tank version; the IV an all-purpose close-support aircraft. Details of these as well as of the Canadian Hurricanes will be given later.

'Normandie' Squadron, which fought with the Red Air Force from the end of 1942 to almost the close of hostilities, but these inevitably were somewhat brief.

Happily, the fall of the Communist regime has resulted in a more satisfactory situation. It is still not possible to give a full account of the Hurricanes' activities because even now much of the necessary information is unavailable. For example, all the records of five Air Regiments that flew Hurricanes – and it will be recalled that each Air Regiment contained at least two, usually three, squadrons – seem to have disappeared completely. Nonetheless, several new sources have become available and a number of fresh details have been revealed.

From such sources, it seems that although Safanov's pilots had expressed a great liking for their new equipment, the Hurricanes were, for a time, the targets of bitter criticism from those responsible for their maintenance, most of the complaints, astonishingly, being directed at the Merlin engines, which were claimed to be 'under-powered'. The reasons for any defects were soon exposed by experts sent out by Rolls-Royce to investigate. In one depot alone, they found more than fifty Merlins allowed to rust in the open. With complete lack of imagination, it had not occurred to the Russians that these had been intended for use in rather different climatic conditions from those for which their own equipment was designed.

A further Russian complaint was that the 0.303-inch Browning machine-guns carried by the Hurricane IIAs and IIBs lacked penetrative power. This the Russians remedied by adapting their Hurricanes in various ways. Some were fitted with two 20 mm cannon and two 12•7 mm machine guns. A few even retained four of their Brownings in addition to these new weapons. Others had their Brownings replaced by 0.5-inch machine-guns taken from American Tomahawks, Kitty-hawks and Airacobras, which the Russians liked less than the Hurricanes because they were not so reliable. Since these adaptations undoubtedly increased the Hurricanes' hitting power, it is perhaps a pity that the RAF did not consider making similar changes to their own Hurricanes. No complaints of course are recorded about the 1,000-plus Hurricane IICs received by the Russians with their armament of four 20 mm cannon.

Once such problems had been rectified, the Hurricanes showed their worth in all manner of situations. They did good work in their interceptor role, especially those of the 78th Air Regiment, one of whose pilots, Major Sgibnev, was credited with eleven victories while flying Hurricanes: the most by any Russian pilot. They guarded targets on the sea as well as on land. Convoy PQ 16, whose ordeal by air-attack has already been mentioned, finally reached harbour under cover of Russian Hurricanes on 30 May 1942. That no German aircraft interfered after the Hurricanes' appearance, suggests that these had earned the healthy respect of the Luftwaffe.

Even in late 1942, a Russian Hurricane squadron, commanded by Major Panov, claimed to have destroyed eighty-three German aircraft, of which thirty-one were bombers, in just over two months. Although there was doubtless a good deal of exaggeration in the claim, it does seem that this unit was enjoying a high degree of success. During the same period, it lost ten Hurricanes but only four pilots – this statement can well be accepted for it has already been shown that the majority of Hurricane pilots escaped with their lives even when their machines were brought down.

The Hurricanes were considered 'equally successful on reconnaissance duty', while the Russians made great use of them as escorts to their Ilyushin bombers, in which task their 'excellent' manoeuvrability proved of great advantage. One Hurricane squadron, under Major Gorshkov, which specialised in this type of work during 1942, is reported to have lost only one of its charges to enemy fighters in three months.

As well as helping 'to solve the important problem of ensuring the safety of the Ilyushin', the 'quality of the Hurricane' enabled it to destroy many of the enemy interceptors. Thus, in an issue of the *Soviet War News* of 3 October

1942, is included an account of how six Hurricanes had beaten off eight Messerschmitt Bf 109s trying to disrupt a bomber formation. Not only were the 109s unable to fulfil their intention but three of them fell to the guns of Lieutenant Dobrovolsky and Sergeants Barishnyov and Bunakov. It is only a pity that more of such combats have not been recorded.

By 1943, as was the case with those in the RAF, the Russian Hurricanes were suffering heavy losses in the interceptor role, so tended to be used increasingly for ground-attack duties. The cannon-armed IICs in particular are said to have proved invaluable against German motor-transport, though it would seem logical that the later Marks of Hurricane were even more useful, since they were expressly designed for this purpose.

As well as demonstrating the Hurricanes' versatility, the Russians made good use of their adaptability. Perhaps the most interesting aspect of this was their conversion of several into two-seaters. One of these carried a dorsal gun position, though it is not known whether this proved advantageous. Remembering the sad fate of the Defiant, it would appear unlikely. It may be a confirmation of this opinion that only the one aeroplane was so equipped. On the other hand, a number of two-seat trainers were used, one by the Free French. After the war, Hawkers also produced such a version and since the Hurricane's mixture of strength and docility made it an ideal advanced trainer, it seems highly probable that this adaptation would have proved a singularly happy one.

From these sparse gleams of light on a largely unknown subject, it can be stated that the Hurricane bore at least a not unworthy part in the titanic struggles on the Eastern Front. Certainly, the scope of its involvement could hardly have been foreseen by the members of 151 Wing as they made their way home over the storm-tossed Arctic, thankful that, with their task in Russia completed, they could enjoy a brief respite from the war.

Far away in the blue Pacific, another war was just about to begin.

Chapter Nine

The Rising Sun

'Air raid, Pearl Harbour!' The awesome message was flashed by Rear-Admiral Bellinger commanding the Naval Air Forces in Hawaii at 0758 on 7 December 1941. 'This', he added, 'is no drill.' They were real bombs. The blood was real too.

The steps by which a once loyal ally of the Western powers had become a deadly, ruthless enemy, were long, complicated and tragic – but special mention must be made of the disastrous Washington Conference of 1921. This saw the end, under American pressure, of the twenty-year-old Anglo-Japanese Alliance, which had been regarded in Japan with sentimental pride as the enduring pillar of the country's foreign policy. Its disappearance therefore was, according to Professor Storry in his *History of Modern Japan*:

> A blow received in sorrow and remembered in wrath. The termination of the Alliance removed, perhaps for ever, the possibility of friendly and effective British influence on policy-making in Tokyo, while it did nothing to strengthen the security of the United States; and of course it greatly weakened the whole strategic position of Australia and New Zealand, to say nothing of Hong Kong and other British possessions east of the Bay of Bengal.

Then, in 1929, came the world depression, which, aggravated by the tariff barriers erected by the Western countries, had a crippling effect on Japanese industry. In particular, the fall in the demand for silk in the United States reduced to desperate poverty the majority of Japanese farmers for whom its production was a major secondary employment.

These events served to discredit the country's political rulers and 'big business' interests, all pro-Western in feeling. Increasingly, they were blamed for the nation's problems, especially by the Army whose leaders advocated, as a short cut to prosperity, a blatantly expansionist policy on the Asiatic mainland.

In furtherance of this aim, on 18 September 1931, without the authority or even knowledge of the civilian government in Tokyo, the Kwantung Army, which from its base at Port Arthur on the Yellow Sea guarded Japanese interests in Manchuria, seized the city of Mukden. This was a preliminary to the conversion by 9 March 1932, of the whole of Manchuria into a Japanese puppet-state under the name of Manchukuo. In February and March 1933, a brilliant campaign in almost Arctic conditions added to the new state the province of Jehol, formerly part of China, which country was also compelled to accept a very large demilitarised zone south of the Manchurian border.

These actions naturally brought about steadily worsening relations, which reached flash-point on 7 July 1937, when fighting broke out between Chinese and Japanese troops on manoeuvres near Peking. Unlike earlier 'incidents' in Manchuria, this had not been arranged by the local Japanese military commanders, who tried to reach a settlement. However, on both sides extremists were only too ready for violence. On 2 August, more than 200 Japanese civilians were massacred by Chinese militia in Tungchow, while on the 14th, Chinese aeroplanes attempted to bomb Japanese warships at Shanghai but hit their own city instead. Stung to fury by these actions, the Japanese mounted a major invasion of China, during which the atrocities committed by the Chinese were repaid with interest.

By 1939, with Peking, Nanking, Shanghai, Hankow and Canton in their hands, the Japanese could claim important military gains, but on the political front there was no indication of any willingness to come to terms from the Chinese leader Chiang Kai-Shek, by now receiving mounting aid from the United States, whose attitude to Japanese conquests was becoming ever more hostile. In April 1940, the Pacific Fleet was transferred from San Diego, California, to Pearl Harbour. From mid-1940, undeterred by Japan's signature, on 27 September, of the Tripartite Pact with Germany and Italy, the American Government placed increasing embargoes on the delivery to Japan of strategic materials such as air-frames, aero-engines, aviation fuel, steel and scrap iron. This process reached its culmination in July 1941, following an agreement by the Vichy French authorities to the occupation of bases in the south of Indo-China by the Japanese (who already held similar rights in the north) – an action that threatened one of the main sources of supply to China. On the 26th of that month, President Roosevelt announced the freezing of all Japanese assets in the United States, about £33 million in value. Britain and Holland quickly followed suit. Japan was thus faced, for want of funds, with a complete halt to all her vital imports, in particular her importation of oil.

Since the Japanese economy was already severely strained, it was clear that it would soon collapse were this decision not revoked. In the words of Admiral Nagano, the Chief of Naval Staff, Japan was 'like a fish in a pond from which the water is gradually being drained away'. Either American demands would have to be met or Japan would have to go to war to seize the forbidden raw materials, in the Dutch East Indies, Malaya, Burma and the Philippines.

In practice, therefore, from 26 July war was inevitable. Although the Japanese commercial interests, already horrified by the cost of the 'China Incident', would gladly have witnessed its termination, the Army had no intention of accepting such a massive loss of face. The weak, though well-meaning, Prime Minister, Prince Konoe, to whom the prospect of hostilities with America was a nightmare, suggested a meeting with Roosevelt in the hope of gaining sufficient prestige to force a settlement on the militarists, but despite the advice of the US Ambassador in Tokyo, Joseph Grew, his request was rejected. On 16 October Konoye resigned – to be replaced as Premier by the Army's representative, General Hideki Tojo.

Even the Army would yet have been prepared to evacuate Indo-China in return for a lifting of the embargoes and an unfreezing of the assets but the United States was not willing to make any such compromise. As Professor Morison points out in *The Two-Ocean War*: 'The fundamental reason for America's going to war with Japan was our insistence on the integrity of China.' It is worth reflecting that she had not been prepared to go to war for the sake of the integrity of Western Europe.

It should be noted also that throughout all negotiations, America spoke for Great Britain as well as for herself. Indeed, on 10 November, Winston Churchill expressly promised that 'if the United States should become involved in war with Japan, a British declaration would follow within the hour'. The only possible explanation for the casual attitude of the British authorities to such a vital matter is that they completely underestimated the military power of their potential foes. James Leasor's comment in *Singapore: The Battle that changed the World* that 'the general impression in Malaya was that the Japanese aircraft were made of rice paper and bamboo shoots' may be an exaggeration, but even in official publications the tired old cliché was repeated that these were mere imitations of Western ones, but some five years out of date – much as similar comments would be passed after the war about Japanese cars or motor-cycles, though not by those fortunate enough to own them.

In 1941, such ignorant arrogance proved lethal, for it led to the attitude epitomised by the observation of Air Chief Marshal Sir Robert Brooke-Popham, Commander-in-Chief, Far East, that the obsolete American Buffalo fighters, slow, unmanoeuvrable, and armed with only four machine-guns, were 'quite good enough for Malaya'. In reality, they were unworthy opposition even for Japanese bombers, while any Buffalo pilot who attempted to dog-fight with the Japanese fighters would be fortunate to survive at all.

Illusions first began to disappear with the shockingly effective attack on Pearl Harbour by the airmen from Vice-Admiral Nagumo's six carriers. Though represented as the height of treachery, it was the intention of the Japanese to keep their opening strike within the letter – though certainly not the spirit – of international law by delivering their declaration of war at 1300 hours, Washington time. This was 0730 at Hawaii, so would have given the Americans only about half-an-hour to warn their forces in the target-zone. However, this cunning device deservedly rebounded on its authors since the inefficient Japanese embassy staff in Washington did not, in practice, deliver the message until after the start of the raid.

Not that it should have made much difference, for American cryptographers had broken the Japanese diplomatic code. By 0600 on the 7th, therefore, it was already known in the capital that the Japanese Ambassador had been instructed to deliver his message at an odd hour, which had a special significance for Pearl Harbour. A warning to be on the alert was dispatched accordingly – but not until noon and then only by normal commercial channels. Some hours after the attack it was handed to General Short, Army Commander in Hawaii, by a messenger boy on a bicycle.

Thus, with total surprise achieved, 351 aircraft, in about two hours, sank four battleships, caused a fifth to beach, and badly damaged three more. On the airfields, 188 American aeroplanes were destroyed, 159 damaged. Although the naval base, with its repair facilities and stocks of oil was almost untouched, and the vital American carriers were not present at the time of the attack, the raiders, at a cost of only twenty-nine of their number, had dealt the Pacific Fleet a stunning blow. However, at least it might be thought that they must have dispelled any doubts as to the calibre of the Japanese armed forces.

Yet this was far from the case. On 2 December, Admiral Sir Tom Phillips had reached Singapore with a fleet designated 'Force Z' centred upon the new, powerful battleship *Prince of Wales* and the ageing battle-cruiser *Repulse*. On the evening of the 8th, these left harbour to attack reported Japanese landings in northern Malaya and southern Thailand (or Siam as it was then also known). However, the following afternoon, when sightings by enemy reconnaissance aircraft had prevented any chance of surprise, Phillips decided to turn back.

Early next morning, the Admiral received a report of a fresh landing at Kuantan, on the east coast of Malaya. He at once headed in this direction but, since he preserved radio silence, he made no request for fighter cover. Doubtless, he hoped that his decision would be anticipated, so that protection would be provided in any event – but, unhappily, the correct deductions were not made in Singapore.

It seems at least probable that Phillips was never really concerned about obtaining cover, for he was convinced that warships at sea could always repel any attack from the air. Mr Leasor quotes a quite weird prophecy made during a pre-war dispute on this subject by the then Group Captain Harris (later Bomber Command's redoubtable leader) who informed Phillips:

> 'I had the strangest dream last night, Tom. I saw your future so clearly. War was declared in the east and the Japanese had taken Siam. You'd been promoted Admiral, and you set off in your flagship up the east coast of Malaya to deal with the situation. Suddenly, down comes a rain of bombs and

torpedoes. You look down from your bridge and say, "What a lot of mines we've hit!"'[37]

On the morning of 10 December, shortly after Force Z had discovered that the Kuantan landing was a false alarm, this dream came true. The high-level bombers and torpedo-planes of the Japanese Navy's 22nd Air Flotilla, based at Saigon, struck with merciless efficiency. Both *Prince of Wales* and *Repulse* went to the bottom, Admiral Phillips perishing with his flagship. Just four of the attackers were lost. The effect on the morale of Singapore was catastrophic.

By this time, the original enemy landings had attained complete success. Japan had mustered some high talents for the invasion of Malaya, for the superbly trained 25th Army was entrusted to General Yamashita, whose abilities were sufficiently outstanding to earn him this command in the face of Tojo's personal enmity; while the supporting naval forces were led by Vice-Admiral Ozawa, Japan's finest strategist whose exemplary co-operation with the Army was in sharp contrast to the usual deplorable relations between the two services.

In the early hours of 8 December, Yamashita's men landed at Kota Bharu in the north of Malaya, and at Singora and Patani in southern Thailand. The same day, Bangkok was occupied, whereupon Thai resistance collapsed with a rapidity remarkable even in a war noted for such debacles. The Japanese III Air Group was rushed to Singora, whence it fell on the northern airfields of Malaya. During the next twenty-four hours, it reduced the RAF strength by half.

Disasters on land quickly followed. The enemy, crossing the Kra Isthmus to the west coast of Malaya, where there were far better communications, fell on the 11th Indian Division at Jitra

---

[37]There are half-a-dozen versions of this conversation but the message is the same in each.

on 11 December, their light tanks, although quite unsuitable for actions in Europe, taking heavy toll of the defenders many of whom had never seen one before. By the morning of the 13th, the Allied forces, who had outnumbered the Japanese by more than four to one, were fleeing southward, leaving behind immense stores of food, guns, ammunition and petrol to speed their foes' advance. This retreat also abandoned the main airfields in the north almost intact. Within hours, enemy airmen, flying on captured British fuel, were dropping captured British bombs on the routed ground troops.

Thereafter, the morale of the services, like that of the civilian population, plunged to the depths while that of Yamashita's men was correspondingly exalted. Subsequent events followed a gloomy pattern. The British and Commonwealth soldiers retired in increasing disorder, hotly pursued by the light-equipped, fast-moving Japanese, whose swift out-flanking movements thwarted any attempts to make a stand before the defences could be consolidated. By 11 January, Yamashita had reached Kuala Lumpur, where another vast mass of supplies fell into his hands.

It was at this stage, when the defences were in ruins, despair had gripped every heart, and the advanced radar stations had long been over-run by the Japanese advance, that Hurricanes at last reached Singapore, fifty-one machines and twenty-four pilots arriving in a convoy on 13 January. The Hurricanes were old Mark Is or early IIAs Series 1, which, having been destined originally for the Middle East, were further handicapped by desert air filters. The pilots were newly trained and, although mustered together as 232 Squadron, had been assembled from four different, quite unconnected, units. But even had the force suffered from none of these disadvantages, it would have been too small to be effective. And even if it had been larger, it would have come too late to turn the tide.

At least it may be said of the Hurricane pilots that they tried. Not until the 19th had the aircraft been assembled and armed by ground crews without previous experience of the type, but next day 232 was aloft to engage a force of twenty-seven bombers raiding Singapore. One flight attacked these, claiming to have shot down eight of them without loss in short order.

However, while this was happening, 232's other flight was having the less fortunate experience of being engaged by the escorting Japanese fighters. Three of these were brought down but three Hurricanes were also lost, as were all their pilots, among them the CO, Squadron Leader Landels.

The enemy aircraft were identified on this and on numerous other occasions throughout the next few weeks as Mitsubishi Zeros, the finest of the Japanese fighters, possessing a top speed of 350 mph, an amazing rate of climb, the ability to out-turn a Hurricane and an armament of two 20 mm cannon and two 7•7 mm machine-guns. In fact, being a Navy fighter, the Zero took little part in the Malayan campaign, in which the Hurricanes' main opponents were Japanese Army Air Force machines – Nakajima Ki 27 'Nates' or (as on 20 January) Ki 43 'Oscars'.[38] These had a speed of approximately 310 mph, though they also could out-turn and out-climb the Hawker fighters.

Yet, in some respects, they were inferior to the Hurricanes. Lacking the cannon of the Zero, their fire power was less than that of the steady Hawker fighters. In contrast with the ability of the Hurricanes to take punishment, the Japanese fighters, if hit fairly, which admittedly with such small, agile machines was not easy to do, would literally tear to pieces. Also, once experience had been gained, the Hurricane pilots learned that the best way to deal with their enemies was to attack them from above, then

---

[38]The Allies avoided difficulties caused by the complicated Japanese system of aircraft classification, plus the problem of pronunciation, by giving each type an arbitrary code-name; the bombers having ladies' names; the fighters men's names. The exception was Mitsubishi's famous fighter, which was almost always known as the Zero, or – to RAF pilots – the Navy Nought, not by its official code name of Zeke'.

continue downwards, since a Hurricane could always out-dive its rivals.

Indeed, the main problems faced by 232 arose less from the performance of its Hurricanes than from the inexperience of its pilots, coupled with a complete lack of information on either the virtues or the defects of the Japanese aircraft. Many fine men lost their lives needlessly while combat knowledge was acquired. Once it had been, 232 gained successes, Sergeant Allen being credited with seven 'kills' and Sergeant Dovell with six.

The squadron also attacked the advancing Japanese armies with good results, according to Yamashita's Director of Military Operations, Colonel Tsuji, who reported:

> The Hurricanes flying low over the rubber forest were a serious challenge. Their intrepid pilots continually machine-gunned our roads, shooting up our motor transport and blocking traffic ... Until then our mobile corps had been advancing on the paved roads in broad daylight taking no precautions against enemy air raids. While the Hurricanes were flying even single cars moved off the road into the cover of the jungle, and all convoys had to move off the road and get out of sight at the first alarm.

On 26 January, the Japanese landed at Endau on Malaya's east coast. That afternoon, two separate attacks were made on the beachhead by Hudsons, Albacores and Vildebeest bombers, escorted by Hurricanes and Buffalos, but, for all the fighters' efforts, the obsolete biplane Vildebeests, with their top speed of little over 100 mph, fell easy victims to enemy interceptors. Eleven of them were lost, as were two Albacores.

No 232 Squadron was suffering grim casualties also. Only one Hurricane fell during the Endau actions, the pilot, Sergeant Fleming, baling out, later to make his way safely back to the

Allied lines, but by the 28th, the squadron had lost seven-teen Hurricanes in all, while thirteen more were under repair, leaving twenty-one, of which the ground crews could probably manage to keep only half fit for action at any one time. By now, all RAF machines except fighters had retired to Sumatra, while the land forces were fleeing to the dubious safety of Singapore Island. On the 31st, the causeway was blown up; the siege of the fortress had begun.

Also on the 31st, the Hurricanes of 258 Squadron came to the aid of 232, but their experiences merely serve to illuminate the folly of bringing in reinforcements at this stage. Four days earlier, sixteen Mark IIBs of 258 had flown to Java from the aircraft *carrier Indomitable*. Here one suffered mechanical trouble but the others proceeded to Sumatra, where two more were damaged on the inadequate landing grounds. The remainder reached Singapore but their long-range tanks had then to be removed and the guns stripped of their protective grease and re-assembled. Thus, on the 31st, only eight were ready to take on a hostile formation, which outnumbered them by at least ten to one. It may be thought greatly to their credit that they destroyed two of the enemy aircraft for the loss of two Hurricanes and one pilot. Three other Hurricanes were badly damaged, but were able to return to base.

By 3 February, an enemy artillery bombardment of the three northern aerodromes on Singapore Island had forced the with-drawal to Sumatra of all fighters except ten Hurricanes and six Buffalos based at the mud-soaked civilian airport at Kallang. By the 8th, on the night of which the Japanese crossed the Straits in the face of collapsing resistance, only the Hurricanes remained. Next day they were in constant action, destroying or damaging seven enemy machines for the loss of three of their own number and one pilot. On the 10th, the last seven Hurricanes took off for Sumatra. Five days after that, General Percival, the British Commander in Malaya, surrendered with 70,000 troops – the greatest capitulation in British military history.

On the day following the loss of *Prince of Wales* and *Repulse*, Japanese airmen had dropped a large bouquet of flowers to honour the dead. It was also, says Mr Leasor, 'the first wreath for the death of the British Empire'.

–

The Hurricanes did not surrender. A number of Mark Is had been on their way to Singapore by sea when its position became hopeless. They were therefore diverted: thirty to Burma; twenty-four to Java where they were taken over by the Dutch; one lone machine (V7476) as far as Australia where it was used for fast communications work as well as for testing a suit, designed by Professor Cotton of Melbourne University, to reduce the effect on pilots of the gravitational forces experienced at high speeds.

However, the area that saw most activity by the Hurricanes, even before the Japanese triumph at Singapore, was Sumatra. The day after 258 had flown from *Indomitable* as described earlier, thirty-two more Mark IIAs or Mark IIBs left her decks, though one, suffering from engine trouble, had to land back on the carrier – successfully, despite the absence of a hook to catch the arrester wires. These were intended as reinforcements for 232, but they got no further than Sumatra, where, on 3 February, they were joined by the survivors of both 232 and 258, apart from the small detachment at Kallang. Unfortunately, by this time five of the first batch had already been written off in crashes on the wretched East Indian airfields.

In Sumatra there were two such: P1 just north of the capital, Palembang; and, some twenty miles to the south, a secret base called P2 – an irregularly-shaped clearing in otherwise dense jungle with clumps of trees growing in the middle. At neither were there proper facilities, accommodation or transport. Spare parts were all but non-existent, as were tool kits, while the ground personnel lacked any previous training on the maintenance of Hurricanes. When the grim landing surfaces are remembered, it is not very remarkable that fewer than half the Hawker fighters were ever serviceable at once – a factor that should be kept in mind, since otherwise the number physically present appears misleadingly high.

Although on 1 February the Hurricanes were formed into 226 Group under Air Commodore Stanley Vincent, there was no ground-control organisation to back them up. In fact, in the absence of very high frequency radio sets, there was practically no ground-to-air communication and very little between individual aircraft. There was no radar system and the Observer Corps posts were much too widely scattered.

The inevitable consequence was that the Hurricanes rarely received adequate warning. P1 was strafed by enemy fighters on 3 February. Hardly any damage was done, but that was only the first raid. Next day, no fewer than nine Hurricanes were shot down or wrecked in crash-landings, four pilots being killed. On the 7th, the enemy destroyed six Blenheims and three Hurricanes on the ground, shot down another Blenheim coming in to land and bounced the Hurricanes as they got airborne, bringing down three more – though only one pilot died. On neither day did the Japanese suffer any appreciable harm.

Had it not been for the arrival of fresh aircraft, there seems little doubt that the Hurricanes would have been wiped out before the enemy landed on Sumatra. However, on the 10th, the last machines from Singapore flew in, while, on the 13th, Wing Commander Maguire, who was to command the fighters, led eight more Hurricanes from Java. As they approached P1, some enemy bombers escorted by Oscars appeared. Sergeant Scott of 258 was jumped. With smoke pouring from his Hurricane's engine, he pulled up sharply to gain enough height to bale out, which he did safely. By so doing, he also brought down his conqueror, for when the Oscar tried to follow him, both its wings tore off, sending it hurtling into the jungle.

At the same time, improved radio communications having at last enabled adequate warning to be passed by the Observer

Corps, the Hurricanes at the base also took off to engage, shooting down two bombers and a further two fighters for the loss of two more Hurricanes and one pilot.

There was little time for celebration. Next day, a convoy moving on Palembang from Banka Island, which the Japanese had just seized, was attacked by Blenheims and Hudsons, though the damage inflicted appears to have been much less than was estimated at the time. Unfortunately, on orders from Java, all serviceable Hurricanes were detailed to escort this raid in defiance of the wishes of Air Commodore Vincent. Although they provided useful protection, none were left to mount guard over their own bases.

This was regrettable, for the Japanese were about to spring a complete surprise in the form of parachute landings, one at the oil refineries at Pladjoe, just outside Palembang, another close to P1. This necessitated the abandonment of the airfield and the destruction of the large number of Hurricanes grounded there. To make matters worse, a flight of nine replacement aircraft from Java was attacked by Japanese fighters when it was desperately short of fuel and five of the RAF aircraft were written off in crash-landings.

Unaware of the existence of the hidden base at P2, the Japanese were confident that they were now exempt from interference from the air. They were soon to be disillusioned. On the 15th, their convoy arrived at the mouth of the Musi River on which Palembang stands. Here the soldiers transferred to barges in preparation for a move upstream to the capital. But now RAF bombers fell on them; the few remaining Hurricanes, flown alternately by pilots of 232 and 258 in order to keep up continuous pressure, strafed individual landing craft without mercy, causing casualties that ran into hundreds; while as a final act in the day's operations, the Hawker fighters raided Muntok airfield on Banka Island, strafing a number of enemy machines that they caught on the ground.

During these attacks, only one Hurricane pilot was injured. Flying Officer Donahue, an American serving with 258, was hit in the leg, but still managed to bring his aircraft back safely. The ground conditions proved more hostile than the Japanese: two of the fighters were put out of action in landing accidents, though in neither case was the pilot harmed.

Had there been anything like a sizable force of Hurricanes on hand, it is possible that the Japanese would have been stopped in their tracks – but it was not to be. Palembang fell the next day. Although P2 had still not been spotted, its position had now become so precarious that it was thought best to evacuate such aircraft as remained to fight elsewhere. By the 18th all RAF machines, including the last six Hurricanes, had retired to Java.

This, the richest of the East Indies, had already seen Hurricanes. First there had been those from *Indomitable* but they had moved on to Sumatra. Then had come the twenty-four Mark Is, which went to the Dutch Java Air Force. Finally, in early February, forty more IIAs or IIBs arrived in crates on the aircraft transport *Athene*. Eight were hastily assembled but were ordered off to Sumatra as previously recounted. Before that island fell, nine more had also been sent there. A few of the remainder were handed over to the Dutch. The others were joined by the survivors from P2, together being re-named West Group, still under the command of Vincent. Ground crews – but no pilots – from 242 and 605 Squadrons had now also reached Java. Since they were fresh, it was decided that flying personnel should be allocated to these new squadrons from 232 and 258, the tired ground crews of which were to be evacuated. Most of 258's pilots left as well but a few together with some from 488 Squadron[39] joined 605. Those of 232 transferred their allegiance en masse to 242.

The two new squadrons possessed, between them, about twenty-five Hurricanes of which the unprecedented number of eighteen were fit for combat – though this proportion soon

---

[39]This was a New Zealand unit that previously had flown Buffalos in Malaya.

dropped. In contrast to the position in Sumatra, they had excellent control facilities at their base at Tjililitan, near the capital city of Batavia (as it was then known), admirable communications and an improved warning system, with radar stations as well as a network of observers on smaller islands to the north.

Finally, the pilots were now fully knowledgeable about the best techniques to employ against their enemies. Despite the odds against them, they forced the Japanese to mount increasingly large escorts for their bombing raids on Batavia, and claimed to have destroyed or damaged about thirty of the attackers, though by the end of February the number of Hurricanes had declined to twelve.

Some bitterness was felt by the RAF about the Dutch refusal to give up any of the Hurricanes they had received although there was a surplus of British pilots without machines to fly. However, the Dutch also needed Hurricanes desperately since most of their own fighters had been wiped out by a Japanese raid on Surabaya on 3 February. Although it seems unarguable that the more experienced RAF pilots would have made better use of them, the Dutch Hurricanes, based at Ngoro under the command of Lieutenant Anamaet, did put up a gallant defence of eastern Java, being credited with destroying or damaging another thirty Japanese machines for the loss of about twenty Hurricanes in combat, in accidents, or in raids on their airfields.

Yet, as had been the case so often elsewhere, events over which they had no control conspired to thwart the Hurricanes' best efforts. The fall of Sumatra had threatened Java from the west. Previously, the loss of Borneo, Celebes and the small but strategically important Amboina had opened it to invasion from the north. On 19 February, Bali to the east was taken, while, on the same day, the airmen from Vice-Admiral Nagumo's carriers fell on Port Darwin, Australia, from which any reinforcements to Java would be dispatched, where, overwhelming a small force of American Kittyhawks, they sank US destroyer *Peary* and five merchantmen, caused three others to beach and damaged ten more. Ashore they devastated airfields, docks, warehouses, fuel tanks and most of the town. For the time being, Darwin had ceased to be a naval base.

Now the Japanese were ready to grasp the greatest prize. From the captured islands in the north, two invasion forces snaked towards Java. Trying to intercept these, an Allied fleet under the Dutch Rear-Admiral Doorman, its every move watched by enemy float-planes, was routed by a strong Japanese covering group on the 27th, the surviving ships being hunted down individually later. On the 27th also, all hope of fresh fighters arriving was ended when the American aircraft transport *Langley*, heading for the port of Tjilatjap with thirty-two Kittyhawks on board, was sunk by shore-based bombers.

During the night of the 28th, the troops from the convoys poured ashore; one force at Bantam Bay in the west of Java, with a detachment at Eritanwetan 100 miles east of Batavia; the other near Rembang in the north-east of the island. Next day, the RAF Hurricanes, twelve in number, attacked both the western landings, making low-level strikes on everything that moved on the ground, on the sea, or in the air – for Sergeant Young of 242 shot down a seaplane, while Pilot Officer Fitzherbert of 242 and Sergeant Kelly of 605 between them destroyed or badly damaged three other seaplanes that had already landed in a small cove. At Eritanwetan, they inflicted particularly heavy losses on enemy soldiers coming ashore in barges as well as setting three light tanks on fire. Almost equal damage was caused by the Dutch Hurricanes at Rembang.

Such actions could hamper the Japanese but only counter-attacks on the ground could throw them back and the demoralised land forces scarcely put up a struggle. As the Dutch Hurricanes landed back at Ngoro after their triumphant foray, they were spotted by some enemy fighters, this time really the deadly Zeros, which strafed them so heavily that those that were not destroyed were too badly damaged to get airborne again before later raids eliminated them. On 2 March, the RAF

Hurricanes, now contracted into 242 Squadron under Squadron Leader Brooker, withdrew to Andir near Bandoeng. From here, in the face of constant air attacks, they continued to give what cover was possible to the retreating ground troops.

The collapse of these was swift. Batavia fell on 5 March, on which day also Nagumo's carriers raided Tjilatjap, completely wrecking the harbour together with the twenty-three merchant vessels therein. By now, 242 had only five Hurricanes left. By the 7th, there remained only two. These the squadron was ordered to destroy on the ground next day. Almost all the airmen were captured. On the 9th, the Dutch commander, General ter Poorten, ordered the final surrender.

It cannot really be claimed that the Hurricanes had done much more than hinder the enemy, while such was the feebleness of resistance in other quarters that the Japanese who had reckoned on taking six months to over-run the East Indies, achieved their aim in half that time. Nonetheless, the casualties inflicted by the Hurricanes, on the ground as well as in the air, were of significance, for already the quality of replacements was perceptibly lower as the wastage of Japan's front-line troops began. Certainly, the Hurricanes' achievements were such as to make it a matter of immense regret that really large numbers of them were not available from the beginning. Had they been, the Japanese might have paid an exorbitant price for their capture of Sumatra; and Java perhaps they might not have captured at all.

–

As it was in Malaya, as it was in the East Indies, so it was in Burma. The Hurricanes performed some splendid feats but they were too few in number, were backed by an inadequate warning system and were faced with the need for constant withdrawals as powerful Japanese thrusts from Thailand bit into the out-numbered defenders on the ground.

When Japan entered the war, the Allied fighter strength in Burma consisted of No 67 Squadron with sixteen of the useless Buffalos and the twenty-one Tomahawks of the American Volunteer Group commanded by Colonel Chennault – the famous 'Flying Tigers'. However, by mid-January, thirty elderly Hurricane Is had arrived in crates, while, not long afterwards, 17 Squadron, with Hurricane II As, made its appearance. This squadron also took over some of the Mark Is while others undoubtedly went to 67, although no official confirmation of this exists – which perhaps signifies little as the squadron records from January to May 1942 have ceased to be.

The Hurricanes, Tomahawks and Buffalos formed part of 221 Group under Air Vice-Marshal Stevenson, who had the unenviable task of opposing the 500 Japanese aircraft assisting their invasion forces. It is scarcely bearable to have to report that, once more, the Allied fighters not only faced enormous odds but suffered from every possible disadvantage. Most of the Hurricanes possessed neither auxiliary fuel tanks nor tropical filters – this last factor improved their performance rather than the reverse but gave the Merlin engines a very short life. There were hardly any AA defences. The only radar station – east of the capital, Rangoon – was so out of date that only once did it send marginally quicker warning of the approach of a raid than that from the Observer Corps, though this gallant body, lacking wireless equipment, was dependent on a highly inefficient civil telephone system.

Yet Stevenson did not wait on the defensive. Using the landing grounds at Moulmein, Tavoy and Mergui in the long Burma appendix, on the Kra Peninsula, his fighters raided the enemy's forward bases in Thailand, while his Blenheims struck at Bangkok. Fifty-eight Japanese aircraft were believed to have been destroyed or damaged in these attacks.

Unfortunately, they imposed no apparent restriction on the Japanese Army, which soon removed the nuisance of the advanced British airfields by the simple method of capturing them. Tavoy fell on 19 January, cutting off the garrison in Mergui, which was therefore withdrawn by sea. Moulmein was over-run on the 30th. Now enemy fighters were stationed at these bases, from which they covered the Mitsubishi 'Sally' bombers attacking Rangoon.

These escorted raids began on 23 January, continuing for six days. From their aerodrome at Mingaladon near the capital, the Hurricanes, Tomahawks and Buffalos met the challenge with undiminished firmness. The lack of warning meant that they had to climb away from their base before diving on the hostile formations; then, using the speed gained, climb again to repeat the attack. These tactics, similar to those used by the Hurricanes in Java, resulted in the destruction of, or damage to, about fifty enemy aircraft. The Japanese also came over at night but the Hurricanes again showed their mettle with several interceptions, one of which, on the 27th, resulted in a certain 'kill' when Squadron Leader Stone, 17's CO, brought down a Sally in flames close to his own airfield.

Towards the end of this period, another Hurricane squadron, No 135 equipped with Mark IIAs, joined the fighter contingent. Its pilots' first successes came on the 29th, when Squadron Leader Carey and Pilot Officer Storey, who at the time were attached to No 17, each shot down a Nate fighter. 135's first mission as a squadron came on 3 February, when it provided an escort for a force of Blenheims. The squadron took part in several such raids though it certainly did not neglect its interceptor duties; Storey destroyed two more Nates on the 6th.

Although checked in the air, the Japanese continued to make steady progress on the ground. In contrast to events in Java, the British and Commonwealth leaders in Burma were so determined not to give up any position uncontested, that their troops were constantly outflanked. The culmination of a series of misfortunes came on 23 February, when a Japanese threat to the vital bridge over the Sittang River necessitated its demolition, trapping most of 17th Indian Division, which all but ceased to exist as a fighting formation.

Encouraged by these portents, the Japanese Air Force renewed major raids on Rangoon on 24 February, keeping up its efforts until the 27th. Again, the defenders broke up the attacks, permitting badly-needed reinforcements to reach the port unmolested. Yet the fighters' achievements could only be temporary, for it was now clear that nothing could stop the Japanese soldiers from reaching Rangoon. It was finally abandoned by the Allies on 7 March.

To date, the record of the Hurricanes had been admirable. The estimated number of enemy aircraft destroyed or damaged had risen to 150. Of these the Tomahawks had claimed the larger proportion but it is not unfair to sound a note of caution by pointing out that in the European theatre the claims of American pilots – for no more sinister reason than that their enthusiasm was less restrained – have proved to have been considerably more exaggerated than those of the RAF. On the other hand, as has been seen, the details of the British part in this campaign are incomplete, so although there may well have been errors of duplication, these are probably rectified by successes of which no mention whatever has been made. Thus, it can be said with some confidence that the Hurricanes had probably shot down or damaged about sixty hostile aeroplanes for the loss of only twenty-two machines in combat – the pilots of most of which were saved – though several others had been written-off as a result of various mishaps on the ground. The ablest airman undoubtedly was the CO of 135 Squadron, Squadron Leader Carey, who was credited with the destruction of at least ten of the enemy, including three fighters, usually reported as Oscars but apparently Nates, in one day on 26 February, on which date also he was promoted to wing commander.

Now problems began to accumulate. As the Japanese closed in on Rangoon, Stevenson withdrew his command, three Buffalos, four Tomahawks and about twenty Hurricanes, north-ward to a strip cut out of paddy fields at Zigon. This was the only landing ground from which the retreat could be covered, but the surface was so bad that many Hurricanes lost their tail-wheels, being temporarily re-fitted with bamboo tailskids, while it is reported that one of them flew in combat with one of its four longerons, which were the backbone of the fuselage and usually steel tubes, made entirely out of bamboo.

By 12 March the fighters had been forced out of this undesirable residence. The Hurricanes of 135 – together with the aged Mark Is flown by 67, of which only half-a-dozen now remained – retired to the island of Akyab off the north-west coast of Burma where they formed part of 'Akwing', while the American Volunteer Group and No 17, now known as 'Burwing', made for the civil airport at Magwe where they found a few reserves of Tomahawks and Hurricane IIAs, which had been held back for the defence of Central Burma, as well as nine Blenheim bombers, which would soon join them in action.

On 20 March, a reconnaissance Hurricane reported that some fifty Japanese warplanes were based at Mingaladon. Next day, the Blenheims blasted the installations at this former RAF base, while ten escorting Hurricanes strafed the machines on the ground, destroying sixteen of them. When the enemy fighters attempted to intervene, the Hurricanes pounced on them as they climbed, shooting down another nine. The gunners in the intrepid Blenheims downed two more.

Retribution followed only too quickly. That same afternoon, the Japanese sent in wave after wave of aircraft, about 230 in all, to make a series of raids on Magwe extending over twenty-four hours, at the end of which it was a pitted, blazing ruin and eight Hurricanes, six Tomahawks and three Blenheims were total wrecks. Of the few Hurricanes that got into the air, two more were so badly damaged that they crash-landed, though they did claim to have destroyed or damaged four of the attackers.

Only eleven Hurricanes now remained, more than half of them too badly damaged to be fit for combat. These struggled off to Akyab but on 23, 24, and 27 March enemy bombers fell on this base also. Seven more Hurricanes perished on the ground.

This all but ended the Hurricanes' resistance. The remnants of the squadrons left Burma altogether. 67 went to Alipore, India, where it re-equipped first with Hurricane IIBs, then with IICs; 135 to Calcutta to receive IIBs; 17 to Loiwing, just over the border in China, where it had a few last clashes with Japanese fighters before making its way to Jessore, India, also to take on IIBs.

Deprived of fighter protection, the ground forces continued the longest, most horrific retreat made by a British army, accompanied by growing thousands of terrified, hunger-stricken, disease-riddled refugees. There is no need to recall their sufferings here, suffice to mention some of the later mile-stones on their path of travail: the blowing-up, on 16 April, of the oil-storage tanks at Yenangyaung to deny them to the enemy; the evacuation, on 1 May, of the once-lovely city of Mandalay, now laid in ruin and ashes by bombs; finally, in mid-May, the last struggle over the Indian frontier through the full fury of the monsoon, which did at least bog down the pursuing Japanese. It would be three years before the vastly superior numbers of the Allies, backed by total command of the air, could recover Burma from its new conquerors.

To protect the flank of their advancing armies, the Japanese Navy struck into the Bay of Bengal. Vice-Admiral Ozawa led a raiding group that sank more than 90,000 tons of merchant shipping and so disrupted sea traffic that it virtually closed most of the ports in eastern India, including Calcutta, for about three weeks. Even more formidable were the vessels commanded by Vice-Admiral Nagumo. Although one of the carriers of the Pearl Harbour force, the *Kaga*, had had to return to Japan to refit, all the others were there: the flagship *Akagi*, *Soryu*, *Hiryu*, *Shokaku* and *Zuikaku*. So were their élite air-crews, the pride of the Japanese fleet, under their leader, Commander Mitsuo Fuchida.

Nagumo's target was Ceylon, but his mission was a some-what muddled one. He was to strike at the great port of Colombo, the Royal Navy base at Trincomalee and any units of the British fleet that he might encounter but the Japanese had no intention of following up with a landing in Ceylon. Yet the conquest of the island could have brought rich strategic gains by severing the convoy routes in the Bay of Bengal, thus necessitating overland supply of the Allied forces in Bengal and Assam, as well as threatening those carrying oil from the Middle East and reinforcements to General Auchinleck in the Western Desert. For this reason, there was a strong school of thought at the Naval General Staff which did favour mounting an invasion. Had Nagumo not experienced any real resistance, it is quite likely that this view would have prevailed in the future.

The stiffest resistance would be provided by Hurricanes, for Admiral Sir Geoffrey Layton, the Commander-in-Chief, had ordered that two squadrons of these being carried by HMS *Indomitable* to Java – where they would certainly have been lost to little effect – were to be diverted to Ceylon. When the carrier came within flying distance of the island, the fighters, which had been stored in hangars minus their wings to increase the number that could be embarked, had to be assembled manually in the intense tropical heat. The comments of those entrusted with this task can be imagined. Nonetheless it was duly completed.

On 6 March, twenty-two Mark IIBs of 30 Squadron, under Squadron Leader Chater, flew from the carrier to the civil airport of Ratmalana, south of Colombo, while next day twenty more from 261 Squadron, led by the South African Squadron Leader Lewis, took off for the aerodrome at China Bay near Trincomalee, though one flown by Sergeant Whittaker that developed a glycol leak had to return to *Indomitable* on which it landed safely, despite the absence of arrester gear. To complete the build-up of Hurricanes, on 22 March, nine IIBs and five Mark Is were transferred from India to Colombo, where a well-camouflaged airfield had been constructed for them on a racecourse, though its short runway necessitated the IIBs carrying only eight machine-guns in order to reduce weight. They officially formed a revived 258 Squadron, under the command of the Rhodesian Squadron Leader Fletcher, on the 30th.

Meanwhile, the British Eastern Fleet, under Admiral Sir James Somerville, prepared to receive the impending assault, of which it had been warned by Intelligence reports, though consisting as it did of elderly battleships and new but inadequately equipped aircraft carriers, it was clearly no match for Nagumo. In any event, the information Somerville had received had indicated that the raid would be launched on 1 April. When the evening of the 2nd came with no sign of this, a strong feeling grew that it had been a false alarm – perhaps the Japanese, learning of the British concentration had called off their operation.

Somerville therefore retired to Addu Atoll, a secret base in the Maldive Islands, 600 miles to the south-west. However, two of his heavy cruisers, were ordered to Ceylon: *Dorsetshire* to complete an interrupted refit, *Cornwall* to escort a troop convoy – while the small Fleet carrier *Hermes*, with a destroyer, HMAS *Vampire*, headed for Trincomalee to prepare for a proposed attack on Vichy French Madagascar. The main fleet reached Addu Atoll in the late afternoon of 4 April – just in time to learn that a Catalina flying-boat had at last sighted the enemy steaming straight for Ceylon.

The subsequent onslaught by Fuchida's warriors took place on 5 and 9 April, being directed on each occasion against two different targets – the detached warships at sea and the ports. On the 5th, which was Easter Sunday, Japanese dive-bombers sank *Dorsetshire* and *Cornwall* in some twenty minutes. On the 9th, an attack on *Hermes* carried out 'perfectly, relentlessly and quite fearlessly', sent her to the bottom in about the same length of time; to be followed, after a valiant defence, By her escort, the *Vampire*.[40]

The raids on Ceylon were not so satisfactory for the Japanese. That on Easter Sunday, carried out by thirty-six 'Val' dive bombers, fifty-three 'Kate' high-level bombers and thirty-six Zeros, was aimed at Colombo. As these splendid aircraft reached the harbour they sighted six old biplane Swordfish, pitiful representatives of Britain's inadequate Fleet Air Arm. All were massacred in short order. Four out of six Fulmar fighters that tried to engage also perished, after shooting down only one dive-bomber.

Once again, radar proved inefficient, giving no warning whatsoever. The Hurricanes of 30 Squadron at Ratmalana all got airborne safely though they were savagely shaken by bomb-blasts as they did so. They were then jumped by Zeros before they could gain height. Luckily, the Japanese did not know of the racecourse airfield, so 258 took off without interference: first the IIBs led by Squadron Leader Fletcher; then the Mark Is under Flight Lieutenant Sharp from New Zealand – all climbing steeply before turning to attack the bombers diving on the harbour.

Tragically, however, the Hurricane pilots made the fatal mistake of trying to dog-fight with the highly manoeuvrable Zeros. No 30 lost eight machines shot down or crash-landed, five pilots being killed. 258 lost nine aircraft and five pilots. They destroyed only one Zero and five Vals. Nine other Japanese aircraft were damaged. Fortunately, the Japanese bombing was much less deadly than at Darwin: damage to installations was not great: an old World War I destroyer and an armed merchant

---

[40]However, on this occasion before the dive-bombers could leave the area, they were attacked by eight Fulmars of 806 Squadron from Ratmalana, which shot down four of them and damaged two more at the cost of two of their own number.

cruiser were sunk; three of the twenty-one merchantmen in harbour were damaged.

On 9 April, Fuchida was back, at the head of ninety-one Kates, escorted by thirty-eight Zeros, this time to strike at Trincomalee. Radar did give the Hurricanes of 261 time to get into a favourable position but since there were only sixteen of them, three having been written off in accidents over the previous few days, they were hopelessly outnumbered. Ten were shot down or crash-landed, though, amazingly, only two pilots died. They destroyed a Zero (two more were shot down by AA fire) and two Kates, and damaged ten more bombers. There was hardly any shipping present but the enemy inflicted heavy damage on the port facilities, though again scarcely approaching the desolation of Darwin or Tjilatjap.

Nagumo now retired from the Indian Ocean, well pleased with events, but had he been able to foresee the future his delight must have evaporated. Perhaps because of the opposition he had met, no change of plan was made to allow a landing on Ceylon, so his raid in practice brought no strategic benefit. On the contrary, it had wasted several weeks when his carriers could have been better used in countering the American build-up in the Pacific, to which they now belatedly returned – only to meet, first with a severe set-back in the Battle of the Coral Sea on 7-8 May, then with a major defeat at the Battle of Midway on 4 June, which abruptly terminated the period of Japanese expansion. The threat to Ceylon was never to be renewed.

The actions in Ceylon marked the end of the period in which the Hurricane had fought with all odds against it. Though it would still see combats against superior numbers at isolated bases such as Malta, in most theatres henceforth, it would form part of air forces and accompany land forces both of which were larger than those of its opponents. While time had been gained for Britain to build up her resources and for her powerful allies, Russia and the United States, to marshal their full strengths, other aeroplanes, some more modern but most more elderly, had played a part, but the backbone of the defence throughout had been the Hawker Hurricane.

It could now take part in offensive campaigns against its enemies, which, in the fullness of time, would lead to their utter destruction.

Chapter Ten

The Sea Hurricane

From the huge, important island of Ceylon, it appears fitting to turn to another island, tiny in size but vast in strategic value: – Malta GC.

The duty of Malta was to be a permanent threat to the security of the German troops in North Africa. The Allied armies had to be supplied by vessels travelling more than 14,000 miles round the Cape of Good Hope, whereas Rommel's transports had only to cross a few hundred miles of the central Mediterranean, but it was the German commander whose audacious plans were constantly fettered by a lack of equipment of every kind.

The reason for this was that, directly in the path of the Axis lifeline, lay the fortress of Malta, whose unceasing activity not only crippled Rommel's movements but decimated the Italian merchant marine. That the ability of Malta to maintain this offensive was the most crucial factor in the whole strategic situation is confirmed by Churchill's statement that: 'The heroic defence of the island ... formed the keystone of the prolonged struggle for the maintenance of our position in Egypt and the Middle East'; while Captain Donald Macintyre in his book *The Battle for the Mediterranean*, concludes that:

> The contest revolved largely round the ability of the British to build up and preserve the little island of Malta as an offensive base in the midst of

waters dominated by enemy air and sea power ...
On the outcome depended, absolutely, the success
or failure of the campaigns in North Africa and
hence, it can be said, of the whole war.

Nor were the enemy unaware of the effectiveness of Malta.
Rommel would later claim that:

With Malta in our hands, the British would have
had little chance of exercising any further control
over convoy traffic in the central Mediterranean...
It has the lives of many thousands of German and
Italian soldiers on its conscience.

The Italian Official History sadly remarks: 'Malta was the rock
upon which our hopes in the Mediterranean foundered.'

It was natural, therefore, that the Axis should try to eliminate
the island. The first major attempt had ended at the beginning
of May 1941, with the coming of the Hurricane IIs, followed
by the transfer of Fliegerkorps X to the Balkans. The lull
that followed enabled Air Vice-Marshal Maynard to group the
recent arrivals together as a new squadron, 185, on 12 May. Later
that month, 261 Squadron, exhausted and with morale low
after its recent losses, was transferred to the Middle East where
it disbanded.[41] However, on 21st May, forty-eight Hurricanes
from 249 Squadron – a mixture of Mark Is and Mark IIAs – flew
off *Furious* and *Ark Royal*, all except one reaching Malta safely
despite appearing in the middle of a raid by Italian bombers.

Thus, when Maynard left the island at the end of May,
his fighters totalled seventy-five, five times the figure at the
beginning of the year, based not only at Takali, Hal Far and
Luqa but at the recently built Safi strip. During Maynard's

---

[41]As mentioned earlier, 261 was re-formed at Habbaniyah, Iraq on 12 July.

tenure of command, his Hurricanes had been credited with the destruction of sixty-two German and fifteen Italian aircraft for the loss of thirty-two in combat – though many others had been destroyed on the ground.

In contrast, his successor, Air Vice-Marshal Lloyd, was able to declare that: 'You wouldn't have known there was a war on.' It was a pardonable exaggeration, for, after the ravages of the Luftwaffe, the raids by the Regia Aeronautica in June were singularly ineffective. The Hurricanes now encountered the Italians' latest fighters, the Macchi MC200s, which had a speed of 312 mph but were armed only with two 12•7 mm machine-guns. However, they inflicted several losses on these, while all the time Malta's strength continued to grow.

On 15 June, forty-seven more Hurricanes left *Victorious* and *Ark Royal*, forty-three reaching Malta safely. Similar operations on the 27th and 30th brought a further sixty-four fighters to the island-base, though some of these were staged on to Egypt. Among the new arrivals was an entire squadron, No 46. Strangely, though, this was disbanded on 28 June, only to be re-formed immediately as a new squadron, No 126.

Moreover, the Hurricanes received in late June included four-cannon Mark IICs, which went to both 126 and 185 Squadrons. On 30 June, 126 engaged some Macchi MC 200s, two of which it shot down without loss. 185 had its first fight on cannon-armed Hurricanes on 4 July, routing thirty more Macchis, of which it destroyed two and damaged three, again without loss.

The next few months saw only desultory enemy raids. 185 claimed eight more Italian warplanes during July. 126's best efforts came later: they claimed to have downed four Macchi MC200s on 19 August and three more on 4 September. Of course, the Hurricanes had their losses as well; none sadder than that of Pilot Officer Barnwell of 185, who on 14 October, just after receiving a DFC for destroying five Italian machines in one week, was killed when his parachute failed to open.

In the meantime, the Hurricanes were also flying at night, 1435 Flight having been specially formed for this purpose in July. It was commanded by a full squadron leader, George Powell-Shedden, a former flight commander in Bader's 242 Squadron during the Battle of Britain, who showed the way with two night victories during his period of service at Malta.

During the night of 25/26 July, the Italians attempted their only seaborne assault on the island, their target being a convoy at anchor in Grand Harbour, Valetta. The attackers were two 'chariots' or 'human torpedoes' and eight one-man explosive motor-boats (or EMBs), supported by two MAS-boats – the equivalents of the German E-boats.[42] They were, however, detected by radar, so found Malta's searchlights and shore-batteries ready to receive them.

It had been planned that a chariot piloted by Lieutenant Tesei, the designer of this weapon, would blast a way through the obstructions in a passage between the harbour mole and the shore. When he disappeared without trace, two of the EMBs made a suicidal attack on the passage but their gallant act merely brought down the footbridge spanning it. The defenders then opened fire on the other EMBs sinking every one. As the light increased, a number of Macchi MC200s arrived but were met by the Hurricanes, which shot down at least two, probably damaged others and drove the rest away.

Then the RAF fighters concentrated on the MAS-boats, which were out of range of the harbour defences. They damaged one of them so badly that it was abandoned by its crew and subsequently blew up and sank. Flight Lieutenant Lefevre of 126, a native of the Channel Islands, attacked the other MAS-boat repeatedly with such determination that he killed eight Italian seamen and the shaken survivors hoisted a white flag

---

[42] The initials stood for *Motoscafo Armato Silurante* – Italian for Motor Torpedo Boat.

before abandoning this vessel as well. Only one Hurricane was lost. Pilot Officer Winton of 185 baled out. He was able to swim to and board the MAS-boat, from which he was rescued by a Swordfish with floats, retaining the Italian naval ensign as a souvenir. The MAS-boat was later towed into Grand Harbour.

Even before this, the Hurricanes were already striking at Italian targets. The scale of attacks on Malta had so declined that pilots could complain of inactivity – a luxury denied to their successors as to their predecessors. Thus, only part of the fighter force was needed for defence, leaving the remainder free for intruder sorties. The first such took place on 7 July when Squadron Leader Mould, who as a young pilot officer had claimed the Hurricane's first victim in France and now commanded No 185, led two other 185 pilots and Squadron Leader Rabagliati of 126 to shoot up the seaplane base at Syracuse. All returned safely, after having destroyed six flying-boats, damaged four more, and, as they reported gleefully, 'severely shaken everyone in the neighbourhood of Syracuse'.

The seaplane base received further attention on 17 October, when it was bombed by six Blenheims, the Hurricanes providing the escort; but the fighters flew numerous other missions over Sicily, by day or night, their targets including aerodromes, railway stations and transport of all sorts. The cannon of the IICs did especial damage to hostile aircraft on the ground. Although inevitably there were losses on such raids – including Mould who was killed in action on 1 October – there is no doubt that in general they were very rewarding.

Yet they were perhaps the smallest part of the offensive that Malta was now launching. Its bomber forces had also been built up. From the beginning of June, scarcely an Axis convoy to North Africa had escaped attack. By August, Blenheims, Wellingtons and old Swordfish, were lashing at Italian merchant ships by day and night, at sea or in harbour, quite regardless of losses. Malta's submarines also enjoyed many successes, while the establishment there of a surface unit, Force K, completed the enemy's misery.

In September, 28 per cent of all supplies sent to Rommel failed to reach him. In October the proportion lost was 21 per cent. In November, it rose to a staggering 63 per cent – this just at the time when the Allied armies were commencing a major attack. It must therefore be a matter for wonder that General Auchinleck, the chief beneficiary of Malta's efforts, showed neither appreciation nor gratitude for these. At one time he even stated that the retention of the island was not absolutely necessary for his plans. It was well that the Navy and Air Force commanders, with the full support of Churchill, had greater strategic insight.

Unfortunately, so had Hitler. By now he had noticed that when Malta was under pressure, the German cause in North Africa prospered, but when Malta was strong, Rommel was invariably in difficulty. On 2 December, Fliegerkorps II, previously conducting operations on the Moscow front, was ordered to Sicily, joining with Fliegerkorps X in the Balkans to form Luftflotte 2. To command this, there came from Russia an old enemy, Field-Marshal Kesselring, who could also call on Luftwaffe forces in Libya if necessary. Including the Regia Aeronautica, this gave him control of 2,000 warplanes.

Hitler's orders to his subordinate were admirably precise. He was to 'ensure safe lines of communication' with North Africa, for which 'the suppression of Malta' was declared to be 'particularly important'. In addition, he was to prevent any further supplies from reaching the island. On 22 December, Kesselring set about these tasks to the utmost of his very considerable ability.

It was just as well that prior to this more Hurricanes had got to Malta. During September, forty-nine had reached the island at various times from *Ark Royal* and *Furious*. On 12th November twenty-one Mark IIBs of 242 Squadron and sixteen more from 605 had flown off *Argus* and *Ark Royal* – the last service the noble '*Ark*' would render, for she was sunk by a U-boat when returning to Gibraltar. All the Hurricanes from 605 and all except three from 242 had arrived safely, thereby causing some confusion for those interested in squadron histories as it will be remembered that the ground crews of both these units had been shipped to the East Indies. However, in practice, the pilots of 242 and 605 were incorporated into 126 and 185 Squadrons respectively, though for administrative reasons they retained their old squadron identities until 18 March 1942.[43]

Thus, there were in reality only the equivalent of three squadrons of Hurricanes ready to meet Kesselring's assault. His Junkers Ju 88s were now escorted by the improved Messerschmitt Bf 109Fs, which the Hurricanes, with their slow rate of climb, found difficult to engage. Yet during the last week of December 1941 and throughout January 1942, they were still able to break up the enemy formations at least sufficiently to reduce the effectiveness of their bombing.

By February, Kesselring's offensive was in full swing. In that month there were 222 raids on Malta's airfields alone, while heavy rain made matters worse except at the well-drained Luqa. Only tremendous efforts by ground-crews, civilians and soldiers from the garrison, all working round the clock, kept the runways in operation. From them the Hurricanes continued to take off to exact their toll from the attackers. On the 15th, 126 Squadron out-fought the 109s, destroying two of them. On the 23rd, 185 claimed to have brought down three more. However, in the prevailing conditions it became increasingly difficult to keep the fighters in action. At the beginning of February, twenty-eight Hurricanes remained serviceable. By the 15th, the number was down to eleven.

On 7 March, fifteen fresh fighters flew to Malta from the carrier *Eagle*, but this time they were not Hurricanes but Spitfires, the first such to operate outside the British Isles – a

---

[43]This was even more confusing since, as already recorded, other 242 and 605 Squadrons were in action in Java during part of the period in question.

measure this of the unsupported burden that the Hurricanes had carried for so long. Sixteen more Spitfires arrived on the 21st and on the 29th. 249 Squadron, which had been credited with about twenty victories with Hurricanes, now converted to Spitfires, followed in early April by 126, whose Hurricanes had been credited with the destruction of thirty-four of the enemy for the loss of ten of their own number.

With their higher rate of climb, the Spitfires undoubtedly were better able to tackle the raiders but it was shortage of aircraft rather than any deficiencies in type that provided Malta's greatest problem. In this connection, the Spitfires' great disadvantage was that being less robust than the Hurricanes, they could be put out of action more easily. By mid-April, virtually all of them had been rendered unserviceable, mostly by attacks on their airfields.

Nor could the Spitfires reverse the steadily growing German air superiority. The raids in March doubled those in February. The attacks in April, in which more than 6,700 tons of bombs were dropped, almost doubled those in March. By now the majority of Malta's bombers had been withdrawn to less exposed bases. All surface warships followed them. Even the gallant submarines, which had been stationed at the island throughout the worst of the previous assaults, were forced to depart.

Yet by mid-March the number of Hurricanes serviceable had increased to about thirty. They still equipped 185 Squadron, which on the 21st gained one of the most spectacular victories of the campaign when four Hurricanes, led by Flight Lieutenant Mortimer-Rose, attacked eight Messerschmitt Bf 110s that were strafing Takali airfield, claiming to have shot down six of them. Also, on 27 March, ten Hurricane IICs of 229 Squadron flew in from North Africa, as did fifteen more during April. This unit also took over Hurricanes that had previously belonged to 126 or 249. It gained its initial success on 1 April.

Thus, whatever may have been stated elsewhere, it was the Hurricanes that upheld Malta's defence during April, on the

15th of which month the island was honoured with the George Gross but in which it received its harshest punishment. As in the Far East, the Hawker fighters would climb away from the raiders in order to gain the necessary height. The odds against them were quite awe-inspiring: eight to one was considered fairly reasonable.

To help the Hurricanes, it was proposed to send in a really huge force of Spitfires. Through the generosity of President Roosevelt, the American fleet carrier *Wasp* was made available for this purpose. With her broad lifts and long flight-deck, she could carry great numbers of these fighters. On 20 April, she flew off forty-seven, of which all except one arrived – but the enemy were quick to pounce on them before they could even commence operations. By the evening of the following day, only seventeen remained serviceable. Within three days every single one had been grounded.

This catastrophe turned the rate of exchange sharply against the RAF. In April, thirty-seven enemy aircraft were brought down or damaged – though much greater claims were made at the time – but the destruction during the month of twenty-three Spitfires, plus fifty-seven more crippled, was a tragedy for the defenders. In addition, eighteen Hurricanes had been destroyed with thirty more damaged. 229 Squadron, the pilots of which had had little combat experience, had suffered heavy losses of men and aircraft, and by 9 May the defence of the island was in effect entrusted to No 185 with a notional strength of eighteen Hurricanes – though by no means all of these were available at any one time.

Yet as had happened in the Battle of Britain, the Hurricanes had kept up their resistance for just long enough. 'Hitler', as the *RAF Official History* unkindly remarks, 'with that improvidence characteristic of the master-plotters of war, was short of aircraft.' Demands from the Russian front, from North Africa where Rommel was poised to re-conquer Cyrenaica, had become too great to be resisted. On 30 April, instead of the usual murderous raids by the Luftwaffe, Malta was attacked only by Italian bombers, flying at great heights. The main part of Kesselring's forces had left for other theatres.

It was a gross miscalculation. This comparative lull gave Malta the opportunity to restore her shattered defences.[44] On 9 May, forty-seven Spitfires flew off *Wasp*, seventeen off *Eagle*. Sixty-one of them reached Malta where some scouting 109s were beaten off by Hurricanes, backed by a determined gun barrage. So splendid were the arrangements for their reception that many were ready for action within five minutes of touching down.

The result was that when next day the Luftwaffe tried to sink the fast minelayer *Welshman*, which had reached Grand Harbour with a precious cargo of ammunition, it was repulsed by superior numbers. Of the fifty RAF fighters that went up, thirteen were still Hurricanes but 185 now also began to re-equip. This move is usually recorded in rather slighting fashion, but when one reads, for instance, of how the Australian Pilot Officer Boyd, one of the squadron's finest airmen, 'at last exchanged his tired Hurricane for a Spitfire', it is worth remembering that this particular pilot had, when flying the Hawker fighter, destroyed four enemy aircraft, including two 109s, and damaged many others. Another of 185's most noted pilots in the latter part of the fighting was Sergeant Tweedale, also an Australian, who shot down four enemy machines in April. However, it is sad to report that both died in action not long after taking over their Spitfires.

By the beginning of June, the remaining Hurricanes of 229 Squadron had flown back to Egypt while 185 had entirely converted to Spitfires. In July, the night fighter flight, No 1435,

---

[44]Kesselring had wrongly advised Hitler that Malta had now been neutralised – but it seems that the Führer would have had to transfer his aircraft in any event in view of commitments elsewhere.

followed suit. Yet the Hurricane's connection with Malta had by no means ended. Nor had Malta's ordeal. The weight of Kesselring's direct assault had been eased but, in the words of the *RAF Official History*, 'all too close ahead' loomed 'the day when, failing the arrival of a convoy, the last reserves of fuel, food and ammunition would be exhausted'.

Which seems an appropriate moment for that splendid convoy-protector, the Sea Hurricane, to make its entrance.

–

The adventures of the Sea Hurricane Mark IAs, which were launched from merchantmen to ward off Focke-Wulf Fw 200s, have already been recalled but, apart from those on the fighter catapult ships, they did not operate from vessels of the Royal Navy, nor, except at the start of their service in a few cases, were they manned by naval pilots, nor could they return to their ships. Therefore, it may be argued that the first true Sea Hurricane was the Mark IB, which was designed specifically to operate from aircraft carriers.

No exact total of the Sea Hurricanes that served with the Royal Navy exists, because they were not new aeroplanes off the production lines but conversions of existing machines for new purposes; but it was probably between 500 and 550. The first such to come from General Aircraft Limited, which carried out the bulk of this work, were the 400 Mark IBs. These were adapted from RAF Mark Is, by definition therefore having Merlin III engines and an armament of eight machine-guns, but in order to fit them for their duties afloat they also carried catapult spools and an arrester hook below the fuselage to catch the crosswires on a carrier's deck.

The majority of such aircraft commenced operations only in 1942, but an advance guard had made its presence felt much earlier. On 30 July 1941, a force of Albacores escorted by Fulmars from carriers *Victorious* and *Furious* attacked the ports of Kirkenes in northern Norway and Petsamo in Finland, at the request of the Russians who believed that the strikes would disrupt the enemy's communications. Left behind on *Furious* to provide fighter protection was 'A' Flight of 880 Squadron – the first Sea Hurricane IBs in service.

The raids achieved meagre results at the heavy cost of twelve Albacores and four Fulmars. However, on the 31st a much worse occurrence was avoided. A Domier Do 18 flying-boat was sighted, shadowing the carriers as they headed away from the danger-zone. An air-attack seemed imminent but this possibility was removed when two of the Sea Hurricanes, flown by Lieutenant-Commander Judd and Sub-Lieutenant Howarth, were hastily sent up to deal with the enemy scout. After a short chase they brought it down, thereby gaining the first victory scored by the Sea Hurricane.

As already indicated, the IBs formed the bulk of the Sea Hurricane conversions. Until late 1942 there was scarcely a single machine aboard a carrier that was not a IB, while some (minus their hooks) were transferred for use on CAM-ships. From early 1942 the Navy boosted the IB's Merlin III so as to lift its speed to 315 mph at 7,500 feet, though at the cost of a considerable reduction to the length of life of the engine.

The IBs were followed by a bewildering series of different types of Sea Hurricanes. These included the ICs, a highly appropriate designation, for apart from the addition of cata-pult spools and arrester hook, these had Mark I fuselages with standard Merlin III engines but were adapted to carry the four-cannon wing of the IIC – though this combination greatly restricted their speed. In any case, probably because of the demands for cannon elsewhere, few such conversions were made and it seems that – with one exception, which will be mentioned later – these did not enter squadron service though they may have been used for trials and evaluation.

Then there were the 'Hooked Hurricane IIs', a name awarded to about sixty RAF Mark IIAs or IIBs, which were adapted by the simple means of adding a hook to them. These were first delivered to Fleet Air Arm squadrons in September 1942. The following month, the Navy obtained thirty sets of IIC wings, which it fitted to the earlier aircraft.

Finally, in early 1943 about sixty Sea Hurricane IICs were converted from RAF IICs to carry full Fleet Air Arm equipment, including naval radio – unlike the Hooked Hurricane IIs, which retained the RAF set – and arrester hook but not catapult spools. Their Merlin XXs gave a top speed of 318 mph, they had a service ceiling of 34,500 feet and could climb to 20,000 feet in nine minutes – abilities which, although no doubt they declined under operational conditions, may entitle them to rank as the finest of the Fleet Air Arm Hurricanes. They were also the last, since a proposed adaptation with folding wings, for more convenient use on carriers, was abandoned without ever going into production.

Ultimately, Sea Hurricanes equipped sixteen Fleet Air Arm units, though not all of these were operational at the same time, as well as forming part of the equipment of at least that number of second-line units such as training establishments. Indeed, in 1942 they were considered the most suitable machines available for all naval fighter training. Sea Hurricanes flew from a total of six fleet carriers and seven escort carriers. Perhaps the high-point of their use came in July 1942, when there were about 300 serving with the Royal Navy. Of these, seventy-four were embarked or about to embark in carriers, about the same number were operating from shore bases, while the remainder were used for training duties or held in reserve.

In the Indian Ocean, Sea Hurricanes of 880 Squadron were flying from HMS *Indomitable* at the time when the Japanese made their thrust towards Ceylon.[45] Though 880 did not see combat in this instance, in the following month – 5 May to be

---

[45]It was personnel of this unit who had the task of assembling the Hurricanes of 30 and 261 Squadrons before they were flown off to assist in the island's defence, as previously recounted. By way of payment, 880 'unofficially'

exact – it was to take part in the first Allied amphibious landings – an indication of greater events to follow.

The object of this operation was to seize the naval base of Diego Suarez in the north of Madagascar, to ensure that it was not occupied by the Japanese instead. It has since been criticised by Captain Liddell Hart among others, but, at the time, information gained from the breaking of enemy codes suggested that the Japanese did intend to exert major pressures in the Indian Ocean – as the heavy attacks on Ceylon seemed to confirm. There was, of course, no way of telling that the promises Japan was making to Germany were meant to boost Axis solidarity rather than be a genuine guide to future plans. On the other hand, it was appreciated only too well that the Vichy Government would no more resist Japanese aggression in Madagascar than it had in Indo-China. Since the island commanded the shipping routes to the Middle East, the Allies can hardly be blamed for taking not the slightest risk of its falling into enemy hands.

Hence the attack on Diego Suarez. Fighter cover was provided by *Indomitable* with the Sea Hurricanes of 880 Squadron and *Illustrious* with two squadrons of Grumman Martlets, the Fleet Air Arm version of the Wildcat, which was then the US Navy's main carrier-borne interceptor.

As it happened, only the Martlets saw air-combat, the Sea Hurricanes concentrating on targets on the ground. At dawn, on 5 May, they attacked the airfield protecting the harbour, causing considerable damage as well as destroying at least three Morane 406 fighters in their hangars. Subsequently they turned their attention to gun positions, petrol lorries, and even a sloop, which was firing on British troops advancing on the naval base.

---

retained the fighter, which had to return to the carrier. It was (equally 'unofficially') hooked and otherwise modified from standard spares held on board.

By the 7th, the port had fallen at minimal cost, though it was not until 5 November that the whole of the island had been taken over by the Allies, after a desultory campaign in which malaria claimed the bulk of the casualties. For the Sea Hurricanes, perhaps the greatest significance of the operation was as a practice for their part in a much more vital landing on Vichy French territory, the invasion of Algeria – but the triumphs and tragedies of Operation Torch may be postponed, since they can best be appreciated in the context of the wider struggle against the Axis armies in North Africa.

Instead, it is convenient to return to the Arctic Ocean, over the freezing waters of which the Sea Hurricane had made its first 'kill' back in July 1941. Almost a year later, on 4 July 1942, Convoy PQ17, making for Russia, was ordered to scatter by the First Sea Lord, Admiral Sir Dudley Pound, who, as a result of incomplete Intelligence, erroneously considered it was threatened by a German surface-force. While the escorts moved to engage an enemy that never materialised, the merchantmen proceeded individually, robbed of their protection against U-boats, deprived of the concentrated fire power that had previously defied air-attacks. Eleven reached Archangel. Twenty-three freighters, a tanker and a rescue ship were sunk. 430 tanks, 210 aircraft, 3,350 vehicles were lost.

The Germans were jubilant, especially the airmen of Luftflotte 5, who not only exaggerated their considerable achievements but mistakenly believed that they had been responsible for breaking the convoy's formation. It was decided therefore that the main effort in future would come from the air. Reinforcements hurried to join KG26 with its Heinkel He 111s and Junkers Ju 88s adapted to carry two torpedoes each, and KG30 with its orthodox Ju 88 bombers. Its strength increased to ninety-two torpedo-planes and 133 long-range bombers, Luftflotte 5 was certain it could annihilate the next Russian convoy to sail.

PQ18 did not set out until September, but for the Luftwaffe, whose Intelligence had obtained detailed information of the

Allies' plans, it was well worth waiting for, since it contained forty merchant vessels, a larger number than ever before, flying the flags of Britain, the United States, Russia and Panama. The defence of these was entrusted to Rear-Admiral Burnett, who commanded the cruiser *Scylla*, eighteen destroyers and, most important of all, the American-built escort carrier *Avenger* with three Swordfish for anti-submarine duties, six Sea Hurricane IBs from 802 Squadron, six others from 883 and six more kept dismantled in her hangar as reserves. It was a good idea to have replacements incidentally, for apart from the probability of combat-losses, one Sea Hurricane went overboard in rough weather while *Avenger* was proceeding to the rendezvous with the convoy at Iceland.

During the passage of PQ18, U-boats made several attacks but the Swordfish, combining splendidly with the surface escorts, thwarted the majority of these. Only three merchantmen were sunk by the submarines, which paid dearly with the loss of three of their own number. Yet it was from the newly strengthened Luftflotte 5, urged to make a 'special effort' against the convoy by Reichsmarschall Göring in person, that the major threat would come.

Naturally, it was *Avenger*'s Sea Hurricanes that were at the heart of the resistance to the German airmen. They were to be in almost constant action, at first experiencing tragic disappointments but persisting to gain an ultimate triumph.

The unhappy phase of the conflict began at 1304 on 12 September, when four Sea Hurricane IBs were launched to intercept a three-engine Blohm & Voss BV 138 flying-boat. They drove this away but were quite unable to finish it off, mainly because they were not armed with cannon, for the eight Brownings that had inflicted such losses in 1940 were nothing like so effective now that the Germans were providing their aircraft with far more protective armour.[46]

On the 13th, other shadowers, up to five in number, kept PQ18 under observation, disappearing into cloud whenever *Avenger's* fighters approached too close. The first air attack did not come until about 1500, when half-a-dozen Junkers Ju 88s from KG30 bombed ineffectively through the cloud-cover. Five of the Sea Hurricanes helped to harry these, though all returned to base undamaged, but while they were thus engaged and while others were re-fuelling after their ineffectual pursuits of the reconnaissance machines, a terrifying formation of more than forty torpedo-carrying Heinkel He 111s and Junkers Ju 88s from KG26 appeared, to sweep down upon the convoy.

The attack was timed to perfection. There was no chance for the fighters already in the air to intervene. Nor could Commander Colthurst, *Avenger's* skipper, hope to get off those remaining on the carrier's decks. The raiders were opposed only by the convoy's gunfire, which was quite unable to prevent them from dropping their torpedoes simultaneously with classic precision. Orders were given for the merchantmen to turn parallel to the course of the deadly weapons, thereby reducing the target afforded, but many of them responded much too late. No fewer than eight were sunk, an ammunition ship disintegrating in a horrific detonation.

This ghastly catastrophe put all other misfortunes into the shade – but the day had one extra misery in store for the Sea Hurricanes. That evening, four of them attacked a Heinkel He 115 float-plane. Once again, the shadower got away but this time it also shot down one of the fighters in flames, killing Lieutenant Taylor, the CO of 802.

---

[46]By a rather grim irony, there were a large number of cannon-armed Hurricanes with the convoy, but since they were in crates on the merchant vessels for delivery to the Red Air Force, they were not of much value to the Fleet Air Arm.

It says much for the spirit of the escorts that the misfortunes of the 13th, far from stunning them, inspired them with a determination that the mistakes made would never be repeated. In particular, Commander Colthurst appreciated that in future he would have to maintain a continuous combat air patrol over the convoy, with his interceptors landing to refuel in strict rotation – also, that he would have to direct them only against really large formations, although this might allow reconnaissance aircraft or even small bombing raids to pass unchallenged. His task was made easier by Göring, who, eager to gain the prestige of sinking a carrier as well as to deprive PQ18 of its fighter protection, ordered that *Avenger* was to be the priority target for future attacks.

Thus, when at 1235 on 14 September the raids were renewed by twenty Junkers Ju 88s carrying torpedoes, the majority went for the carrier which, according to Admiral Burnett, was 'peeling off Hurricanes whilst streaking across the front of the convoy'. Six fighters got among the Junkers, shooting down two of them, damaging several more beyond repair and so disrupting their formation that no torpedoes found their mark. Hard on the heels of this raid came another, again aimed at *Avenger*, by Ju 88s armed with bombs. Luckily the Sea Hurricanes that were still airborne were able to engage these also, to such good effect that the bombers inflicted no damage.

At 1340, another bombing attack began but this time was opposed only by the guns, since Colthurst rightly suspected that it was a feint to draw away his fighters. In consequence, when twenty-two torpedo-carrying Heinkel He 111s from KG26 made their appearance there were ten Sea Hurricanes in the air to meet them. A few of the Heinkels launched their weapons at the convoy where another ammunition ship simply vanished leaving behind only 'a vast column of fire and smoke many thousands of feet high'. However, most courageously tried to strike at *Avenger* although this necessitated their flying low over the merchantmen and their escorts in the teeth of a savage defence. Equally undaunted were the Fleet Air Arm pilots who, in their determination to engage, followed their targets through the full fury of the barrage.

Once more the combination of fighters and guns was deadly. The carrier avoided all torpedoes. Five Heinkels were shot down. Nine more reached base so badly damaged that they never flew again. Three of the Sea Hurricanes also fell, all victims of their own ships' gunfire, but mercifully in every case a destroyer was quickly on hand to pick up the pilot.

The next day saw only sporadic raids by Junkers Ju 88s, bombing through the cloud cover. Badgered by flak and by fighters, they made no hits but lost three of their number. On the 16th, in increasingly bad weather, even these limited attacks died away.

As was related elsewhere, that afternoon, Burnett transferred most of his forces, including *Avenger*, to guard QP14, a convoy on the return trip from Russia. Even with the carrier gone, the Luftwaffe encompassed the destruction of but one more ship, while an even older Sea Hurricane, the Catafighter of Flying Officer Burr from *Empire Morn*, added to the enemy's losses. Twenty-seven of the merchant vessels survived to deliver their cargoes. According to the War Diaries of the Luftwaffe – which was unlikely to exaggerate the extent of its failure – these 'represented hundreds of modern tanks and aircraft, thousands of road vehicles and a mass of other war and industrial materials – enough to equip a whole new army for the front'.

The numbers of enemy aircraft brought down were, as usual, overestimated, yet the Diary of Luftflotte 5 admits that it lost thirty-three torpedo-planes, six bombers and two long-range reconnaissance machines destroyed or damaged beyond repair. In addition, many of its most experienced crews perished. The guns claimed the majority of this total but the Sea Hurricanes certainly destroyed three Heinkels and two Junkers. The number that they damaged is not known for certain but the most conservative estimate puts this at seventeen. How many of these fell easy victims to the ships' gunners because they could not manoeuvre properly, or reached their airfields only to be written off, will never be recorded. In any case, it matters little, for there is no question that the interceptors' main value lay not in the downing of individual aeroplanes, but in the breaking-up of enemy formations. The havoc caused by the systematic attack on the 13th, which was not engaged by Hurricanes, contrasts sharply with the almost entire failure of later raids where the guns and the fighters both provided opposition.

Of *Avenger*'s, Sea Hurricanes, one had been lost in bad weather earlier. Four fell in combat, though all their pilots except Lieutenant Taylor survived. Seven others were damaged in action or in accidents, but it is a tribute to the reliability of the type, as well as to the dedicated attention of *Avenger*'s hangar crews, that by the 17th, all injuries had been made good, so that, having brought its reserve aircraft to an airworthy condition, the carrier had thirteen Sea Hurricanes ready for action, one more than when the battle for PQ18 had started. However, as it transpired, they were not needed for the defence of QP14, since this convoy experienced no air-raids, though U-boats inflicted several losses. The Luftwaffe had accepted that the price it would have to pay was too high.

Indeed, the passage of PQ18 was decisive. Thereafter, the Germans transferred the bulk of their torpedo-bombers to the Mediterranean, where if they had to ditch the crews would at least have a fair chance of survival without freezing to death in short order. Never again would the Arctic convoys face anything like the same weight of attack from the air.

Thus, although Sea Hurricanes continued to protect merchant shipping en route to Russia throughout 1943, they saw no combat, though they faced all the many dangers arising from sorties in bad weather, culminating in landing on a heaving, pitching flight-deck. During the year also, they tended to be used increasingly from shore bases, being gradually replaced by Martlets − which by this time had reverted to their US Navy name of Wildcats – on board carriers. Thus, in January 1943, for example, the Sea Hurricanes of 804 Squadron were operating over convoys off the east coast of Britain, being joined in this task later by those from 895 and 897 Squadrons; while in April, the territory covered by the Fleet Air Arm was extended to Tanga, Tanganyika, where 877 Squadron was formed with ex-RAF Hurricanes for the defence of East Africa and the western Indian Ocean.

By the spring of 1944 only three squadrons of Sea Hurricanes remained, all equipped with Mark IICs. These were all serving afloat: 824 on HMS *Striker*, 825 on HMS *Vindex* and 835 on HMS *Nairana* – their duties being to provide fighter cover by day and night, as well as monotonous but vital anti-submarine patrols. On 26 May, 1944 Sea Hurricanes from 835 made the last 'kills' for the type by shooting down two Junkers Ju 290 transports over the Bay of Biscay; but it was 825 that in August 1944, shortly before re-equipping with Wildcats, flew the last Sea Hurricane sorties during the successful passage of yet another Russian convoy.

–

Yet, as has already been hinted, the finest service rendered by the Sea Hurricane was its defence of the convoys to Malta – which in defiance of chronology therefore, may form a fitting conclusion to its exploits.

The reduction in the weight of Kesselring's attacks at the end of April 1942, not only enabled fighter reinforcements to be flown to Malta, but led to its resurgence as an offensive base. By the middle of May, the Wellingtons and the torpedo-carrying Beauforts had renewed their raids on the lifeline of the luckless Axis forces in North Africa, though it was not until July that submarines would re-join them. Once more the island had become the key to the control of the Mediterranean.

However, if it was to continue as such, or even survive at all, it had to receive essential supplies. The greatest need was for food. Next came oil, which provided the power for all public utilities from the flour mills, via the pumps drawing water from the deep wells, to the cranes serving the harbour. Equally vital was kerosene, which on the tree-less island supplied the means of all light and heat. Finally, there was the requirement of military stores such as aviation fuel and ammunition.

For all these commodities, Malta depended on convoys arriving safely. It was therefore a source of terrible anxiety that lately they had not done so. The tale of troubles really began in January 1942, when, following a series of confused encounters, another victory by Rommel gave the Luftwaffe airfields in Cyrenaica from which British ships bound for Malta could be attacked, while the RAF retired to bases further east from which such ships could not be given adequate fighter protection.

The consequences of this change of fortune followed swiftly. On 14 February, three merchantmen making for Malta were attacked in 'Bomb Alley' – the area between Crete and Cyrenaica outside the range of fighters from North Africa or Malta. Two were sunk, the third so badly damaged she had to return to Alexandria. No provisions reached the island. Next month, four more vessels set out from the Eastern Medi-terranean. Although the covering force, under Rear-Admiral Vian, beat off a far stronger Italian fleet, the Luftwaffe again mauled the convoy. Only two ships reached Grand Harbour – on 23 March. For three days, Malta's fighters kept the dive-bombers away from them but on the 26th, both were sunk at their moorings. Of the 26,000 tons of supplies that had left Egypt, only 5,000 had been unloaded.

Now the scanty rations issued to troops and civilians alike were cut to the absolute minimum, although already unpleasant skin diseases caused by malnutrition had begun to appear. The shortage of oil was also critical – the Governor, Viscount Gort, who succeeded General Dobbie on 7 May, had to tour Valetta on a bicycle. It was decided therefore that in June, two more convoys would set out simultaneously from Gibraltar and

Alexandria for the relief of the island. They were code-named 'Harpoon' and 'Vigorous' respectively.

To guard the six freighters in the Harpoon convoy were the carriers *Eagle* and *Argus*, the latter carrying five Fulmars, the former twelve Sea Hurricanes of 801 Squadron, four Sea Hurricanes of 813 Squadron and four more Fulmars. However, it is important to note that it was not intended that they should accompany the merchantmen all the way, but only as far as the Narrows – the waters between Sicily and Cape Bon, Tunisia. Here the main covering forces would turn back. The convoy would pass through the Narrows at night, then make a final dash for Malta. It was appreciated that losses would occur on this last stage but it was hoped that the protection of the Fleet Air Arm would have provided a sufficiently good start to ensure that the bulk of the supplies got through.

On 13 June, the convoy came under constant observation by shadowers, one of which was shot down at dusk by the Hurricanes of 813. Next day German and Italian aircraft from Sardinia and Sicily – almost 150 torpedo-planes, dive-bombers, high-level bombers and fighter-bombers, escorted by more than 100 fighters – made a series of attacks, but, despite the heavy odds against them, the mere handful of naval fighters so hampered the raiders as to prevent them from sinking more than one of the vital merchantmen. The Sea Hurricanes destroyed ten enemy machines, the Fulmars four. Several more fell to AA fire. Three Sea Hurricanes and four Fulmars were lost.

Most of the escort now retired, though the Hurricanes of 801 made a further 'kill' during the carriers' return to Gibraltar next day. However, 15 June was much less satisfactory for the convoy. An attempted intervention by an Italian cruiser-force was beaten off by British destroyers, but the defence against air-attacks was much inferior in the absence of the interceptors – thereby emphasising their value on the previous day. Three more merchantmen, including a precious tanker, went down but in the early hours of the 16th, the survivors, *Troilus* and

Orari, reached Malta safely. Their 15,000 tons of flour and ammunition provided at least a breathing space for the battered fortress.

However, it was little more than this, for while the convoy that was protected, in part at least, by Sea Hurricanes was a costly success, the one that was not was a costly failure. For a brief period, the eleven freighters from Alexandria, though unaccompanied by aircraft carriers, did have cover from Hurricanes, and also Kittyhawks, all operating at long range from bases in the Western Desert. These drove off a number of threatening attacks, probably inflicting some losses in the process. But as the merchant vessels moved westward, the Hurricanes and Kittyhawks disappeared – and the Junkers Ju 87s and Ju 88s returned.

The convoy had reached Bomb Alley where, during the afternoon of 14 June, it was persistently assaulted. Though surprisingly little damage was done, the escorting warships began to use up their anti-aircraft ammunition at an alarming rate. Meanwhile, a strong Italian force built around two battle-ships had been sighted, moving to intercept. It was attacked by the RAF without much effect but conflicting reports resulted in the convoy steering backwards and forwards in Bomb Alley awaiting clarification of the position. By the evening of the 15th, two merchantmen and two destroyers had been sunk by bombs and a third destroyer by E-boats.

Although it was now known that the Italian fleet had turned away, Rear-Admiral Vian, whose ships had already expended more than two-thirds of their ammunition, rightly appreciated that a continued advance in the face of uninterrupted air-attacks would soon leave them quite helpless. He therefore returned to Alexandria, losing cruiser *Hermione* to a U-boat on the way. The attempt to supply Malta from the east had collapsed miserably.

It was therefore clear that the next convoy to the island, code-named 'Pedestal', would have to come from the west, where carrier-based fighters could mount guard during at least some of the journey. It was even considered sending the heavy fleet units all the way through. Though this suggestion was not adopted in view of the losses certain to occur in the Narrows where the movements of such vessels would be much restricted, its rejection must have caused some heart-ache. Pedestal was planned for August. If it did not succeed then there would be no time to organise another convoy before the food, oil and kerosene in Malta ran out – this would happen by 7 September, after which the fortress would have no option but to surrender.

Thus, a vital duty was laid upon the crews of the thirteen big, fast freighters, two of them American, which began to assemble at the mouth of the Clyde in June and July, loaded with 85,000 tons of flour, ammunition, and aviation fuel in cans. They had to be accompanied by a tanker but no British tanker could make the 14 knots needed to keep up with them. Fortunately, with typical generosity, the Americans again came to Britain's aid with the very finest of all the fine ships to make the 'Malta Run'. On 21 June, the tanker *Ohio* steamed into the Clyde. She was taken over by the Ministry of War Transport, loaded with 11,000 tons of oil and kerosene and placed in the nominal ownership of the British Eagle Oil and Shipping Company, which entrusted a hand-picked British crew, under Captain Dudley Mason, with the task of carrying this crucial, vulnerable cargo to Malta.

In order to provide an adequate escort for Pedestal, Vice-Admiral Sir Neville Syfret, the South African officer who had succeeded Somerville in command of Force H at Gibraltar, was given battleships *Nelson* and *Rodney*, seven cruisers and twenty-six destroyers. More important still, however, he also had three aircraft carriers, the first time that so many British carriers had operated as a squadron: *Indomitable*, embarking twenty-two Sea Hurricanes of 800 and 880 Squadrons plus nine Martlets; *Victorious*, with five Sea Hurricanes of 885 and sixteen Fulmars; and *Eagle*, which had sixteen Sea Hurricanes of 801 and 813 ready for action with four more unassembled as possible reserves.

Axis Intelligence was well aware of the details of the convoy as of its importance. On aerodromes in Sardinia and Sicily, the Luftwaffe and the Regia Aeronautica had mustered about 500 aircraft – fighters, fighter-bombers, dive-bombers, medium bombers and torpedo-bombers – plus reconnaissance machines. Their plans were carefully co-ordinated and included the use of new weapons: the *motobomba*, a torpedo which, after coming down by parachute, took a zigzag course through the water; and a radio-controlled Savoia Marchetti SM79 packed with explosive. However, as it transpired, the former proved quite ineffective, while the SM79 ultimately crashed in Algeria, much to the indignation of the Vichy French authorities.

More orthodox methods were more dangerous. On 11 August, the Sea Hurricanes had a few clashes with shadowers, Lieutenant-Commander Judd, the CO of 880, shooting down a Ju 88 in flames, but although there was one bombing-raid just before sunset, this was little more than a reconnaissance in force for the massive operations planned for the 12th. However, also on the 11th, at 1315, a U-boat sank the gallant old *Eagle*. Four of the Sea Hurricanes that were aloft at the time landed on the other carriers but the remainder went down with her.

The 12th began with a victory for the Fleet Air Arm. Just after 0900, nineteen Junkers Ju 88s from Sardinia, escorted by sixteen Bf 109s, opened the day's proceedings. The Sea Hurricanes quickly got among the bombers, dispersing them so thoroughly that only four reached the convoy, where their attacks were ineffective. Nine Ju 88s were brought down by the fighters or by AA fire. Two Sea Hurricanes were also lost.

At 1240, a series of major raids from the Sardinian bases commenced, but the defending fighters carried out their primary task of breaking up the attacks so effectively that the enemy managed only to cripple one freighter – which was afterwards sunk while proceeding independently. By 1345, the action seemed to be over. Suddenly, two machines roared in low upon *Victorious*. The carrier did not fire, thinking they were

Hurricanes coming in to land, but they were, in fact, Reggiane Re 2001 fighter-bombers, which bore a considerable resemblance to the Hawker aircraft. The Italians achieved complete surprise, but their light fragmentation bombs had little effect on the carrier's armoured deck, though splinters killed six men. Both planes escaped unharmed as, it must be admitted, they thoroughly deserved to do.

That concluded the strikes from Sardinia, but those from Sicily were yet to come. They commenced at about 1800, continuing throughout the next hour. *Indomitable* was badly damaged by two bombs; destroyer *Foresight* was sunk by a torpedo; yet not one of the merchant ships was hit.

At about 1900, Vice-Admiral Syfret turned back for Gibraltar with his weary but triumphant carriers, leaving Rear-Admiral Burrough, with four cruisers and eleven destroyers, to conduct the convoy the rest of the way to Malta. David Brown, after checking with Axis records for his book *Carrier Fighters*, has concluded that the naval airmen brought down thirty-five enemy machines during this one day. Others fell to the AA gunners, while it would appear that many more were damaged, of which seven made force-landings in Tunisia. Of the 'kills', the Martlets gained four, the Fulmars seven, the Sea Hurricanes twenty-four.

Many of the Sea Hurricane pilots achieved considerable personal success on 12 August, even allowing for duplicated claims. Lieutenant Ritchie of 800 took part in the destruction of two Italian Stukas; while his CO, Lieutenant-Commander Bruen, helped to bring down two Italian torpedo-bombers and a Junkers Ju 88. Sub-Lieutenant Thomson of 800, having just shot down one of a group of four Stukas and driven away the others, which jettisoned their bombs, was attacked from above by about eight Bf 109s – but he out-turned them, destroyed one and returned safely.

However, the best performance of all was that of Lieutenant Richard Cork, a member of *Indomitable*'s 880 Squadron. Cork was a veteran Hurricane pilot who had been one of the naval airmen seconded to Fighter Command at the start of the Battle of Britain, during which he served with Bader's 242 Squadron, winning the DFC – later amended to the DSC by order of the Admiralty. He added a DSO to this after the Pedestal convoy.

On 12 August 1942, Cork flew at first in Hurricane Z4642, which had been received on board *Indomitable* as a standard Sea Hurricane IB but had been "unofficially" converted to a IC by providing it with the four-cannon wing. In this he shot down two Junkers Ju 88s, a Messerschmitt Bf 110 and an Italian bomber usually reported as a Savoia Marchetti SM 79 but apparently a Cant Z1007. In the course of these engagements, his aircraft suffered damage and, while it was being repaired, Cork took off again in a standard Sea Hurricane IB, Z7093, to shoot down a genuine SM 79. He was then attacked by two Reggiane Re 2001s but his Hurricane, though badly hit in radiator and rudder, brought him safely back to his carrier.

The Fleet Air Arm's losses in combat were eight aircraft, four of which were Sea Hurricanes. Four of the pilots survived. At least five more machines were destroyed by ditching due to loss of fuel, or by being pushed into the sea as too badly damaged for repair, a harsh but necessary decision since the injuries to *Indomitable* had forced her fighters to land on a badly overcrowded *Victorious*.

These casualties were not suffered in vain. Thanks to the Fleet Air Arm pilots, even the most determined of attacks had sunk only one merchant ship. The 'good start' had been achieved.

It was just as well. During the night and early the following day the convoy endured terrible sufferings under assaults by U-boats, motor torpedo-boats and aircraft. Cruisers *Manchester* and *Cairo* were lost. The destruction of the merchantmen was so ghastly that had it not been for the triumphs on the 12th, it seems certain that none would have got through. As it was, on the evening of the 13th, three of the freighters, *Port Chalmers*,

Rochester Castle and *Melbourne Star*, arrived at Malta, where they were joined next day by the damaged *Brisbane Star*, which had been proceeding independently.

One of the island's requirements had been met but, some seventy miles to the west, a small group of escort vessels was still battling the enemy, the sea and growing fatigue in order to provide the remaining necessities. The *Ohio* had been reduced to a sickening travesty of the fast, handsome, graceful tanker that had set out on Operation Pedestal. Totally disabled, dead in the water, a vast hole from a torpedo hit gaping in her side, struck by at least one bomb, racked by about twenty near-misses, one of which had split her stern open, with the remains of a Stuka and a Ju 88 littering her decks, it seemed impossible for her to survive. But the shipwrights of Chester, Pennsylvania had achieved perfection and Captain Mason, her crew and her rescuers were all utterly determined that her 11,000 tons of oil and kerosene would get to Malta.

And, in the end, their efforts were crowned with success. With destroyers *Penn* and *Bramham* lashed on either side of *Ohio* to help keep her afloat, destroyer *Ledbury* secured to her stern and minesweeper *Rye* towing, the slow procession moved towards the island, while the tanker's buckled plates groaned and she sank ever lower until the seas washed over her main deck. All that day and all that night they struggled on and, at 0800 on 15 August, it seemed that the entire population of Malta was there to roar a welcome as the unconquerable *Ohio* finally entered Grand Harbour.

Fittingly, Captain Mason was awarded the George Cross, the same decoration as the island that he had done so much to save. For saved it was – soon, indeed, it was striking out harder than ever against the enemy's supply-lines – until such time that great events in North Africa ensured the permanent retention of the air-bases necessary to provide its convoys with fighter-cover all the way through.

In those great events, the Hurricane, as so often, would have a major role to play.

Chapter Eleven

Fighter-Bomber and Tank-Buster

On 15 November 1941, Mr Churchill signalled to General Auchinleck, whose offensive, code-named Operation 'Crusader', was planned for the early hours of the 18th:

> Now is the time to strike the hardest blow yet struck for final victory, home and freedom. The Desert Army may add a page to history which will rank with Blenheim and with Waterloo.

Certainly, only such an achievement could justify the receipt by Auchinleck's command, which since September had borne the immortal title of Eighth Army, and by its supporting Western Desert Air Force – a name officially bestowed on 9 October, though, in practice, the word 'Western' was rarely used – of the reinforcements that had been designated originally for the Far East. So much was thus diverted that Captain Liddell Hart has claimed that Rommel 'indirectly produced the fall of Singapore'. The Allies had 710 gun-armed tanks as against 174 German and 146 Italian. Despite repeated assertions to the contrary, the majority were slightly superior to the German tanks and all were vastly superior to those of the Italians. Moreover, Auchinleck had some 500 reserves while Rommel had none whatever, for his supply lines were then under maximum pressure from the forces based in Malta.

Because of Eighth Army's superiority, it was planned that while XIII Corps pinned down the frontier positions at Bardia, Sollum and Halfaya, XXX Corps with the bulk of the armour, would seek out and destroy the German Panzer Divisions, after which the relief of Tobruk, the re-conquest of Cyrenaica, the advance to Tripoli, would follow as a matter of course. It is therefore astonishing that Auchinleck should have appointed, as his Army Commander, Lieutenant-General Sir Alan Cunningham, who had done so well in East Africa, but who was quite inexperienced in armoured warfare.

Air Vice-Marshal Coningham's Western Desert Air Force also outnumbered its enemy. Of forty fighter squadrons in the Middle East, twenty-five flew Hurricanes, the bulk of which were Mark IIs, though some of these units were resting, re-equipping, working-up or operating in quieter areas, among them such famous ones as 6 and 73, as well as 335, which formed with Greek personnel on 10 October at Aqir, Palestine, but did not see action until February 1942. However, that still left thirteen Hurricane squadrons in the front line, including the Royal Navy Fighter Unit, under Lieutenant-Commander Alan Black, which had been created in August from the pilots of 803 and 806 Squadrons, flying Hurricanes, and 805 with Martlets. With a strength of eighteen Hurricanes and nine Martlets, it was roughly the size of two RAF units, like them providing fighter cover for and close support to the Allied soldiers.[47]

Support, of course, took many forms, a particularly close relationship with the ground troops being that of the Tac

---

[47]Although the Hurricane IIs were of course superior to the Mark Is, their performance also was reduced by their tropical filters, as well as by the wear and tear of their airframes caused by wind-driven sand. This had a particularly bad effect on the canopy Perspex, which, in time, would become almost opaque. The Royal Navy Fighter Unit even tried flying without canopies, as did several RAF pilots despite the buffeting they received from the wind. The Germans did not suffer from this handicap as their Perspex was of a higher quality.

R (Tactical Reconnaissance) Hurricane Squadrons, 208, 451 RAAF, and, prior to February 1942 when it moved to Iran on interceptor duties, 237 with Rhodesian personnel; all of which were the 'eyes' of Eighth Army. On 20 November the garrison in Tobruk was ordered to break out the next day, whereupon four of 451's machines were flown into the perimeter, whence, in addition to their normal duties, they provided liaison with the main forces by delivering messages too detailed or too secret for dispatch by wireless.

The same day saw Hurricanes operating in a new role in the desert when six of them from 80 Squadron – which at this time still flew Mark Is, receiving a mixture of IIAs and IIBs only in January – attacked motor vehicles with four 40 lb bombs under each wing. Similar weapons had been used previously by some of the Malta Hurricanes in their raids on Sicily commencing on 20 September. However, while the mounting assaults on Malta prevented development of this idea there, the fighter–bomber version in the Desert – the 'Hurribomber' as it became known – was to have a long, distinguished career.

It was also to carry larger weapons. On 18 April 1941, an old Mark I, P2989, had flown with two 250 lb bombs, a direct hit from which could destroy a tank. It was decided to adapt Mark IIBs to carry these, the first such, Z2326, making its trial flight in May. Experience showed that to allow enough space for the bomb racks and associated wiring, it was best to omit one gun on each wing leaving the Hurribomber with ten – though, in practice, many of the fighter-bomber squadrons flew with all twelve guns installed, presumably for better protection after use of the bombs.

These reduced the speed of the Mark IIB by some 20 mph, but they had hardly any other effect on its handling character-istics or manoeuvrability. Mark IICs were also adapted to carry the bombs but the speed of a IIC in the Desert, with its tropical filter and while carrying bombs, was reduced to little more than 275 mph.

Such aircraft would appear in the Desert later but for Operation Crusader, 80's Hurribombers were the only ones available, supplemented by the other Hurricane squadrons in the low-level strafing of Axis troops or vehicles. However, their effectiveness was somewhat restricted by the difficulty of telling friend from foe amid the ever-changing complications of the land-battles.

For the British and Commonwealth forces, all went well at first. They gained complete surprise, partly because the evening before Crusader opened, the heaviest rainstorm of the year flooded the Axis airfields but largely missed those of the RAF.

In consequence, the Hurricanes were able to operate almost without hindrance. On the 18th, 33 Squadron destroyed a number of Italian motor vehicles south of Tobruk, then strafed enemy airfields with less effect. The pilots saw combat only with three Fiat CR 42 biplanes, all of which they shot down. For four days, the Hurricanes, joined on the 20th by 80's fighter-bombers, raided aerodromes, transports and troop positions, though suffering a number of losses to AA fire.

Already by the 20th, however, the Luftwaffe was trying desperately to re-enter the fight. Attacks by Junkers Ju 87s were thwarted by the Hurricanes, which forced them to jettison their bombs well away from any worthwhile targets. On the 20th, the Royal Navy Fighter Unit completely routed a Stuka formation, Sub-Lieutenant Charlton of 803 shooting down three before his Hurricane was hit and so badly damaged that he had to crash-land. Next day, 33 and 80 destroyed or damaged a total of eight enemy aircraft, losing only one Hurricane each. On the 23rd, 229 downed three Stukas and damaged eight others, for the loss of two Hurricanes and one pilot. 238 also destroyed three Stukas, damaging four more, while Flight Lieutenant Forsyth shot down a Fiat G50 fighter; but it lost four aircraft and three pilots, with a fifth machine crash-landing.

Allied soldiers had also destroyed Axis aeroplanes on the 19th, when they captured the important airfield of Sidi Rezegh.

Unfortunately, Eighth Army now made its almost traditional error of scattering its armoured formations, which the Afrika Korps could therefore attack in succession. By the evening of the 23rd, Sidi Rezegh had been re-taken and the break-out from Tobruk checked. Thereupon, over-estimating the extent of his successes, Rommel conceived a daring if reckless plan – he would thrust for the frontier, destroying the forces engaged with his garrisons there and wrecking the British lines of communication in the process.

By the following evening it seemed that he had made the right decision. He had reached the frontier, causing widespread panic, while General Cunningham, who had been visiting XXX Corps headquarters, had escaped capture only by dashing to a waiting Blenheim, which took off under fire. It was small wonder that Cunningham felt the situation was lost, but happily, Auchinleck, not being so closely involved, was able to take a more realistic view – he ordered that the offensive must continue. On the 26th, Cunningham was relieved from command – though it should be added that he filled several important posts with distinction on his return to Britain.

He was replaced by Auchinleck's Deputy Chief of Staff, Major-General Neil Ritchie. This was an equally strange choice for Ritchie lacked experience in desert operations, was junior to his Corps commanders and had a confusing relationship with Auchinleck, who spent much time at his HQ, giving 'advice', which it was difficult to interpret as other than orders.

Yet, at the moment, this scarcely mattered, for it was now Rommel who lost his grip, his forces becoming increasingly dispersed. Furthermore, Axis accounts admit that immense damage was done on the 25th and 26th by the Desert Air Force, particularly No 80's Hurribombers. Their 40 lb bombs could inflict little harm on a tank but they proved deadly against the supporting motorised infantry or petrol and ammunition lorries, while they even caused some gruesome casualties among tank crews who were decapitated when they opened their hatches to see what was happening, or blown to pieces when caught by surprise outside their vehicles.

Meanwhile, behind Rommel's back, on the night of the 26th, the New Zealand Division linked up with the Tobruk garrison. Thereupon the Panzers hurried back from the frontier to commence a series of violent fights that, by 1 December, had again isolated Tobruk but had failed to gain any decisive advantage.

During these desperate days, the Hurricanes saw every possible type of service. No 2 Photographic Reconnaissance Unit[48] – with its royal-blue coloured Mark Is, minus guns but provided with cameras and extra fuel tanks – made flights of more than 1,100 miles, lasting over five hours, at heights of over 33,000 feet, bringing back vital information of enemy movements far behind the lines. The Tac R squadrons supplied similar details of more immediate importance and it was on 29 November, that Flying Officer Cotton of 208, in Hurricane I, Z4063, won a DFC for an especially fine piece of airmanship.

While operating over the enemy airfield at El Adem, Cotton was intercepted by two Messerschmitt Bf 109s. They pursued him for nearly half-an-hour but he evaded their attacks by a series of very tight, fast turns at ground level. His Hurricane was hit repeatedly but kept flying, enabling him to fire several bursts at the Messerschmitts, though without apparent results. Eventually, with their ammunition exhausted, the 109s abandoned the chase but the Hurricane's hard-used engine now gave out, necessitating a force-landing only a few miles from Cotton's base. The aircraft which had suffered only minor damage was later salvaged and repaired and the pilot was able to deliver his valuable information to XXX Corps headquarters.

The ground-attack Hurricanes continued their assaults on anything that moved or anything on an enemy airfield. 33 made

---

[48]The Hurricanes of this unit had previously carried out secret reconnaissance missions over Iraq, Syria and Iran prior to the campaigns in those countries.

a speciality of raiding Agedabia, destroying or damaging four-teen machines there on 2 December, while Sergeant Challis, making a really low-level strike, hit a Fiat CR 42 with his wingtip, 'damaging both aircraft', though not enough in the Hurricane's case to prevent it returning to base. Three days later, the squadron was back, destroying or damaging ten more Axis aeroplanes. On the 8th, Agedabia was visited again, six enemy aircraft being hit – probably there were now few left to strafe. Flying Officer Charles, damaged by flak, force-landed, but Flight Lieutenant Gould touched down in the desert to pick him up safely – though, sad to say, Charles was killed in action eight days afterwards.

Further aloft, the Hurricanes were engaged on interceptor duties. On 1 December, the Royal Navy Fighter Unit had its biggest combat with Junkers Ju 88s, escorted by Fiat G50s. It forced the bombers to jettison their weapons, then shot down one of them, plus two Fiats, while Sub-Lieutenant Dennison out-turned another Fiat to such good effect that it stalled and crashed. It was a sad loss to their RAF comrades when the Fleet Air Arm pilots reverted to their 'official' task of shipping protection at the end of the month, though they did score a few more victories in this role; even more so when, late in February, they replaced their Hurricanes with Fulmars prior to leaving for the Far East.

The RAF and Fleet Air Arm pilots were not the only ones flying Hurricanes, for the part played by No 1 Squadron, South African Air Force should not be forgotten. On 7 December, it encountered a force of 109s and Macchi MC 202s – superb fighters with a top speed of 370 mph and very high manoeuv-rability though armed only with two 12•7 mm and two 7•7 mm machine-guns. Outnumbered by more than two to one, the South Africans still shot down two 109s and a Macchi but lost two Hurricanes, one pilot being killed, the other becoming a prisoner of war. A third Hurricane, badly damaged, crash-landed, while two more, also hard hit, were able to get back to their airfield.

To emphasise the different nationalities flying Hurricanes, the most successful airman during this period was Pilot Officer Lance Wade of 33 Squadron, an American citizen from Texas. He had gained at least six victories when on 5 December, he attacked a low-flying Savoia Marchetti SM 79 so fiercely that when it blew up, the Hurricane was caught in the blast. Wade force-landed about twenty-five miles from Eighth Army's forward positions. Sergeant Wooler came down to assist but so damaged his own machine that both pilots had to start back on foot. They ultimately rejoined their squadron, Wade at least receiving balm for his blisters in the news that he had been awarded a DFC.

This trick of landing to rescue a colleague was used fairly frequently by the Hurricane pilots. On 1 December, 1 SAAF and 274 Squadrons, while escorting a Blenheim raid, were attacked by hostile fighters. They shot down three 109s, damaging two more, but lost three Hurricanes, though only one pilot died. Another Hurricane, flown by Lieutenant Hoffe of the South African Air Force, had to land in 'no man's land' with a damaged engine. Spotting this, Flight Lieutenant Tracey of 274 touched down beside him and took off again with Hoffe sitting in his lap. Similarly, on 25 December, 94 and 260 were strafing enemy vehicles when Sergeant McKay of the latter unit was shot down by flak – whereupon Wing Commander Mayers landed his Hurricane behind the enemy lines to carry McKay to safety.

However, the Hurricanes jealously guarded their privileges. On 7 December, 274 Squadron lost three of its own aircraft in combat but shot down two MC 202s, a G50 and a Ju 87. At least five other enemy machines were damaged, including a second Stuka, which force-landed. Another Ju 87 went down to assist but was observed by Sergeant Parbury who promptly, if somewhat unchivalrously, destroyed both the dive-bombers.

That same day, Rommel learned that he would receive no substantial reinforcements for at least a month, which meant that a continuation of the struggle must lead to his destruction. With considerable moral courage therefore – abandoning his frontier garrisons, which capitulated by mid-January, abandoning the investment of Tobruk – he fell back to a new line stretching south from Gazala.

The Allied forces followed up cautiously but, by 17 December, Rommel was again retiring. Assisted by the timely arrival of twenty-two tanks at Benghazi on the 19th shortly before it was evacuated, as well as twenty-three more at Tripoli, he twice turned to maul his pursuers who were quite unable to prevent his reaching the bottle-neck of El Agheila by 6 January 1942.

During these events, the Hurricanes frequently protected Blenheims or Marylands attacking the retreating Axis army – at some cost to themselves, for when tied down guarding the bombers they were very vulnerable to the higher-flying Messerschmitts. Thus, on 17 December, 1 SAAF lost four machines, three airmen being killed, but shot down only two 109s in compensation. However, a week later, Sergeant Maxwell of 94, with his Hurricane already badly damaged, counter-attacked a 109, the pilot of which, Oberleutnant Graf von Kageneck, JG 27's top-scoring ace, reached his base only to die of his wounds.

The fighter-bombers of 80 Squadron also raided the enemy columns, though suffering losses to hostile fighters or AA fire. On 12 December, four of the squadron's aircraft were downed – all the pilots survived but as prisoners of war. However, on 8 December, while six Hurribombers were striking at the coast road, leaving it strewn with blazing transports over almost a mile, four more were able to beat off an attack by twelve enemy machines without loss. Flying Officers Reynolds and Mason each shot down a 109. Sergeant Whyte damaged two Macchi MC 202s.

Next day saw the finest individual performance of all by 80's CO, Squadron Leader Michael Stephens. After leading another bomb attack on German vehicles, he was engaged by a Macchi MC 202.[49]

Wounded in both feet, with his Hurricane on fire from a hit in the petrol tank, he was about to take to his parachute when the Italian overshot him. In spite of the intense heat, Stephens scrambled back into his seat, took quick aim and brought his foe down. He then baled out, beating out his burning clothes during his descent. He was rescued by some Polish troops who confirmed his victory and brought him to Tobruk hospital where, a month later, he learned that for his 'great courage and devotion to duty' he had been awarded the DSO.

Crusader may have been far from a complete victory, as well as proving only a very temporary one, but in one respect its achievement was immense. On the landing grounds of Cyrenaica, the Allies discovered 458 wrecked enemy aeroplanes, half of them German, plus many more is various resting places in the desert. A large proportion of these bore witness to the high effectiveness of the bombs, cannon-shells or bullets used by those expert airfield-attackers, the Hurricane squadrons.

From this time onwards, the Desert Air Force enjoyed superiority. This may seem an extravagant claim in view of the splendid abilities of the Messerschmitt Bf 109Fs and the Macchi MC 202s, but their losses had been such that they could no longer indulge in dog-fights. Instead, they used their superior speed to dive onto machines at a disadvantage, whether straggling, recovering from strafing attacks or guarding Allied bombers.

It must be conceded that they often struck heavy blows in such circumstances. Thus, on 9 January 1942, 229 was bounced when on bomber escort, losing four Hurricanes but only one

---

[49]This aircraft has frequently been reported as a Messerschmitt Bf 109, which Stephens originally identified it as being. It seems, however, more probable that it was a Macchi and Stephens himself would later agree.

pilot. More humiliating was an assault by only two 109s on 94 Squadron on the 14th. They shot down four Hurricanes, though again only one pilot was killed, damaging three others. Yet the 109s seldom managed to get through to the really important targets, the bombers, which were never prevented from keeping up regular, effective strikes on the Axis land forces.

Furthermore, the 109s were rarely able to keep the Hurricanes from breaking up the raids by their own bombers, though they occasionally pounced on the Allied fighters to some purpose afterwards, before these could re-gain height. The next few months seemed to provide little more than a series of British retreats, with choked lines of communication, miles of fleeing vehicles, bottle-necks where hundreds of lorries were almost stationary. Determined air-attacks could have turned retirement into rout. Yet such was the protection afforded by the RAF that this was never seriously threatened.

The first retreat came on 21 January. Kesselring's assault on Malta had forced this base to look to its own defence rather than the disruption of the Libyan supply lines. With sizable reinforcements at last reaching his command, now re-christened Panzerarmee Afrika, Rommel bounced back with astounding speed, completely fooling Auchinleck who had believed that he was 'hard pressed' to maintain his present position.

Thus, the enemy achieved complete surprise, made worse by the forward RAF base at Antelat being flooded by heavy rains. On the 22nd, the fighters there, Hurricanes and the new Kittyhawks – similar but superior to Tomahawks – had to make a rapid evacuation. The state of the ground necessitated man-handling each aircraft to the only area not waterlogged. Two Hurricanes and four Kittyhawks, lacking such essential items as propellers, had to be destroyed but the last of the remaining machines got off just as German shells began falling on the airfield.

However, the Axis air force still achieved little. On the 24th, 274, meeting thirty Junkers Ju 87s, shot down four plus three escorting Italian fighters. One Hurricane crash-landed but the pilot was unhurt. Next day, the faster Junkers Ju 88s took a hand but at least four fell to various Hurricane squadrons. Two days later 274 was again in good form, destroying four enemy aircraft, after which four Hurricanes landed in the desert out of fuel – but without harming any of the airmen.

Unhappily, the old, sad story of disasters on the ground was already being repeated. By the 25th the British and Commonwealth forces were in full flight, while the arrival of Auchinleck at Ritchie's HQ made matters worse by starting a series of conflicting instructions that had the traditional consequences.

On the 26th came a gleam of hope when five squadrons of Hurricanes, using bombs, cannon and machine-guns, formed the backbone of a series of strafing attacks, which destroyed or damaged about 120 enemy vehicles. Then the weather broke completely, though between storms the Hurricanes continued to strike at Axis transports despite suffering a number of casualties to flak. On the 28th, Eighth Army fell back to Rommel's old line stretching from Gazala to Bir Hacheim. That splendid Hurricane squadron, No 73, which had landed at Mechili with its IIBs and IICs to re-join the fighting, had to retire rapidly to El Adem. By 5 February, with most of Rommel's transports immobilised by lack of petrol – a circumstance of which the RAF fighters took full advantage, destroying or damaging another 100 – the front stabilised. The German recapture of the Cyrenaican 'bulge' enabled their bombers to dominate the seas east of Malta, whose ordeal by hunger now began. The British were left to wonder how they had let the same disaster happen two years running. Somehow it did not quite measure up to Blenheim or Waterloo.

The next three-and-a-half months were a genuine lull during which, for once, the Hurricanes saw little action, though since such valuable aeroplanes could not be allowed to remain idle, they were subjected to much re-organisation, the results of some of which have already been noted. 229 Squadron retired to the Canal Zone in February, after which it moved on to Malta. No 30, having shot down three SM 79s attacking a convoy on 15 February, was ordered off next day to Ceylon to oppose Nagumo's naval airmen. 94 converted to Kittyhawks but, after suffering heavy casualties, reequipped with Hurricane IICs in May, for patrol duties in the Eastern Mediterranean. 213 – which had had most of its aircraft in Cyprus, where it had flown a number of sorties without result due to lack of proper radio communications – re-assembled in Egypt to protect the Fleet at Alexandria, gaining several victories over reconnaissance Junkers Ju 88s. Its place in Cyprus was taken by a detachment from No 127.[50]

At the front, although fighter Hurricanes escorted bombing raids on enemy positions and fighter-bomber Hurricanes attacked targets of opportunity, the heaviest duties were those of the Tac R Squadrons, 208 and the newly formed No 40 SAAF. Flying alone, often with only Mark Is, these were favourite victims for the 109s, to such an extent that, by mid-May, other Hurricanes were giving them cover. However, on 12 February, Pilot Officer Montagu of 208 drove off two attacking 109s, badly damaging one of them in the process. He was then hit by 'friendly' AA fire but Hurricane P2646 made a landing in very rough country without injury to its pilot.

The Axis air forces made few attacks at this time, though 12 February also saw an especially savage fight in which 274 destroyed four enemy machines but lost four Hurricanes, with two pilots killed. After that, the pressure eased though there were still a number of encounters, some more satisfying in retrospect than at the time – as on 26 March when a Hurricane of 80 Squadron flown by Flying Officer Mason returned safely having been damaged not only by 109s but by the Tobruk AA guns.

---

[50]It may be recalled that 127 had disbanded at Habbaniyah in July 1941. It was reformed in February 1942 with Hurricane Is, receiving IIBs in June.

On the night of 25/26 May, the lull ended abruptly when Junkers Ju 88s struck at the fighter bases, at least three being downed by the Hurricanes of 33 and 73 Squadrons. It was the prelude to violent events. Next morning, Rommel launched his offensive on the Gazala Line.

Since British Intelligence had broken the German cypher, so could read the signals between Rommel and his masters in Berlin, the Allied commanders were well aware of the impending conflict which they awaited with considerable confidence, having a heavy superiority in tanks, artillery and reserves. No wonder General Sir William Jackson remarks in *The North African Campaign 1940-43* that:

> If the Battles of Gazala had been fought by the British as they should have been ... Auchinleck and Ritchie would be remembered as the men who turned the tide of British fortunes in the Second World War. Instead both were to lose their commands as defeated generals before the summer was out – and rightly so.

Yet it all began so well. Rommel, with his three crack divisions, 15th Panzer, 21st Panzer and the mechanised infantry of 90th Light, swung south of the fortified position, or box, at Bir Hacheim, held by General Koenig's Free French. However, since an attack on this strong-point by his Italian troops failed completely, he was left with a long supply line, especially vulnerable to assault from the air.

In theory this should not have mattered greatly, since it will be remembered that Kesselring, at the cost of abandoning sustained raids on Malta, had sent large reinforcements of aircraft to Africa, so that the numbers on each side were now approximately equal. However, in practice, the Luftwaffe showed much less aggression than the RAF – for example, when the Hurricanes of 274 sighted a formation of Junkers Ju 87s escorted by 109s on the 27th, the Stukas unloaded their bombs at random and departed – though not before one had been shot down by Flight Sergeant Neil.

By next day Coningham had concentrated his main effort on Rommel's supply route. This led to some risks for the Hurricanes – 80 Squadron was bounced on the 29th, losing four machines and two men – but the bombs, cannon-shells and bullets that they and the Kittyhawks – now also adapted as fighter-bombers – poured onto the Axis vehicles destroyed or damaged about 300 of these. The Panzer divisions – already shaken to find that their tanks, contrary to popular myth, were badly outgunned by Eighth Army's new American Grants – were further crippled by the absence of supplies, particularly of petrol. By 30 May Rommel had been forced to withdraw to an area among the British minefields, aptly, if unimaginatively, named the 'Cauldron'.

It seemed that the 'Desert Fox' was trapped at last but now control of the battle began to disintegrate. This was mainly due to the odd command system. The unlucky Ritchie was bombarded by a stream of orders from Auchinleck, most sound, some unrealistic, a few mutually contradictory. Ritchie was also junior to both his corps commanders, Lieutenant-Generals Gott and Norrie, so inevitably perhaps, came to rely on their advice rather than giving them a definite lead.

In consequence, Rommel was allowed to clear a lifeline through the minefields to link-up with the Italians, then beat off a sequence of badly co-ordinated attacks with his anti-tank guns. On 2 June, he was even able to dispatch '90th Light' to join in the fighting around Bir Hacheim.

It was in defence of the gallant Free French garrison that the Hurricanes flew most of their sorties in the next few days. The fighter versions gave cover while the Hurribombers struck at the investing forces, having an especially good day on the 3rd, when they hit about sixty enemy vehicles. Moreover, it was at this time that an entirely new form of Hurricane made its first appearance.

Since 1939, British manufacturers had been working on the development of 40 mm armour-piercing cannon for use against tanks. Since these were relatively light, it was not surprising that the Air Ministry had considered that they might be carried by aircraft. In May 1941, Hawkers were asked to investigate the possibility of fitting such weapons under each wing of the Hurricane II. On 18th September, Seth-Smith took up a IIB, Z2326, armed with two Vickers Type 'S' 40 mm anti-tank guns. The trials were satisfactory.

Thus, there came into being the Hurricane IID, known throughout the RAF as the 'tank-buster' or more rarely as the 'Hurribuster' or even affectionately as the 'Tin-opener'. Like other Mark IIs, it had a Merlin XX engine but, unlike them, was not a fighter but a pure ground-attack aeroplane. It retained two Browning 0.303-inch machine-guns but these normally fired tracer ammunition as an aid for sighting the two 40 mm cannon, either Vickers Type 'S' or Rolls-Royce 'BF' belt feed – though the former was much preferred, if only because it carried fifteen rounds per gun, compared with twelve on the Rolls-Royce version. Top speed, fully loaded, was 304 mph but this declined to 288 mph when the IID was fitted with a tropical filter, as most were.

Of the RAF squadrons in Britain, only No 184, from December 1942, flew IIDs – but saw no combat with them. The great majority were shipped overseas, the first unit to be so equipped being 6 Squadron, then stationed in Egypt under Wing Commander Porteous, which received its new machines in May 1942. Practices against a captured German tank soon showed that although the recoil of the guns threw the Hurricane's nose down so that fresh aim had to be taken, it was still possible to fire the next salvo almost at once. According to the South African Squadron Leader Weston-Burt, who became No 6's second-in-command in June – later taking over as CO in January 1943 – if the IID fired its first pair of 40 mms at 1,000 yards, 'two more pairs could be got away accurately before breaking off the attack'. 'It is no exaggeration,' he adds, 'to say that any good pilot could guarantee to hit his target with one or more pairs on each attack.'

Certainly, No 6 soon proved this. Its first mission, on 7 June, was abortive but next day it engaged a German column, destroying a number of vehicles including three half-tracked troop-carriers and four of Rommel's vital tanks.

By this time, the enemy was concentrating overwhelming numbers against Bir Hacheim. On the 9th, Hurricanes dropped supply canisters to the defenders. Next day, 73 and 213 broke up raids by Stukas before they could find their targets. But now continuous artillery bombardments were rendering the fortress untenable. That night, the great majority of the garrison broke out, having done much to restore the reputation of the French armed forces.

With his supply route at last safeguarded, Rommel, though still outnumbered by two to one, was again confident of success. He thrust for the sea, threatening to cut off the British 50th and the South African 1st Divisions in the northern part of the Gazala line. As they poured back along the coastal road, their lorries, bunched helplessly together, provided an exceptionally vulnerable target. Yet during the whole of this withdrawal, only six Allied soldiers died as a result of attacks by enemy aeroplanes.

In part, this was due to the pre-occupation of the German bombers with the 'Vigorous' convoy for Malta, described earlier, but it also resulted in no small measure from the extremely effective interceptions made by the Desert Air Force. These sometimes cost the interceptors dear – on 17 June, 73 Squadron lost four Hurricane IICs, two pilots baling out but Squadron Leader Ward and Pilot Officer Woolley being killed – but they did manage to break up even large Axis formations.

On 12 June, the Hurricanes of 33, 73, 213 and 274 Squadrons met huge numbers of Ju 87s and Ju 88s guarded by 109s and MC 202s in the El Adem area. As in the Battle of Britain, there is no doubt that on this and other occasions where large numbers of aircraft were involved, the RAF pilots overestimated the harm they had inflicted. They also apparently made mistakes in aircraft recognition. On the other hand, it will be remembered that in the Battle of Britain there had been instances of enemy casualties being reported through different channels and enemy records undeniably proving incomplete. Bearing this in mind, the very small Axis losses admitted on 12 June do appear somewhat doubtful, particularly since this was potentially an especially dangerous day for the Allied divisions in the north of the Gazala Line and heavy losses of Axis airmen would provide a good reason for their otherwise inexplicable immunity.

Therefore, while making due allowance for exaggerated or duplicated claims, the RAF pilots' beliefs are worth recording if only to show the comparative fortunes of individual squadrons and the extent to which each was involved. 33 was the unlucky one, losing three aircraft and two pilots and claiming only to have damaged five enemy fighters. 213 believed it had shot down two enemy fighters for the loss of two Hurricanes, the pilot of one of which baled out; 274 that it had downed two enemy fighters without loss. While the other units fought the escorts, 73 got among the bombers, claiming five Stukas and one Junkers Ju 88 destroyed and four more Stukas damaged, again without loss.

Of 274's claims, both were made by Sergeant James Dodds. On the 17th, this Scotsman shot down a pair of Macchis, which brought his total of victories in the desert to fourteen. As with many of the finest Hurricane pilots, this figure was by no means exaggerated. On 4 December 1941, for example, he had attacked a Messerschmitt Bf 110, which he last saw diving to earth with both engines on fire. He claimed only a 'probable', but German records show that the 110 (from ZG26) did in fact crash, killing both crewmembers.

It is especially interesting that Dodds was credited with the destruction often of ten 109s or Macchi MC 202s, which had performances superior to that of his Hurricane. However, perhaps too much should not be made of this, for when all tribute has been paid to the pilot's skill, it should be recalled that the Hurricane's 'inferior' abilities had already showed more on paper than in practice in half-a-dozen different skies.

This contention is supported by some observations of the Canadian Wing Commander Keefer, who was a flight lieutenant with 274 during 1942. Though admitting the 109 was 'faster, had a better climb and much better altitude performance', which constantly enabled it to attack with the advantage of height, he states also that

> The old 'Hurri' provided some considerable comfort in its ruggedness and extreme manoeuvrability. I certainly had the feeling that with this ruggedness and manoeuvrability no one could get me as long as I could see him coming.[51]

For the Army, the Hurricanes gave even more comfort by its ground attack duties. Dodds, for instance, gained no more victories after the 17th, because most of 274's future operations were attacks on Axis vehicles with each aircraft carrying two 250 lb bombs. No 6 was making even better use of its IIDs' 40 mm cannon, destroying five tanks, five lorries and an anti-tank gun on the 15th. In the course of this action, Flight Lieutenant Hillier struck a tank, knocking off his machine's tailwheel and the bottom half of the rudder–but it returned to base despite its injuries.[52] On the 18th, the squadron destroyed, appropriately, eighteen enemy vehicles, including several tanks. By the end of

---

[51]Quoted in *Fighters Over the Desert* by Christopher Shores and Hans Ring.
[52]Hillier became the first anti-tank 'ace' with a score of at least nine destroyed. He was killed in an accident on 6 September, when, by a sad irony, he was demonstrating his technique to leading Army officers.

the month, it had claimed a total of twenty-six tanks, thirty-one armoured troop carriers and large numbers of other useful modes of transport.

The most recent activities of the Desert Air Force had been the more admirable in that it had had to retreat from its forward airfields – but this now had a disastrous consequence. Tobruk, which had again been cut off, could no longer be covered by fighters. Delighted by this rare immunity, Kesselring mustered every available Stuka for a crushing onslaught on its defences. Captain Schmidt, who saw this, gives a clear reason for the dive-bombers' effectiveness: 'They flew without interference, for the RAF had been driven off Gambut airfield, and the Luftwaffe had no *Huren-kähne* to harass them.'

Under cover of this bombardment, the Axis soldiers burst through the Tobruk perimeter. At dawn on the 21st, the South African General Klopper surrendered with 32,000 British and Commonwealth troops. Rommel had gained vast quantities of supplies, petrol, vehicles, guns, ammunition – and his field marshal's baton. He can be forgiven for having allowed success to turn his head.

For at this moment of greatest triumph, Rommel made a fatal decision. It had been intended that, once Tobruk had fallen, he would pause on the frontier while his supply lines were guaranteed by the conquest of Malta. Now, forgetting his own warning that 'without Malta the Axis will end by losing control of North Africa', heedless of the protests of that fine airman, Kesselring, the new field marshal won the consent of another great gambler, Hitler, to dash for the Suez Canal, leaving the island fortress still unsubdued.

It was a catastrophic error. Even after the fall of Tobruk, Eighth Army still had many more men, tanks, guns and supplies than Rommel. It did not seem to occur to him that his booty offered false promises; that without spares, his captured vehicles would rapidly become unserviceable; that once his captured ammunition had been fired, his captured guns would be useless.

Nor did he appear to be aware that forces from Malta were again starting to menace his supply lines, which would become more difficult to maintain with every mile he moved into Egypt; nor that Kesselring was having to withdraw the bulk of his units to Sicily to counter Malta s renewed threats, an action that would prevent the Luftwaffe from giving full support or protection in the desert.

In contrast, the British decisions were at least based on reality. Ritchie, on the advice of Gott, rightly determined to retire to Mersa Matruh, thereby aggravating Rommel's supply problem. Then, on 25 June, Auchinleck dismissed Ritchie, taking personal control of Eighth Army. This undoubtedly improved the chaotic command situation – but in fairness to Ritchie this had not been of his making. It should be remembered that he later did very well in the campaign in North-West Europe. Yet the most important factor of all came from outside the Army. 'It was the RAF,' states General Jackson, 'which was the real British fighting force at this time. Its morale was very high as it rose to the occasion of saving Eighth Army.'

The Hurricanes still formed the bulk of the RAF in the Middle East, their strength being increased by Tedder's decision to withdraw numbers of them from Operational Training Units. These provided machines for pilots of squadrons equipped with aircraft less easy to keep serviceable than Hurricanes, who would otherwise have been grounded – such airmen being attached temporarily to 80, 213 and 274. Of course, not every Hurricane combat was satisfactory – 238 lost five aircraft on 28 June, only one pilot baling out – but it was two Hurricane squadrons, 73 with IIBs and IICs and 213 with IICs, that were the highest scorers during this campaign. On the 26th, for example, 213 reversed the usual roles by bouncing a group of 109s, shooting down five without loss. Later that same day, it intercepted a Stuka formation, destroying three, damaging four, again without loss.

At the same time, the strikes by Hurribombers on 'soft-skinned' vehicles were wringing loud protests from the luckless

German soldiers. The 'tank-busters' of No 6 were also making their contributions. While Eighth Army was almost immune from air attack, the momentum of Panzerarmee Afrika was petering out under constant pressure. There can be little doubt that a determined stand at Mersa Matruh would have held Rommel but since Auchinleck had decided on a delaying action only, the defence was far from resolute – by 29 June all Allied forces were retreating to El Alamein.

Here was a situation unique in the desert. Forty miles from the coast was the Qattara Depression, a mass of quicksands through which no armoured vehicles could possibly pass. North of this, 'boxes' had been prepared, particularly around the rail-halt at El Alamein, where the original defences had been ordered by Wavell. Auchinleck had received reinforcements, including the New Zealand Division, which earlier Churchill had urged in vain should be brought from Syria for the Gazala battles. Many of his other units were also fairly fresh: 1st South African Division had been resting in the Alamein defences for about a week. Rommel, in contrast, had been reduced to commanding only fifty-five tanks, plus thirty of the useless Italian ones. He had only 1,500 motorised infantry, most of whom, having been in continuous action for three weeks, were in desperate need of rest. Furthermore, most of his air forces were lagging far behind his advance.

Thus, when Rommel moved upon the Alamein-positions on 1 July, he was gambling on a miracle that did not transpire – nor did he deserve one for he made no attempt at reconnaissance. His Panzer Divisions encountered a wholly unexpected box, which they only captured at the cost of a wasted day and eighteen tanks lost – while '90th Light', running into the defences at Alamein, was routed by the artillery of the South African and 1st Armoured Divisions.

By the 3rd, the Germans had at least managed to get their air force up. Terrific combats took place all day, with the Hurricanes in the thick of the fighting. This time 213, probably worn out by its previous efforts, was the unlucky one, losing five aircraft and three pilots. 73 did much better with a score of three Stukas and three 109s for the loss of one Hurricane, the pilot of which was killed, plus another crash-landed.

However, the day's finest achievement was that of the Hurricanes of 1 SAAF when defending their countrymen at Alamein from a Stuka attack. While 274 kept away some escorting 109s, the South Africans, led by Major Le Mesurier, tore into the Junkers. Hopelessly trapped, the dive-bombers jettisoned their weapons before seeking safety in flight but the Hurricanes pursued them for miles, claiming thirteen destroyed plus one 109. One Hurricane force-landed, the pilot being unhurt. Although some of the Stukas crashed well away from the battle area, Eighth Army was able to confirm the certain destruction of nine from the ground. Air Vice-Marshal Coningham hastened to send a congratulatory telegram.

Next day, 1 SAAF was less successful, downing a 109 but losing two of its own machines, crash-landed – though the pilots escaped with wounds. However, it did prevent the Messerschmitts from interfering with 80 Squadron, which was attacking another hapless Stuka formation. Four of the Junkers fell to the Hurricanes' guns. One of the Hawker fighters was also shot down but Pilot Officer Hill baled out. Two more were badly damaged but returned safely. One of the dive-bombers was shot down by Flight Lieutenant Sowrey who then increased the enemy's losses by colliding with a fifth Stuka. Both machines crashed but Sowrey escaped by parachute.

By evening, with only twenty-six of his tanks intact, Rommel was forced onto the defensive. Auchinleck on the other hand was urging his troops 'to destroy the enemy as far east as possible and not let him get away as a force in being ... Eighth Army will attack and destroy the enemy in his present position'. It must have seemed to the British commander that nothing could possibly prevent this. He had gained further reinforcements, including perhaps the finest of all divisions, 9th Australian. His strength in tanks increased to more than 200 by 10 July, almost 400 by the 20th. That of Rommel varied from under thirty to the stupendous total of about fifty. Eighth Army had a short, easily defended supply line; Rommel's was becoming ever more tenuous – exactly as Kesselring had warned.

The Axis situation in the air was nearly as bad. This was partly because the German airmen had succumbed to an ace complex. They were concerned mainly with building up high personal scores regardless of other considerations. This attitude was seen at its very worst on 16 August, when five 109 pilots claimed a total of seven Hurricanes and five 'Curtiss fighters' – the general name for Kittyhawks or Tomahawks. That the Allies suffered no such losses is not surprising for another pilot reported having seen these men firing their guns into the sand in mock combat[53] – a fact that shows his own honesty but also leaves no doubt as to the way in which purely personal achievement was considered all-important among German fighter squadrons at the time, even though few pilots went to such discreditable extremes.

Quite apart from this attitude, the enemy air forces simply did not have enough machines. To disguise their weakness, they reverted to long-range attacks on Cairo, mainly at night; 73 Squadron, now defending the Canal Zone, claimed to have destroyed or damaged twenty-three raiders during July for the loss of six Hurricanes. In the combat area, shortages were so dire that, even at this stage of the war, the Hurricanes frequently encountered biplane CR 42s trying to attack the

---

[53]It may be instructive to contrast this incident with a later one on 14 October. Three pilots of 1 SAAF, Capt. Viljoen and Lieutenants Smith and Gilson, engaged some 109s, one of which was seen to crash. At the 'de-briefing', each gave the credit for this to his fellows. Finally, they agreed to share the victory between them.

Allied land-forces – a fact affording much satisfaction to 1 SAAF, which shot down five of these on 13 July.

Yet with these immense advantages, Auchinleck failed utterly to repulse his enemy from Egypt. The same old mistakes were made: no co-ordination between different arms; no concentration of the armour, which furthermore was dashed vainly against hostile artillery. On the 22nd and 23rd, the Allies lost a total of 118 tanks, the Germans only three.

As a crowning disaster, co-operation between Army and Air Force reached its lowest level – probably because Auchinleck had moved his HQ into the desert, miles away from that of Coningham. The Hurricanes kept up their fight all through July. There were triumphs like that of 1 SAAF, which, on the 23rd, shot down six Stukas plus a 109, though losing two aircraft destroyed, two badly damaged, one pilot killed and one wounded. There were tragedies like the death, on the 17th, of the 238's popular CO, Squadron Leader Barclay. There were fine feats of airmanship, such as that of Squadron Leader Stewart of 208 who, on the 27th, crash-landed without injury in a Tac R Hurricane, the tail unit of which had been almost shot away by a 109 – an achievement, curiously enough, almost duplicated exactly a month later by Lieutenant Salmon of 1 SAAF, save that his tail-plane had been hit by flak. Yet such actions took place, as it were, in a vacuum, unconnected with the needs of the fighting on land.

It was a grim waste of lives and equipment, but at least it showed clearly that for victory to come the integration of land and air forces was an essential pre-requisite. The lesson was well learned by a new army commander whose proud boast it would be that Eighth Army and Desert Air Force together constituted 'one fighting machine'. 'Therein' he added, 'lies our great strength.'

By the end of July morale was at its lowest, with the commanders of the Dominion divisions protesting, the Air Force loud in its complaints, a 'most intense distrust, almost hatred' between the different arms, and rumours of further withdrawals persisting – quite untrue but readily explicable when defence positions were being erected at Cairo and Alexandria; when strongpoints in the Alamein 'line' were being sighted with a view to the possibility of quick retreats; when potential evacuation routes were being marked out. This was the situation that confronted Winston Churchill, who now reached Egypt accompanied by Wavell, Field-Marshal Smuts and the Chief of the Imperial General Staff, General Sir Alan Brooke.

They did not take long to realise that immediate, drastic changes were needed. Auchinleck was replaced as Commander-in-Chief by General Sir Harold Alexander, who had achieved so much at Dunkirk and at least more than anyone else in Burma. Eighth Army was entrusted to Gott. On 10 August, this splendid officer was flying to Cairo in a Bombay transport. Because of an over-heated engine, it had climbed to 500 feet instead of proceeding at ground level as was usual. It was sighted by two 109s, which shot it down. Gott was killed. A new, controversial figure was ordered out from Britain to take his place.

'Montgomery,' states General J.F.C. Fuller in *The Decisive Battles of the Western World*,

> was a man of dynamic personality and of supreme
> self-confidence. Known to his officers and men as
> 'Monty', he was a past-master in showmanship and
> publicity; audacious in his utterances and cautious
> in his actions. Though at times smiled at by his
> officers, his neo-Napoleonic personal messages …
> electrified his men. He was the right man in the
> right place at the right moment; for after its severe

defeat the Eighth Army needed a new dynamo and Montgomery supplied it.

However, none of these points would have had much relevance had he been unable to win battles, for Rommel was already planning a new offensive at the end of August. For this, Montgomery was well prepared, not, as it has been fashionable to relate, because he took over a wonderful, visionary plan from his predecessors but because his foe's intentions were fully revealed to him by the ability of British Intelligence to decipher the German signals. Yet, as Liddell Hart has pointed out: 'The strength of the two sides was nearer to an even balance than it was either before or later.' From which he concludes, with good reason, that Rommel 'still had a possibility of victory – and might have achieved it if his opponents had faltered or fumbled as they had done on several previous occasions when their advantage had seemed more sure'.

Happily, this time past mistakes were not repeated. By seeing that his plans were widely known, then keeping firm control of the situation, Montgomery ensured a proper co-ordination of the different branches of his command. By reversing the previous mania for dispersal, he was able to concentrate his armour, concentrate his artillery and bring up to the battlefield soldiers who had previously been employed uselessly guarding the cities of Egypt.

He also altered the earlier intentions of fighting a mobile action, which allowed him to scrap the cumbersome box system, strengthen his minefields and locate his tanks in prepared defensive positions, backed by anti-tank guns. This had the additional benefit of helping the RAF to distinguish friend from foe more easily than was possible in a fluid battle.

For the final change made by Montgomery was that he brought relations between the Army and the RAF closer than ever before. They became indeed 'one fighting machine', a fact symbolised by the prompt return of Eighth Army's HQ alongside that of the Desert Air Force.

'Beyond this, however,' reports the *RAF Official History*,

> he [Montgomery] also brought to his post a
> remarkably keen, clear and vigorous appreciation
> of the part that could and should be played by air
> forces in a land battle. Commanders like Auchin-
> leck and Ritchie had never been anything but
> highly co-operative; but Montgomery insisted that
> goodwill was translated at all stages into practical
> action. If air cooperation was the gospel in the
> GOC's caravan, it would also be the gospel all the
> way from base to front line.

The two services, thus newly united in a common purpose, won their spurs when the Battle of Alam Halfa opened on the night of 30 August. Junkers Ju 88s attacked Allied landing-grounds doing much damage among dummies carefully provided for their attention. No 73 was the first Hurricane squadron in action, one of the night raiders being shot down by Squadron Leader Johnston, another by Warrant Officer Joyce – who by the end of the year had destroyed nine enemy aircraft, six at night, all while flying Hurricanes.

Rommel then attempted to wheel south of the British defences as at Gazala – but with rather different results. The Germans got through the minefields with heavy casualties, only to spend three days dashing their tanks against well-sited positions, just as the Allies had done in earlier encounters. By 3 September, the Axis forces were falling back. The danger was over. The enemy advance on Egypt would never be resumed.

Throughout the fighting, the RAF remained in continuous action, by night as by day. There were at this time more Hurricanes in the Middle East than any other aircraft – about 250 of them. Also, about 75 per cent were serviceable, a higher proportion than that of any other type. They therefore naturally formed the backbone of the air support, paying an inevitable price for this. In the four days to 3 September, the Desert

Air Force, in addition to thirteen bombers, lost forty-three fighters, a good proportion to flak. Seventeen of these were Hurricanes but eight of the pilots escaped with their lives. Another seventeen Hurricanes suffered heavy damage but were made battleworthy again by the magnificent British salvage organisation.

In return, the Hurricanes made repeated strikes against enemy transports, especially the vital petrol lorries. Rommel lost about 375 'soft-skinned' vehicles during the battle, a large proportion to the Hurribombers. The 'tank-busting' IIDs of No 6 were also seen to good effect, destroying at least nine of their favourite targets.

On the other hand, the interceptor Hurricanes prevented the Luftwaffe from doing any harm to Eighth Army. On the morning of 31 August, the Hurricanes of 213, which had previously been strafing enemy fighters on the ground, broke up a formation of Stukas, destroying three, forcing the rest to jettison their bombs futilely, losing only one Hurricane whose pilot baled out. A similar success was gained by the Hurricanes of 80 Squadron that evening. In gathering darkness, they were able to claim only two Stukas damaged but again the dive-bombers fled, dropping their weapons anywhere but on their targets.

On 1 September the tale was similar. 213 again dispersed an enemy raid, this time of Ju 88s, shooting down three. Only then did the escorting 109s give battle, with the advantage of height destroying three Hurricanes, two of whose pilots were killed – badly damaging two more. Next day the 109s again showed more concern with personal scores than with their bombers, by engaging the Hurricanes of 127 and 274 only after they had scattered another Stuka formation, which lost six machines while the rest dropped their bombs at random. Nor, this time, did the 109s even achieve any successes. Also on the 2nd, Lance Wade, now a flight lieutenant, shot down a 109 – his twelfth victory and officially 33 Squadron's 200th, – a claim undoubtedly exaggerated but by no means vastly so.

To complete Rommel's ill-fortune, Montgomery refused to pursue. He has been criticised for this, but there can be few doubts that if he had, the German anti-tank gunners would have had a field-day just as in the 'Cauldron'. However, the Hurricanes gave no rest, especially the IIBs of a new Squadron, No 7 SAAF. This had been bounced on 3 September, losing three Hurricanes shot down – though only one pilot died – with another crash-landed. On the 4th, it gained vengeance by a highly effective attack with 'sticky-bombs', which would not glance off the armoured sides of tanks – though so low did the South Africans fly that two of them force-landed, damaged by splinters from their own weapons. On the 6th, however, 7 SAAF was again attacked from above, losing five aircraft and three pilots including the CO, Major Whelehan. Shortly afterwards, it converted to IIDs for further anti-tank operations.

The victory at Alam Halfa was a turning-point. At last, a commander had said what would happen and made good his words in action. The whole of Eighth Army made ready in good heart for the fresh offensive that it knew would have to follow.

During the time of preparation, every attempt was made to deceive the enemy as to the place and date of attack – with what success can be seen from the fact that when it came, Rommel was on sick leave in Austria, having temporarily entrusted his command to General Stumme. This happy result owed much to the RAF's ability to prevent adequate German reconnaissance. In between times the Hurricanes made ground-attacks, an especially heavy one coming on 9 October from 33, 213, 238 and 1 SAAF, all under Wing Commander John Darwen. It was a costly effort, for although some damage was done, eight Hurricanes failed to return, most falling victims to flak. The wing commander was among those brought down but was able to get back to the Allied lines – he would have a sweet revenge later.

The Eighth Army gained its revenge for past misfortunes with the Battle (or Second Battle as it is now fashionable to call it) of El Alamein, which began on the night of 23-24 October. Montgomery had heavy odds on his side, especially in tanks – though less than those enjoyed by Auchinleck in July – but the Germans did have certain assets that helped to balance the scales. Since his flanks rested on impassable obstacles, Rommel was able to bar his front with half-a-million mines in two belts about two miles apart, between which were areas called 'Devil's Gardens' containing every form of booby-trap that ingenuity could devise. Behind this sinister barrier, any force, let alone Panzerarmee Afrika, could take fearful toll of any number of attackers.

What is more, if Rommel could but hold back his opponent, he would achieve a success of incalculable strategic value. Montgomery was very short of time – he had to take the Cyrenaican airfields by mid-November in order to protect a convoy to Malta, again in dire straits. Also, Operation Torch, the planned Allied invasion of Vichy French North Africa, was due to take place on 8 November. It was felt, with some reason, that the reception of the landing forces would depend very much on whether Eighth Army had previously gained a decisive victory.

In support of, almost part of, Eighth Army were, as always, Coningham's airmen. Though they were now, proportionally, less dominant than before, there were still twenty-two squadrons of Hurricanes in the Middle East, of these, fifteen took part in the battle in several widely differing roles.

At 2140 on 23rd October 1942 a tremendous artillery barrage heralded the Eighth Army's greatest effort. Twenty minutes later the infantry moved up while the sappers began their nerve-racking task of clearing the mines. Overhead, the first squadron in action was No 73, its night-fighter Hurricane IICs strafing troops, vehicles, supply dumps. 73 was active at night throughout the battle, destroying an occasional German raider but principally hitting targets of opportunity on the ground, often well behind enemy lines.

Also on the first night, six squadrons of Wellingtons bombed the Axis positions. Then, as the 24th dawned, the medium bombers, the RAF Baltimores, the South African Bostons, took over, while the Hurricane and Kittyhawk fighters or fighter-bombers struck at everything German or Italian.

The Hurricane squadrons usually operated in pairs, the first such in action on the 24th being the IICs of 1 SAAF and 33, which made four joint patrols during the day. Also flying IICs were 213 and 238, the majority of their machines carrying only two cannon in order to increase speed and range. Close support was provided by the Hurribomber IIBs and IICs of 80 and the Hurribomber IIC of 274 in both of which squadrons each aircraft now carried two 250 lb bombs. They were given fighter cover on their missions by the IIBs and IICs of 127 and 335, one of whose pilots, Flight Sergeant Soufrilas, gained the Greek squadron's first victory by shooting down a 109 on 2 November.

However, the most effective ground attack aircraft were the Hurricane IIDs of 6 and 7 SAAF Squadrons. These made their first appearance on the 24th, hitting at least sixteen tanks, Squadron Leader Weston-Burt of No 6 getting three of them. Thereafter they were constantly in action on subsequent days, claiming tanks, lorries, petrol vehicles and half-tracks. 208 and 40 SAAF, in contrast, flew IIBs on reconnaissance missions, to which they added the old-fashioned but important task of 'spotting' fall of shot for the artillery. Finally, to complete the muster of Hurricanes, the rear areas were guarded by 94 Squadron with IICs armed with only two cannon, joined by the IIBs and IICs of 417, a newly formed Canadian unit that had scored its first victory on 26 September.

October 24 was a day of crucial tension for both armies. The bombardment had so shattered the German communications that Stumme decided to go forward in person to check the situation. He ran into Australian troops and was found dead sometime later – which forced Rommel to make a hasty return flight from Austria to resume command. On the Allied side, the tank commanders, alarmed by the extent of the minefields, believed that the offensive should be abandoned to avoid further useless loss of life.[54] Montgomery was woken up in the early hours of the 25th to listen to such pleas. He refused adamantly to alter his plan, making it quite clear that, if necessary, he would appoint new leaders who were prepared to carry out his orders.

This was perhaps the moment that Montgomery really won the battle. During the 25th, his troops began to maul the German infantry. Attempts by the Axis armour to assist brought only a series of heavy tank losses – especially to the anti-tank guns of the Rifle Brigade in its celebrated defence of the position code-named 'Snipe' on the 27th. And by the night of 28/29 October, the Australians were pushing steadily forward in the north, thereby pinning down Rommel's reserves.

Above the grappling armies, their air forces also strove to decide the conflict. The Hurribombers and anti-tank IIDs kept up their pressure. Flying Officer Houle of 213, catching a Stuka formation at dusk on the 26th, shot down two, damaged three others. Next day a strongly escorted dive-bomber raid in support of the German counter-attack was scattered by 33 and 213, though three Hurricanes were lost in the process. Then on the 30th, 33 Squadron, this time accompanied by 238, broke up another body of Stukas.

On 2 November, Montgomery launched his planned breakout, code-named 'Supercharge'. This brought dive-bomber strikes on the British armour, but once more the Hurricanes intervened. 33 and 238 again drove off one group of Ju 87s, which released their bombs on their own troops, while 1 SAAF and 213 utterly routed another, shooting down six Stukas without loss.

---

[54]It is amusing to note that once the battle was safely won, everyone was quite certain that it could not have been lost. At the time several high officers thought otherwise. There were also grave anxieties in London. Furthermore, almost to the last Rommel believed that if he could maintain his resistance just a little longer, his enemy must call off the attack.

The Hurricanes continued in good form next day in the ground attack as well as the fighter role; No 6 making a particularly impressive strike, in which Squadron Leader Weston-Burt alone hit two tanks and four transports. In the early morning, 33 and 238 yet again forced a Stuka formation to bomb its own soldiers, bringing down two of the dive-bombers without loss for good measure. About 1230, an even bigger raid was encountered by another pair of Hurricane squadrons. No 127, which was providing top cover, was jumped by 109s, losing no fewer than six aircraft, but while the enemy escort was thus engaged, the Hurricanes of No 80 ripped into the Stukas without interference – enemy records confirm that they shot down nine for the loss of one Hurricane force-landed.

On 4 November, Eighth Army finally broke the Axis resistance, capturing General von Thoma, the Panzer commander. For the first time in the Desert War, Rommel's troops fell back not in good order but in full flight, while the Hurricanes helped to savage their retreat regardless of casualties – 238 Squadron, for example, lost four aircraft (but only one pilot) during the day. Eighth Army had suffered 13,500 casualties, under 8 per cent of the forces involved, of whom about 4,000 were killed. A hundred guns were wrecked. It has been widely stated that 500 tanks were destroyed but most of these, having suffered only minor damage to tracks or suspension in the minefields, were quickly back in action – total losses were only 150. Axis casualties were more than 50,000 men, including 30,000 prisoners, 10,700 of them German. More than 1,000 guns were total losses, as were 450 tanks, while the Italians abandoned seventy-five more during the retreat. Most of the soldiers who did escape had lost their fighting equipment.

That the Hurricanes had played a large part in the victory is undisputed. Their air battles have already been recounted. Their actions against the enemy army show few such spectacular individual exploits but, ultimately, they resulted in the destruction of thirty-nine tanks, forty-two guns, 212 lorries or troop carriers, twenty-six petrol vehicles, more than 200 other forms of transport and four small ammunition or fuel dumps. Thirty-seven Hurricanes had been shot down, a good proportion by flak. Fourteen others had been damaged. Twenty-five Hurricane pilots had lost their lives.

The pursuit was undoubtedly the least satisfactory part of the battle. It got off to a slow start, mainly because everyone was exhausted after twelve gruelling days. Thereafter, it proved 'a dull and measured affair' – though remembering all those exciting times when the Allies had rushed into the muzzles of anti-tank guns, this may not have been too disadvantageous. It seems also that the ability of the RAF to check the retreat was over-estimated. Certainly, complaints were made by some Army officers, though these appear a little unfair when it is remembered that the Air Force also had not been idle during the fighting.

In any case, no one could have included the Hurricanes in any criticisms. The 'tank-busting' IIDs in particular were still highly effective – as witness the fact that special reports on them were sent to Rommel and Kesselring; the latter even came out to inspect a tank that had received the full force of a number of 40 mm shells.

Even more devastating was the action of Wing Commander John Darwen's 243 Wing, which contained thirty-six Hurricane IICs, from 213 Squadron, led by Squadron Leader Olver, and 238, under Squadron Leader Marples. This flew to Landing Ground 125, about 140 miles behind the enemy lines, on 13 November, its task, code-named Operation 'Chocolate', being to savage the retreating Axis troops from the rear. Until the 16th, when, with its secret base located, the Hurricanes flew back to the airfield at Fuka, they performed their duties to the limit, totally destroying about 130 vehicles together with the wretched troops manning them, crippling another 170 or so. They also strafed hostile aerodromes, destroying fifteen enemy machines on the ground, plus two more in the air.

The price paid was two Hurricanes from 213 brought down by AA fire and six more landed in the desert having got lost on the 14th. All were undamaged. Fuel was flown out to them next day, after which they returned safely to LG 125. Squadron Leader Olver also had cause to be grateful for the Hurricane's reliability. On the 13th, he hit a telegraph pole, damaging his aircraft's tail. Next day it was again damaged, this time by debris from an enemy aircraft, which blew up as Olver was strafing it on the ground. Both times the squadron leader got back unhurt.

Such activities helped to ensure that the advance from Alamein, unlike previous ones, never lost its momentum. Tobruk was recaptured on 13 November; Benghazi on the 20th. With enemy bombers driven from the Cyrenaican 'bulge', a convoy code-named 'Stoneage', numbering four merchantmen, the British *Denbighshire*, the Dutch *Bantam* and the American *Mormacmoon* and *Robin Locksley*, was able to reach Malta from Egypt on the night of 19th/20th, finally raising the siege of the island. El Agheila fell by 17 December. On 23 January 1943, three months to the day since the bombardment at Alamein, Eighth Army crowned its 1,400-mile advance with the capture of Tripoli.

At the same time, the Tac R Hurricanes of 40 Squadron SAAF and the night-fighter Hurricanes of No 73 moved up to occupy the Tripolitanian airfields.

–

Before dawn on 8 November, while the Hurricanes in Egypt were preparing to hasten Rommel's withdrawal, eighteen IICs from 43 Squadron took off from Gibraltar to head for Maison Blanche aerodrome near Algiers. They were led by Wing Commander Pedley, who must have blessed the Hurricane's reliability, having only once previously flown the type, and Squadron Leader Rook who, his men reckoned, did well to get into a Hurricane at all since he was believed to be the tallest pilot in the RAF. It must have been a long, anxious flight for if the airfield proved to be in hostile hands, the Hurricanes would not be able to land – nor, even with their long-range tanks, could they have got back to Gibraltar.

Mercifully, when the Hurricanes reached Maison Blanche at about 0900, the troops holding it were American, so 43's distinction of being the first Allied squadron to land in Algeria was achieved without any problems. That evening, Pedley sighted a Junkers Ju 88. He claimed only to have damaged it but enemy records indicate that it failed to return to its base. Next day in encounters with unescorted bombers, 43 Squadron destroyed or damaged a Heinkel He 111 and two more Ju 88s without loss.

The men who had captured Maison Blanche were only a few of those engaged, under the over-all command of General Eisenhower, in Operation Torch which comprised landings at Casablanca, Oran and Algiers. The first was an all-American affair but Sea Hurricanes from 802 and 883 Squadrons on the escort carrier HMS *Avenger* covered the Algiers landing, while others from 804 and 891 on escort carrier HMS *Dasher* and 800 on escort carrier HMS *Biter* performed a similar role at Oran; – all incidentally wearing the US white star, since it was feared that the similarity between the British and French roundels might lead to misidentification with tragic results.

Resistance at Algiers was sparse, but it was a different story at Oran where the Sea Hurricanes from *Biter* earned especial praise. They escorted Albacores in a bombing attack on La Senia airfield, where twenty-five machines were destroyed in their hangars, while twenty-two more were later found abandoned in the open, though some of these had already been unserviceable prior to the raid. 800's strafing caused some of these casualties and the squadron did further damage in the air when Vichy French Dewoitine D520 fighters appeared. Though it could not prevent these from shooting down three of the slow Albacores – a fourth was claimed by AA fire – it did succeed in destroying five of the hostile fighters without loss.

By the 10th, all landings were secure though eleven Sea Hurricanes had been lost, the great majority due to the inexperience of their pilots. Thus, on the 8th, six members of 804 Squadron could not find their way back to *Dasher* in the prevailing thick early morning haze. One baled out while the others crash-landed on a race-course near the coast. The airmen were unhurt but all the Hurricanes except one were written off.

The success of Torch now decided the Vichy authorities to throw in their lot with the Allies. Unfortunately, they also made no opposition to Axis intervention in Tunisia. Although taken completely by surprise, Hitler acted with his usual speed. By the 9th, the advance guards of his land and air forces were reaching Tunisia, while, two days later, ten German divisions invaded the unoccupied part of France, which they quickly over-ran.

The German presence in Tunisia made an instant Allied advance to the east imperative. On the 11th, troops landed at Bougie, 120 miles along the coast from Algiers, protected by the Sea Hurricanes from *Avenger*.[55] Next morning, the Hurricanes of 43 escorted twenty-seven US Dakotas, which dropped two companies of British paratroopers on the airfield at Bone, a further 175 miles east – just in time to thwart a planned German airborne attack, since Junkers Ju 52s arrived only a few minutes later to witness the maddening sight of the last Allied parachutists landing. That same evening 43 was in action over Bougie, destroying four enemy aircraft without loss. It escorted further transport aircraft to Bone next day. On the 15th, it joined with 253, also flying Hurricane IICs, to guard twenty Dakotas carrying American paratroopers to Youks-les-Bains,

---

[55]This was the last service rendered by the gallant *Avenger*, which had so nobly defended PQ18. Four days later she was sunk by a U-boat, six Sea Hurricanes from 802, six from 883 and three 'spare' aircraft going to the bottom with her, as well as all of her crew but ten, the senior of whom was a Petty Officer.

well inland. Finally, on the 29th, it covered a parachute landing at Depienne, only about twenty miles from Tunis.

Yet it proved impossible for the Allies to advance those last few miles. By the end of November, 15,000 German and 9,000 Italian soldiers had reached Tunisia, many by air but the majority by sea. On 8 December, General von Arnim arrived to take command. A day earlier the weather had broken, turning Allied communications into a sea of mud, but not interfering with Axis shipping, which was soon to give von Arnim five more divisions including 10th Panzer. By the end of December stalemate was complete – it lasted until mid-February 1943.

During the winter months, the Hurricane pilots had cause to be thankful for those wide, strong undercarriages that made their aircraft easier to operate in bad weather conditions than most other fighters. There were now six Hurricane squadrons in North Africa, mainly occupied in protecting convoys, on one of which missions on 15 February, Flight Lieutenant Mason and Sergeant Whittaker of 253 attacked a Cant Z501 flying-boat, which hastily landed on the sea near Algiers, to be captured by a naval patrol vessel. However, 225, equipped with IIBs and IICs, and 241, which originally had IIAs but soon also converted to IIBs and IICs, kept up the Hurricane's reputation for versatility by combining Tac R duties with the use of 250 lb bombs against targets varying from headquarters buildings to transport via gun batteries. On the other hand, they were not immune to bombs either; 225 lost five machines in an enemy raid on Bone airfield during the night of 28 November.

These Hurribombers were soon even more welcome, for, by 15 February, 1943 Rommel's Panzerarmee Afrika had reached the shelter of the Mareth Line on Tunisia's frontier with Tripolitania. On the 19th, the 'Desert Fox' thrust for Kasserine Pass, hoping to break through to Tebessa, the main Allied supply depot in central Tunisia. The Pass was in his hands by the afternoon of the 20th, but by the 22nd he had decided to halt his advance in the face of increasing resistance, which was aided by the fact that on that day bad weather, which had previously hampered the RAF, cleared, allowing 225's Hurribombers to make destructive strikes against Axis vehicles. Three days later the Allies had reoccupied their lost positions but they had suffered such heavy casualties that they could no longer threaten Rommel's line of retreat.

Von Arnim, who had refused to send help during this operation, now, on the 26th, began an attack of his own in northern Tunisia, directed towards another vital communications-centre at Beja. On the 27th, the Hurribombers of 241 destroyed three German tanks, after which for three successive days, 225 and 241 made two or three raids a day with their 250 lb bombs on enemy transport. By 2 March, this offensive also had broken down, much to Rommel's contempt.

However, he was soon to suffer a far worse defeat. During February Eighth Army had been making ready for an assault on the Mareth Line. In an attempt to disrupt his enemy's plans, Rommel attacked at Medenine on 6 March, but warned by intercepted messages, Montgomery had prepared a perfect defensive position. Rommel lost about fifty tanks, mainly to the British artillery, without doing any damage or gaining a yard of ground. It was his last throw in Africa. He returned to Germany three days later, a sick, disillusioned man.

There was one last German advance in the south. On the 10th, General Leclerc's Free French force which had come all the way from French Equatorial Africa across the Sahara to join Eighth Army, was attacked by a strong Panzer group at Ksar Rhilane. The anti-tank Hurricane IIDs of 6 Squadron were called to Leclerc's aid. They duly made three successive strikes, destroying or damaging six tanks, five half-tracks, thirteen armoured vehicles, ten lorries, a gun and a wireless van. Not one Hurricane was lost. The enemy retired – probably the first time that an armoured force had been turned back by air action alone. Messages of congratulation poured in to No 6, from Tedder, from Montgomery, and, presumably with especial fervour, from Leclerc.

The Free French were at Ksar Rhilane to screen an attempt by Montgomery to outflank the Mareth Line, the formidable defences of which had been designed to protect the French from the Italians but now, ironically, protected Germans and Italians from British and French. It was hoped this move would divert the defenders' attentions from the main direct offensive against the Line. This went in on the night of 20 March 1943, while the night-intruder Hurricanes of 73 Squadron, as at Alamein, prowled far ahead looking for targets.

Also as at Alamein, the early break-in proved disappointing. The first troops quickly crossed the Wadi Zigzaou, a considerable obstacle covering the Axis defences, but heavy rain then fell, preventing reinforcements from getting through to consolidate the positions gained. The Germans counter-attacked. By the early hours of the 23rd they had recovered all lost ground.

Montgomery, often blandly described as a cautious, unimaginative general, reacted with superb flexibility, transferring his main weight to his subsidiary outflanking movement – the famous 'left hook'. For this, he enjoyed splendid backing from the Desert Air Force, commanded since February by Air Vice-Marshal Harry Broadhurst. Broadhurst, in turn, had as his spear-head the 'winged tin-openers' of No 6, which had already destroyed nine tanks and eleven other vehicles on the 22nd, and the Hurribombers of 241, which flew south to assist on the 23rd. The culmination of the fighting came on the afternoon of the 26th, when Eighth Army broke through the Tebaga Gap, only four miles wide, packed with enemy guns. This might have proved an impossible task a few months earlier, but not now that co-operation between tanks, artillery, infantry and aircraft had reached its peak. By the next evening the Axis troops were pouring back from the Mareth Line to avoid being cut off, while the IIDs pounced on concentrations of tanks, raising their total of destroyed or damaged during the battle to thirty-two, and 241 dropped its 250 lb bombs on gun positions, lorries or petrol vehicles.

The men of Eighth Army pressed after their retreating foes; forced the defences at Wadi Akarit on 6 April; linked-up with the Allied forces in western Tunisia on the 8th; captured the port of Sfax on the 10th and that of Sousse on the 12th. These victories deprived the Axis of vital harbours, which instead became available to Allied merchantmen as well as light units of the Royal Navy. In addition, the capture of the central Tunisian aerodromes put the whole area remaining under enemy control within easy range of the Allied air forces.

In the face of mounting naval pressure on their seaborne communications and bombing raids on their remaining ports, the Germans made a last desperate effort to supply von Arnim by air. On 15 April, the Hurricanes of 73 shot down two Junkers Ju 52 transports at night but this paled into insignificance compared with the achievements of American units, which reached a culmination on the 18th when four squadrons of Kittyhawks and Warhawks – a version with an improved ceiling – destroyed twenty-four Ju 52s besides causing thirty-five more to crash-land on the coast. However, the Hurricanes were not entirely left out even then, for that evening, 73 attacked the stranded Junkers, finishing off at least five of them.

One final disaster awaited the Luftwaffe commanders. On the 22nd, they sent over more than twenty Messerschmitt Me 323s, huge, six-engine transports, each carrying ten tons of petrol. Intercepted by two South African Kittyhawk squadrons, these were shot down almost to the last machine. That settled the issue. Most of the remaining German aircraft withdrew to Sicily while there was still time, leaving von Arnim almost unsupported to face the final Allied advance.

This was entrusted to General Alexander, who now led all land forces in Tunisia under the overall command of Eisenhower. He planned his major attack in the north but before he could order the transfer of the cream of the Eighth Army's troops, Montgomery voluntarily offered them to him, a gesture showing the mutual understanding that existed between these very different but very fine soldiers. The great offensive was launched in the Medjerda valley on 6 May. The next afternoon, 7th Armoured Division entered Tunis, after which Alexander sealed off the Cape Bon peninsula to prevent a last-ditch stand there.

The doom of the Axis army was now sealed. Admiral Cunningham's sailors wrought a fitting finish to their work in support of the advancing ground troops by preventing any evacuation by sea. A few Junkers Ju 52s tried to get men off by night but the Hurricanes of 73 were ready for them, destroying or damaging four enemy transports after dark on 8 May. Next evening, Sergeant Beard downed two more. On the night of 10/11 May, Warrant Officer Hewitt and Pilot Officer Bretherton completed the list of the Hurricane's victories in Africa by destroying a further pair of Ju 52s.

By this time enemy soldiers were already surrendering in droves. Von Arnim was captured on the 12th. Next day, the Italian commander, Field-Marshal Messe, capitulated with all remaining Axis troops. Estimates of prisoners varied from 150,000 to 250,000. The long conflict in North Africa was over.

As a change from all the tales of devastation, it may be pleasant to conclude by mentioning some examples of the services that the Hurricanes rendered to their pilots. 241 Squadron had reached Maison Blanche from Gibraltar on 30 November 1942. It first saw action at the end of December, after which it was constantly engaged in fighter-bomber raids, interspersed with the odd Tac R mission, until enemy resistance terminated – the ground crews often meeting the Hurricanes at the end of the runway to service, refuel and re-arm them ready for take-off again in eight minutes. In almost five months of this dangerous low-level work, only fifteen Hurricanes were lost and more than half of their pilots survived.

No 6 had been temporarily equipped with IICs, with which it had gained an occasional victory while on convoy patrols, during the winter months. It re-entered the fighting with its 'tank-busting' IIDs at Ksar Rhilane on 10 March 1943. By the end of the Tunisian campaign, it had claimed the destruction of forty-six tanks as well as many other vehicles, but with its long, very low approach run, the IID was a target for every kind of weapon. The squadron lost twenty-five aircraft, twice – on 25 March and 7 April – having six machines fall to flak. Yet during that whole period only four pilots died. The reason for the survival of the other twenty-one can best be illustrated by the experience of Warrant Officer Mercer on 24 March. In an attack on enemy tanks, his starboard wing was so badly damaged that he had to crash-land in open country at 200 mph. He walked away from the wreck, unhurt.

It might be a terror to its foes but to its pilots the Hurricane was a staunch protector, which could shield them from all but the very worst of dangers.

Chapter Twelve

Preparations for Invasions

It was the confidence that the Hurricane thus engendered in its pilots, which led to its selection for a number of experiments, in the conviction that with this aeroplane there was the least chance of fatal consequences. Many of these trials brought about spectacular results, such as the adaptation of the Hurricane to be catapulted from merchantmen, or fly from carriers, or carry bombs or anti-tank cannon, but even those that did not come to fruition at least often provided useful lessons for future reference.

One such unusual development was entrusted to the firm of Hill & Sons Ltd, from whom it took the name of the Hillson FH40 'slipwing' Hurricane. In order to provide extra lift for take-offs from very small airfields, an old Mark I, L1884, which had gone to Canada before the war, subsequently returning to Britain re-registered as No 321, was converted into a biplane by the addition of a top wing of similar span. Once airborne, it was intended that the extra wing, plus the interplane struts which held it in position, would be jettisoned, thereby restoring the inherent advantages of the monoplane. The biplane Hurricane made several flights but the top wing was never jettisoned before the scheme was abandoned in 1943. Previously in 1941, a similar desire for short take-offs at overload weights had led to trials at Farnborough to launch a Hurricane by means of rockets – but again this variant did not see active service.

Even more remarkable was the idea, put forward in January 1941, of mounting a Hurricane on top of a Liberator bomber, which could thus carry its own fighter protection to areas normally outside the range of interceptors. About the same time, four Hurricane IICs were allocated to Flight Refuelling Ltd at Staverton, Gloucestershire, where they made several journeys towed behind Wellingtons by a cable connected to hooks in the outer wing sections – this being another possible means of providing long-range escorts as well as a prospective way of delivering fighters direct from Britain to Malta. Unfortunately, apart from showing that the towing cable snapped if the Hurricane got out of position, these trials quickly demonstrated that the long flight caused such severe icing to the Hawker fighter's engine, that it would refuse to start when the aircraft was released ready for combat. The scheme therefore was abandoned, as was the composite Liberator plan in anticipation of similar problems.

Another strange experiment, code-named 'Sunflower Seed', saw a six-inch tube fitted behind a Hurricane's cockpit with the open ends protruding above and below the fuselage. It was intended that a rocket should be fired upwards through this at enemy aircraft, which were adopting the tactic of dropping bombs on Allied bomber formations from directly overhead. The first such trial caused some damage to the fuselage but after the area near the mouth of the tube had been reinforced by duralumin sheeting, later rockets were launched without harm, other than to the nerves of onlookers. It seems that the idea was never used in action – but as will be seen, these were far from the only occasions on which the Hurricane fired rockets.

Plans were also afoot for employing new powerplants in the Hurricane. By the end of 1941, a new version was proposed with a Merlin 28 engine, which was then being produced in growing numbers in the United States by the Packard Motor Corporation. This was designated the Mark III but, although arrangements were made to introduce it into production lines in British factories if the need arose, in practice, no Mark IIIs ever appeared since the supply of Rolls-Royce Merlins at no time fell below the required level. However, the Packard-Merlins were used in most of the 1,451 Hurricanes that were built by the Canadian Car and Foundry Corporation.

Although the original Mark Is assembled in Canada were fitted with engines shipped from Britain, later machines from the end of 1940 onwards, were powered by those built in America. The first such were known as Mark Xs – Marks I-IX being reserved for future British developments. The Hurricane Xs had a 1,300 hp Merlin 28 engine driving a Hamilton three-blade propeller (without a spinner), giving a performance approximately equal to that of the Rolls-Royce Merlin XX. A good number of these remained in Canada but most were shipped to Britain where their initial armament of eight machine-guns was increased in the majority of cases to one of twelve Brownings, or four Oerlikon or Hispano cannon.

They were followed by the Mark XIs also with Packard-built Merlin 28s, which were very similar to the Mark Xs save that they had increased fire-power, equivalent to that of the British IIBs or IICs, from the outset, as well as certain minor adaptations for use by the Royal Canadian Air Force. Naturally therefore, the majority remained in Canada, where they equipped five squadrons, though a few went to Britain or Russia.

In November 1941 the Hurricane XIIs appeared, powered by a different engine, a 1,300 hp Packard-Merlin 29, and armed with twelve machine-guns, though some were later modified to carry four cannon. They were widely distributed to Britain, Russia or the Far East, while at least one, No 5624, was fitted with a ski-undercarriage and tail-skid for use on snow-covered airfields in Canada.

The final Canadian version was the XIIA, which was an eight machine-gun edition of the Mark XII with the same engine, a Packard-Merlin 29. A number of these arrived in Britain but most were sent to Russia or the Far East during 1943, some again being modified to carry heavier armament.

By mid-1941 the Canadian Hurricanes were reaching Britain at the rate of ten per week, to supplement those being produced by the British factories at an average of twelve a day. Since their performances were virtually indistinguishable from those of the RAF IIAs, IIBs or IICs, depending on the type of armament carried, they were soon serving alongside them in all manner of different operations.

One such was 'Channel Stop', a campaign designed to halt German coastal shipping, thereby throwing ever-greater burdens on the Continental road or rail systems. At first, the Hurricanes flew as escorts to Blenheims but they were soon used to engage the flak-ships, which guarded the convoys in increasing numbers, so as to divert their attention from the bombers; after which it was not long before the Hurricanes, particularly the IICs with their four powerful 20 mm cannon, were sent to attack the enemy vessels on their own.

The first squadron to excel at this hazardous work was 615, commanded by Squadron Leader Gillam, ably seconded by the South African ace Flight Lieutenant Hugo. On 17 September 1941, 615 escorted Blenheims against a convoy defended not only by flak but by Messerschmitt Bf 109s. The RAF lost two bombers and two Hurricanes, the pilots of both of which were killed. The Germans lost one freighter, two flak-ships, two 109s. Thereafter, 615 formed its own striking-force, attacking two minesweepers on 19 September, sinking a third on the 26th. By 27 November, when it turned to other duties, this squadron, though at heavy cost, had been credited with having sunk or damaged between twenty and thirty ships. It had also claimed to have shot down four other 109s (all on 15 October), a Junkers Ju 88 and two Heinkel He 59s, as well as destroying several more seaplanes at their anchorages.

Gillam had little doubt that much of the credit was due to 615's aircraft. He regarded the Hurricane as 'the finest gun platform of them all.' He added:

It also took a staggering amount of punishment and still managed to get home. I have seen pilots bring them back with most of the fin and rudder missing; with a hole in the wing where a Bofors shell had penetrated and taken out the complete ammunition box; indeed any pilot in a Hurricane squadron will recall the extraordinary amount of damage this fighter could absorb and still keep flying.[56]

In October 1941 the Hurricane IIBs of 402[57] and 607 Squadrons also joined in Channel Stop, their weapons being two 250 lb bombs each, which, though not especially accurate against shipping, could do immense damage when they did strike home. On 27 November, for instance, 607 sank a merchantman and two escorting vessels in an attack on a convoy. However, later that day, it lost three aircraft together with their pilots in a raid on ships in Boulogne harbour.

The most perilous missions flown by Hurricanes against targets at sea were those of 12 February 1942, for the enemy force consisted not of freighters but of two battle-cruisers, one heavy cruiser, six destroyers, thirty-four E-boats and numerous flak-ships. Hitler, obsessed with a belief that the Allies were planning an invasion of Norway, had determined to send thither *Scharnhorst, Gneisenau* and *Prinz Eugen*, which for many months had been lurking in Brest, where raids by Bomber and Coastal Commands had caused damage that the Führer feared would become fatal as the scale of attacks mounted. With his usual decisiveness, he brushed aside the protests of his naval advisers, insisting that the warships be brought through the Channel

---

[56]Quoted in *Fight for the Sky* by Douglas Bader.
[57]Formerly No 2 Squadron Royal Canadian Air Force. To avoid confusion, 1 and 2 Canadian Squadrons were re-numbered 401 and 402 respectively on 1 March 1941.

where they could be covered by a massive fighter escort. To achieve surprise, they would leave Brest at night. This meant passing through the Straits of Dover in daylight, but Hitler was confident that his enemies would be too amazed by the boldness of his action to respond with the necessary speed.

This forecast proved distressingly correct. A series of faults in radar sets, both in Coastal Command's reconnaissance Hudsons and ashore – it is worthy of note that these were far from infallible well over a year after the Battle of Britain – meant that the German ships were only reported at 1110, after being sighted accidentally by two Spitfires on a fighter sweep. It was not until about 1245, by which time the enemy vessels had already passed the Straits, that the first attack on them was made by six Swordfish from 825 Squadron. This venture was doomed from its inception. All the old biplanes were lost. 825's leader, Lieutenant-Commander Eugene Esmonde, was awarded a posthumous VC, the Fleet Air Arm's first.

Low cloud accompanied by drizzling rain helped to ensure the failure of later efforts, including those by Hurricanes. 607 Squadron flew two separate missions against the enemy's light forces; it lost four aircraft, all the pilots being killed, although one of them, Flight Sergeant Walker, had previously scored a direct hit on a flak-ship with a 250 lb bomb. No 3, flying Hurricane IICs, attacked E-boats, while No 1, also with IICs, raked the superstructure of three destroyers with cannon-shells – though the damage done could not have been severe – losing two of its machines to AA fire. Sergeant Blair, a South African, went down in flames, but Pilot Officer Marcinkus, who as a Lithuanian was of unusual nationality even for a Hurricane pilot, ditched in the Channel. He was taken prisoner, only to be murdered later along with forty-nine other officers, following the 'Great Escape' from the Stalag Luft III prison camp.

There is no doubt that the safe passage of his ships was a personal triumph for Adolf Hitler, who seems to have been the only person who had believed it possible. In Britain, naturally, humiliation was intense. Yet the story did not end there. It was stated at the time that the battle-cruisers had succeeded where the Armada had failed. The comparison was meaningless for the Spaniards had intended to clear the way for an invasion whereas the Germans were intent only on escape – but there was a strange parallel, which only emerged later. The Armada traversed the Channel with minor losses, but met with calamity thereafter. The same was true of the warships that made the Channel Dash in 1942. Both *Scharnhorst* and *Gneisenau* were heavily damaged by mines in the North Sea. Only *Prinz Eugen* reached Norway to guard against a non-existent invasion – just in time to be crippled by a submarine's torpedo. *Scharnhorst* also was immobilised for months – she achieved nothing thereafter before her destruction by *Duke of York*, mentioned earlier. On the night of 26/27 February, Bomber Command hit *Gneisenau* while under repair in dry-dock. She never sailed again.

Thus, the affair brought about no serious consequences; nor would anything similar re-occur. The Hurribomber squadrons reverted to their normal Channel Stop activities. Indeed, shortly afterwards, the narrowest part of the Straits of Dover as far west as Beachy Head was reserved as a hunting-ground for them alone. For the rest of the year, they continued to inflict losses on enemy shipping, which in turn forced the Germans to use increased numbers of escorts, which were badly needed elsewhere. Thus, on 15 May 1942, 175 Squadron, newly formed with fighter-bomber Hurricane IIBs, attacked three minesweepers, sinking two outright and damaging the third so badly that it went down later; while the similarly-equipped 174 Squadron, which had taken over a high proportion of 607's pilots, so damaged a 4,000-ton freighter on 29th June that it had to be beached.

It will be noticed that new Hurricane squadrons were being created specifically for fighter-bomber duties, while even veteran units were now dealing with the enemy on land or sea rather than in the air. In fact, the Hurricane was fast disappearing as a fighter in north-western Europe. Its speed had reached its limit. Its manoeuvrability, certainly, was as great as ever but this was less valuable for offensive operations into enemy airspace, if only because it would have to stop manoeuvring eventually in order to fly home.

Thus, though by the autumn of 1941, no fewer than fifty-seven squadrons in the United Kingdom flew Hurricane IIs of various sorts, during the last months of the year or early in 1942, large numbers of them began to convert to other types. Most of the day-fighter units received Spitfires but several of those that had been employed on night-operations changed to Beaufighters; 56 became the first squadron to fly the new Hawker Typhoons; 601 the first to be equipped with American Bell Airacobras.

Yet the Hurricanes were still engaged in the destruction of enemy bombers over their own territory – but only after dark when the RAF pilots could better elude their opponents. These night-intruder missions had been started by the Hurricanes of 87 back in March 1941. They were continued by several different squadrons throughout the year, well into 1942, one of the most able pilots at the task being Flight Lieutenant Shaw of No 3, who by 1 May 1942 was credited with the destruction of five enemy aircraft on such missions.

However, the squadron that led the way in these sorties during 1942, was once more No 1, equipped with Hurricane IICs fitted with two 45-gallon drop-tanks. Its CO now was Squadron Leader James Maclachlan, who had flown with 261 in Malta, being one of the survivors of the ill-fated Operation White reinforcement mission. While on the island he had gained two victories before, on 2 February 1941, he was shot down by a 109, baling out with his left arm so shattered by a cannon shell that it had to be amputated above the elbow. Fitted with an artificial limb, he returned to operations, serving with 73 Squadron in North Africa for a time, before returning to Britain to command No 1.

Under his inspirational leadership, No 1 began its intruder sorties on 1 April 1942, continuing until 2 July, when it moved north prior to converting to Typhoons. During this period, four Hurricanes were lost, one to flak, three by accidents, only one pilot surviving – but apart from the damage done to enemy warplanes, the squadron wrecked seventeen railway engines, of which Flight Sergeant Bland destroyed five on the moonlit night of 1 May.

'Train-busting', however, was in the nature of a bonus of No 1's primary objective of savaging the Luftwaffe. One gratifying consequence of its flights was that much of the German night-fighter effort at this time was diverted to deal with them, rather than with the heavy bomber raids now being mounted. On the night of 30 May, No 1 made a number of sorties designed to keep the Luftwaffe busy on the occasion of the first 'thousand bomber raid' on Cologne. Although only one hostile aircraft was encountered, by Warrant Officer Scott who could do no more than damage it, it seems that the squadron succeeded in its diversionary duties to judge from the grateful signals it received, from the Secretary of State for Air, Sir Archibald Sinclair, as well as from Bomber Command's redoubtable chief, Air Marshal Harris.[58]

There were other missions in which the Hurricanes of 1 Squadron attained more visible results. By 2 July they had shot down twenty-two enemy aeroplanes at night, damaging a further thirteen. Maclachlan, flying Hurricane IIC, BD983, had set a fine example, earning a DSO for his leadership, in addition to gaining five of the confirmed 'kills', but the outstanding night intruder pilot was the Czech Flight Lieutenant Karel Kuttel-wascher, who, at the controls of another IIC, BE581, achieved no less than fifteen of No 1's twenty-two victories, including one during the first night, and two during the last night on which the squadron operated. His greatest success came on 4

---

[58] No 3 Squadron was also intruding on this night with its Hurricane IICs but sighted no targets at all.

May, when he caught six Heinkel He 111s orbiting the airfield at St Andre. Joining the circle, he shot down three in flames within four minutes, before the lights on the ground went out, while AA guns began firing wildly. Kuttelwascher retired safely, to be rewarded with a DFC soon afterwards.

It may also be of interest to record that, on 26 June, Pilot Officer Perrin, finding no targets over Holland, flew east as far as Düsseldorf in search of prey. His quest was unsatisfied but his fellow-pilots found it very encouraging that a member of the squadron had penetrated German air-space for the first time since the campaign in France. It seemed an omen, pointing the way to the future re-occupation of France, followed by the final conquest of Germany.

The path for this was already being prepared by the fighter-bomber Hurricanes, which as well as their attacks on shipping previously described, found plenty of use for their 250 lb bombs in missions against occupied Europe. The first such was flown by Squadron Leader Craig and Sergeant Lees of 607 on 30 October 1941, against a transformer station at Tingrey, which was duly damaged as was a bridge at Samur. Next day the squadron bombed another transformer station at Holque, together with barges on a nearby canal. On 1 November, eight Hurribombers from 402 joined in, hitting the airfield at Berck-sur-Mer, a railway bridge near Etaples, railway wagons and gun posts.

Thereafter attacks continued, with new Hurribomber squadrons adding to the weight of the raids as they were formed – for instance, 174's first mission was another strike against Berck-sur-Mer's aerodrome on 28 March 1942, while 175's first was against Maupertus airfield on 16 April. The aircraft chiefly used were Hurricane IIBs, which were ideal for these tasks since they could be manoeuvred violently when flying at very low level, thus affording the anti-aircraft defences very little time in which to bring their guns to bear. The fighter-bombers normally each carried two 250 lb bombs fitted with delayed-action fuses so that they would not blow each other up, but on occasions they were equipped instead with a number of small fragmentation bombs for use against enemy personnel or road vehicles. They could also fly with two Smoke Curtain Installations to screen amphibious operations.

Finally, the Hurricanes – not only IIBs but IICs as well – came to carry two 500 lb bombs, one under each wing, which gave them the same hitting power as the Blenheim or Boston light bombers. These various external stores naturally caused a reduction in the Hurricanes' performance, which became more pronounced as the weight of their loads increased – a IIC with its four heavy cannon and two 500 lb bombs had a speed of only 180 mph. Usually therefore, the fighter-bombers were protected by Spitfires, though sometimes the escorts suffered heavier losses than their charges, as, for example, on 23 November 1941, when a raid in the Calais area saw the destruction of two Hurribombers, the pilot of one of which was rescued by a naval launch, but the loss of five Spitfires, all of whose pilots were killed.

These Hurribomber sorties continued throughout 1942, well into 1943. They covered the whole of the Low Countries, plus a generous area of France as far south as Rheims. Targets attacked successfully varied widely: they included aerodromes, road bridges, rail bridges, road vehicles, rail transport, railway stations, factories, depots, canal barges, transformer stations, troop concentrations, gun positions and distilleries.

The culmination of these flights came on 19 August 1942, in Operation Jubilee, the raid on Dieppe, organised by Vice-Admiral Lord Louis Mountbatten, Chief of Combined Operations, but carried out mainly by Canadian forces. At this stage of the war, there was no question of achieving a permanent seizure of any part of the fortified coastline of northern France. On the contrary, it was intended that the troops involved would be evacuated after doing as much damage as was possible during a single day – indeed, of a single tide.

Unhappily, there were many reasons, no doubt easier to see in retrospect than at the time, that made Dieppe a dubious choice for such an assault. It was heavily defended on both sides of the harbour, with high cliffs dominating the narrow, rocky beaches, which were, in any case, poorly suited for landing craft. A massive sea-wall blocked any advance from the coast. Furthermore, the plan of attack was far too complicated and far too inflexible. Finally, fatally, the weight of firepower from the supporting naval forces was nothing like sufficient.

In consequence, though Commando raids on the flanks were more successful, the main landing was pinned down on the beaches. Hardly any of the limited objectives were gained. Nor was it easy to bring off the survivors under a savage cross-fire. The Canadians lost well over 3,000 men, of whom almost 2,000 were taken prisoner. The air-fighting was not entirely satis-factory either. Air Vice-Marshal Leigh-Mallory's No 11 Group prevented the Luftwaffe from hampering either the army or the navy but at the cost of almost twice as many losses, including more than 60 Spitfires – a result that must be considered disap-pointing, at the least.

Yet in its main task, as a preparation for later, greater events, the raid was invaluable. The very fact that casualties were high ensured that its lessons would be learned. From the bloodstained beaches at Dieppe would arise the tremendous amphibious triumphs in Sicily, Italy and, finally, Normandy.

One aspect of such operations – close air support – had already been demonstrated to good effect. The first aeroplanes over Dieppe were the Boston and Blenheim light bombers, striking at the German batteries or laying smoke-screens, but the first fighter to cross the hostile coast was a Hurricane, flown by Squadron Leader Du Vivier, 43's Belgian CO. The rest of the 'Fighting Cocks' were right behind him,[59] followed shortly afterwards by five other Hurricane squadrons, 3, 32, 87, 245 and 253. All of these were equipped with IICs, though most had some IIBs on strength as well, while No 3 at least, also flew some Canadian imports, Mark Xs modified to carry four cannon. 174 and 175 Squadrons, with their fighter-bomber IIBs, also took part in the raid. Their machines were armed initially with two 500 lb bombs each, though as the day wore on, supplies of these ran out, forcing the Hurricanes to revert to their usual 250 lb ones.

No 43 Squadron had taken off at 0425, reaching the beach-head at first light fifteen minutes later. Thereafter the Hurricanes gave all possible aid to the Canadians as they struggled to advance. Later they provided covering fire to protect the evacuation, which had been completed by about 1330. The naval forces had also to be guarded as they withdrew across the Channel; 32 and 245 taking off for the last Hurricane patrols at about 2010.

Including these defensive tasks, all the fighter Hurricane squadrons flew four missions during the day except 87, whose total was three. The two Hurribomber units made three raids each. Going in so low that Flight Sergeant Brooks of 174 returned with the aerial of a German tank wedged in his radiator, the Hurricanes used bombs, cannon or machine-guns against gun batteries, troop positions, wireless stations, railway sidings, cranes, tanks, lorries and buildings of various kinds; Pilot Officer Stevenson of 175 somewhat unchivalrously bombed a church, but since gunfire was pouring from it there is no doubt that it was a legitimate target. In addition, Flight Sergeant Meredith, also of 175, en route to a successful attack

---

[59]The day's activities marked the end of 43's distinguished career in north-western Europe. Three months later, it moved to North Africa, where, as has been seen, it added to its laurels. It converted to Spitfires in March 1943.

on a gun position, shot down a Heinkel He 111 which flew across his path.

All the Canadian accounts agree that the assistance afforded by the Hurricanes was extremely effective, lamenting only that it could not be continuous. Clearly also, much damage was done to the enemy; aside from the reports of returning Hurricane pilots, the escorting Spitfires confirmed numerous instances of bombs bursting right on the target, of flames springing up, or of the volume of gunfire from positions that had been strafed decreasing drastically. There is no doubt that the Hurricanes' activities gave every justification for pride, being properly acknowledged by the award of fifteen DFCs, and one DFM.

Inevitably, however, the Hurricanes paid for their triumph. Twenty of them were lost. Although five pilots were rescued after various perils – the most fortunate surely being Pilot Officer Baker of 87, whose parachute landed him neatly on the foredeck of a friendly vessel – eleven were killed, while four more were taken prisoner. 174 had the highest loss rate, five of its IIBs failing to return. Two of the pilots survived only to be captured but Sergeant James and Pilot Officer Du Fretay were killed, the latter's Hurricane being seen to explode among some enemy armoured vehicles; while Squadron Leader Fayolle, who like Du Fretay was a Free Frenchman, disappeared over the Channel, never to be seen again. It had been his first sortie as 174's CO.

It was long suspected that Fayolle had fallen victim to a hostile fighter but only recently has it been established that he had in fact collided – perhaps deliberately – with a Focke-Wulf Fw 190, which also came down. No 3's New Zealand leader, Squadron Leader Berry, was always known to have been shot down by a Focke-wulf on his fourth mission of the day, but although the 190s attacked several other Hurricanes, these used their manoeuvrability to such purpose that none suffered more than minor damage. Indeed, Pilot Officer Peters of 175 damaged two Focke-Wulfs, one badly enough to be reckoned as a 'probable'. Of the other Hurricanes brought down, seventeen fell to flak, the remaining one being lost when Flight Lieutenant Connolly of 32 crashed in flames after a collision with his Number 2, Sergeant Stanage. The sergeant was able to get back to base, with three feet of his Hurricane's port wing missing.

Many another pilot returned thanks to the toughness of his aircraft. In fact, there were few Hurricanes that did not have at least a number of holes as mementoes of their sorties. Squadron Leader Mould of 245 crash-landed on the beach at Littlehampton, churning through shore defences but climbing out unhurt. Flight Sergeant Tate, an American member of 253, crash-landed at base minus his starboard flap. Flying Officer Shaw, also of 253, came down safely although his rudder controls had been shot away. But the mission that can best stand as a portrait in miniature of the entire ground-attack exploits of the Hurricane against occupied Europe was the first of the day by No 43.

Twelve Hurricanes, six IIBs and six IICs, took part in this raid. They 'went in', relates the squadron history,[60] 'in line abreast against gun positions on the beaches and in the buildings to the west of the harbour entrance'. The pilots reported numerous strikes on these targets but came under intense, very accurate fire from every sort of AA weapon, no fewer than seven of the fighters being hit. Two of the cannon-firing IICs were shot down. Flight Sergeant Wik from Canada was seen to crash into a field, reported 'missing' and thought to have been killed, but he somehow survived though as a prisoner of war. Pilot Officer Snell baled out, to be rescued by a tank landing craft, which 'set him to earn his keep for the rest of the day by operating the machine-gun on the port side of the bridge'.

---

[60] *43 Squadron; Royal Flying Corps; Royal Air Force: The History of the Fighting Cocks 1916-1966* by J. Beedle – from which all subsequent quotations are taken.

Of the other five damaged aircraft, two suffered only minor injuries to wing or fuselage but the remaining three were seriously mauled. As the squadron history puts it:

> Flying Officer Turkington made a most creditable touch-down with the elevators partly missing and wholly shredded, and Pilot Officer Trenchard-Smith of Australia earned for himself the enduring name of 'Tailless Ted' after returning to the owners a Hurricane on which the top half of the fin and the rudder had vanished and the remnants of the latter hung to the rudder post only by the bottom hinge and the control cables.

Both these aircraft, Z3687 and Z5153, were IIBs, so it emphasises the reliability of Hurricanes in general, to state that 'the supreme exposition of all' was given by BN 234, a IIC flown by Flight Lieutenant 'Freddy' Lister. This machine suffered damage to the port wing detailed as follows:

> The panels above the cannons (were) blown off; a four foot square hole in the trailing edge took in the aileron up to its inboard hinge, and the outermost cannon, wrenched from its by no means slender mounting, was askew, damaging the front spar and buckling the leading edge. But the Hurricane kept flying.

In fact, it flew all the way back to 43's airfield at Tangmere, where, since experiments on the return trip had shown Lister that only at high speed could he maintain level flight, while a glance at the shattered wing convinced him that it would be most unwise to lower the flaps, he drove his fighter onto the ground with the airspeed indicator showing 210 mph. It seems amazing that it never occurred to him that he might bale out. Presumably it was unthinkable that he should abandon such a valuable aircraft, or perhaps, like most Hurricane pilots, he had supreme confidence in the virtues of his 'mount'.

If so it was not misplaced.

> Most aeroplanes of that period, hitting the ground at such speeds, would simply have disintegrated; others, dug in the nose, flipped over on to their backs and crushed the skulls of their unfortunate occupants. The Hurricane, possessed of no such venomous streak, just kept going straight and level, ripping open the green turf beside the runway and spewing out the brown clods to either side until, with the radiator wrenched away, it slid to a halt close to the boundary fence and the sunken road beyond. The air intake had gone, the propeller blades were all sheared at the roots, the spinner was stove in, the bottom cowlings torn. The wheel bay and cannon muzzles were packed with hard driven earth, and the port wing was a scarcely cohesive jumble of twisted metal and torn skin. But from out of it stepped, quite unhurt though a little stunned by his good fortune, one very valuable flight commander.

Lister flew three more sorties that day, receiving the award of the DFC. 'No bad effort after that kind of landing', concludes the squadron history, 'and no slight recommendation for the Hurricane either.'

–

One of the lessons learned from the Dieppe operation was that neither the cannon nor the bombs of a Hurricane were adequate for smashing the concrete emplacements protecting heavy guns. It must be admitted that Sydney Camm had not designed his fighter for this purpose, but the problem was soon to be solved by yet another Hurricane adaptation.

It was concluded that the ideal weapon for the demolition of strong fortifications was the air-launched rocket. Thereupon, the Hurricane took to carrying these, being the first Allied fighter so to do. In fact, the original proposal for this installation had already been made back in October 1941, though it was not until February 1942 that a Hurricane IIA, Z2415, first flew with six rockets, three under each wing; a steel plate being fitted to guard against blast or premature explosion.

From the start, the missiles were somewhat crude, consisting merely of an iron pipe, three-and-a-quarter in diameter, containing the solid fuel, to the front of which was screwed the warhead, either a 25 lb solid armour-piercing shot or a 60 lb high explosive shell, while on the back were bolted the tail fins. Crude, not always accurate, but lethal – especially after the number of rockets carried by a Hurricane had been increased to eight.

They were fired electrically from the cockpit, the launching causing no recoil as the anti-tank cannon had done, though the aircraft was often buffeted as it passed through the displaced air. The pilot had the option of using them in pairs, one from each side, or of sending all eight away in one salvo. The effect of this latter can be judged from the fact that it was considered to be the equivalent of a full broadside from a light cruiser.

While the rockets were used to a limited extent on IIBs or IICs, it was felt that with the range of external stores now being fitted to Hurricanes, it would be advantageous to have a version that could carry any of these. This new Mark, originally known as the IIE, was intended to be employed primarily for close support duties, so had only two Browning machine-guns for aiming purposes. It would be supplemental to, not a replacement for, the twelve gun or four cannon Hurricanes – indeed, many squadrons would operate with a mixture of all three types on their strength.

If its effectiveness against enemy aeroplanes was low, the latest Hurricane could do immense damage to targets on land or sea, for it could be equipped not only with bombs, 40 mm anti-tank guns or Smoke Curtain Installations, but also with the new rockets. Indeed, flexibility proved the key virtue of the type. Because of its intended use in low-level strikes, 350 lb of extra armour plate was provided to protect the front fuselage and radiator, which latter was deepened to give improved ceiling in the tropics.

Finally, it was decided to install a different engine, whereupon the designation IIE was dropped, the aircraft henceforth being called the Hurricane Mark IV – the Mark III, it will be remembered, was the proposed Packard-Merlin version, which had not materialised. The prototype Mark IV, KX405, which was first flown by Philip Lucas at Langley on 14 March 1943, had a Merlin 32 engine with a Rotol four-blade propeller, but subsequent production machines – the first of which, KZ193, was flown by Lucas on 23 March – were powered by a 1,620 bhp Merlin 27 – or, in the case of some later models, a Merlin 24 rated to give the same power – driving a Rotol three-blade constant-speed propeller. Without encumbrances the Hurricane IV had a speed of about 330 mph, a service ceiling of 32,500 feet and a rate of climb much the same as that of a IIC, but naturally when carrying external stores its performance fell away sharply, the speed being reduced to about 200 mph.

Plans were also made for the production of a variant of the Mark IV – in due course to be known, obviously enough, as the Mark V – which would be powered by a ground-boosted Merlin 32 rated to give 1,700 bhp at low level, driving a Rotol four-blade propeller. KZ193, duly altered to these specifications, made its initial flight as a Mark V on 3 April 1943 with Lucas at the controls. It was soon joined by KX405, similarly converted, and then by NL255, the first aircraft built by Hawkers from the start as a Hurricane Mark V.

These Mark Vs, though fitted with a tropical filter and two anti-tank guns, could boast a maximum speed of 326 mph at sea level: nearly as fast as the Hurricane Mark I had been though the aircraft's weight had increased by almost 60 per cent. Sadly, though, a lengthy series of tests showed that the life of the ground-boosted engine was unacceptably short, so the scheme was abandoned and Hawkers converted or re-converted all three Mark Vs to standard Hurricane IVs.

Mark IVs were certainly needed. 524 were issued to the Royal Air Force, seeing service throughout the world. Of the home-based Mark IV units, No 184 made the first strike using rockets, this being against shipping at Flushing on 17 June 1943. Thereafter, it carried out a series of such missions, an especially useful one coming on 29 July, when, for the loss of one aircraft with its pilot, it sank two enemy vessels out of a convoy of five, forced another to beach on fire, and damaged both the remaining pair. The squadron was quickly joined in this task by the Mark IVs of 137 and, from 20 August, 164 – flying, since the Hurricanes, when weighted down by their rockets, were very vulnerable, under the protection of Typhoons.

It may be worth repeating that the Mark IV could carry many weapons. Indeed, one of its most rewarding raids was made on 23 July 1943, by four aircraft from 137 armed with 40 mm cannon. Between them the pilots, led by Flight Lieutenant Bryan, covered a considerable area of Belgium, destroying four railway engines, two barges and an army lorry. However, the rocket was the favourite armament used, 137 joining 164 – together with an escort of Typhoons – on 2 September, to attack lock gates at Zuid Beveland on the Dutch Hansweert Canal. At least one gate was wrecked but 137 lost Flying Officer De Houx, while 164 lost its CO Squadron Leader McKeown – though on the 15th, it showed its resilience by a successful assault on the airfield at Abbeville.

During December 1943 and the early months of 1944, the Hurricanes, particularly those of 184 Squadron, turned their attention to the 'No-Balls' – this being the code-name for the VI launching sites along the French coast. Several raids were made on various of these with rockets or, incidentally, 500 lb bombs, but because of their importance to German hopes, the sites were very strongly defended, with the result that the Hurricanes often suffered heavy losses to flak. These strikes had ceased by March 1944 at which date all the Mark IV Squadrons stationed in Britain had converted to Typhoons, whose later exploits, especially in Normandy, owed a good deal to the experience gathered on the sorties flown by their predecessors.

By this time also, the majority of the squadrons equipped with Hurricane IIs, had also received more modern fighters or fighter-bombers as the case might be. During 1943, they had already tended to be used less on front-line duties than on supporting roles, such as defensive patrols or providing cover for air/sea rescue launches. Among the machines engaged on such work may perhaps be included the twelve Mark Is and 6 Mark IICs which by the end of the war had been sold to Eire for service with the Irish Air Corps at Baldonnel.

It was on similar duties that the last fighter unit to fly Hurricanes in Britain was employed. The Polish squadron, No 309, received Hurricane IICs in April 1944, with which it guarded shipping off the east coast of Scotland against raids by German bombers from Norway, as well as mounting patrols over Edinburgh. Much to the pilots' disappointment, 309 did not sight a single foe on such missions, though perhaps its presence had a deterrent effect. It converted to Mustangs in October 1944.

However, other Hurricanes were still carrying out a variety of secondary tasks well after this date. From December 1944 until the close of the fighting in Europe, the IICs of 521 Squadron, stripped of their cannon, made twice-daily climbs to 24,000 feet from their base at Langham in Norfolk, to report on meteorological conditions. Similarly, 527 Squadron continued to the end of hostilities with its work of providing data for checking the accuracy of radar, searchlights or gun batteries. Hurricanes were also still used for experimental purposes, notably KZ706, a Mark IV, which, in 1945, was engaged in a series of trials in which it carried a large 'Long Tom' rocket with a 500 lb warhead under each wing.

The same strength and reliability that rendered the Hurricane so suitable for experiments, made it an ideal advanced fighter trainer, large numbers being delivered to a total of eighteen Operational Training Units in Britain or abroad. Not that it trained only fighter pilots; from April 1943 until just before D-Day, the IICs of 516 Squadron participated in assault exercises along the west coast of Scotland, laying smoke screens or making mock attacks during the incessant preparations for the return of the Allied armies to northern France.

No 63 Squadron was also involved in preparing for the Normandy landings on 6 June 1944. In March, this squadron replaced the Mustangs it had been using for tactical reconnaissance work with Hurricane IICs and IVs. Officially, it converted to Spitfires at the end of May, but it is clear that it retained Hurricanes on its strength: indeed, one famous Hurricane – LF363, of which more later – is recorded as serving with 63 as late as 30 November 1944.

It seems that on D-Day it was 63's Hurricanes, with which of course its pilots were much more familiar, that, starting at dawn, ranged above the beaches rectifying one of the weaknesses revealed at Dieppe by spotting fall of shot from the naval forces bombarding Hitler's 'Atlantic Wall'. Certainly, this was an entirely appropriate role for Hurricanes as it was undramatic but immensely important and required a high degree of skill in the pilots and great versatility in the aircraft.

During the final advance through the liberated countries of western Europe to the heart of Nazi Germany, the Hurricane saw no front-line fighter or fighter-bomber service – not that any established Hurricane pilot could have felt that a campaign where the overwhelming odds were on the side of the Allied air-forces was other than unnatural. These duties had now been delegated to its successors from the Hawker 'stable', the deadly Typhoon with its eight rockets; the speedy Tempest, which proved Fighter Command's main weapon against the flying-bombs.[61]

Yet Hurricanes still performed important work in less arduous roles. Those of 116 Squadron made most of the flights needed to test the mobile radar sets used by the advancing armies, while other individual aircraft were employed carrying high-priority despatches to and from the Continent. They even found some excitement in the course of this; on 10 June 1944 two such unarmed Hurricane IICs landed on an airfield which turned out to be still in German hands. Luckily Squadron Leader ffrench Beytagh and Squadron Leader Storrar were used to emergency take-offs as a result of their experiences with 73 Squadron in the desert fighting. They both got airborne just in time to avoid capture.

However, not long afterwards came an event with sadder connotations. The last Hurricanes had been issued to the RAF; now, on 12 August 1944, the final production Hurricane was completed at Langley – PZ865, a IIC, which did not join the RAF as it was purchased from the Ministry of Aircraft Production by Hawker Aircraft Limited. Officially, 14,553 Hurricanes had been built, 10,030 by Hawkers at Kingston, Brooklands or Langley, 2,750 by Gloster Aircraft Company Limited at Brockworth, 300 by Austin Motor Company Limited at Longbridge,

---

[61]There was a curious parallel here to the Hurricanes' part in the Battle of Britain, in that the Tempests never received adequate recognition of their successes, though they destroyed 638 V1s in fewer than three months, Squadron Leader Berry of 501 Squadron heading the list with a score of sixty-one. Incidentally the Tempests, with a top speed of about 430 mph, were faster than the early Gloster Meteor jet fighters, which claimed only thirteen flying-bombs.

1,451 by Canadian Car and Foundry Corporation, twenty by Rogozarski at Belgrade, Yugoslavia, and two by Avions Fairey at Brussels, Belgium. For practical purposes, it can be said that there were more Hurricanes even than this, for many served in more than one entirely different version – those that were converted to fly from CAM-ships or from carriers for example.

In any case, PZ865 was 'The Last of the Many' and a few weeks later, with this legend painted on the sides of its fuselage, with a banner above it proclaiming the Hurricane's battle-honours and carrying the flags of the nations whose pilots had flown it, this machine took part in a ceremony held by Hawkers at Langley to bid farewell to the company's most famous aircraft. It was joined, appropriately, by an old Hart, the type whose advent had led to the creation of the Hurricane, as well as by Hawkers' latest fighter, the mighty Tempest.

After the three machines had flown past the spectators, the Hurricane gave a solo exhibition of its skills, being displayed in masterly fashion by an ex-test pilot who had come out of retirement to do so. His name was 'George' Bulman. Nearly nine years earlier, it was he who had been at the controls of the prototype on its first flight. The Hurricanes had achieved a great deal since then.

–

They would achieve much even after the last model came out of the factory, for those already in service continued to play their part in all manner of operations up to the close of hostilities. In north-western Europe, as already seen, they were limited to supporting roles during the final stages, but elsewhere they remained in the frontline until the end.

At the conclusion of the North African campaign in May 1943, the Mediterranean theatre still contained twenty-three RAF or SAAF Hurricane squadrons, among which may be mentioned a second Greek unit, No. 336, which had formed with IICs on 25 February, but was only just becoming fully operational. In addition, about twenty IIBs or IICs were provided to equip two squadrons of the Royal Egyptian Air Force. Previously, during 1942, fourteen IIBs or IICs had been supplied to Turkey to supplement the Mark Is delivered before the war. Other Mark IIs were taken over by the French in North Africa, as were some Canadian-built Hurricanes, at least one of which appeared later in the victory celebrations in Paris.

Iran also acquired Hurricanes – in somewhat unusual circumstances. On 1 December 1942, 74 Squadron, under Squadron Leader Illingworth, arrived at Meherabad in the north of that country with Mark IIBs. Its duty was to patrol the area, which it will be remembered had been taken over by Britain and Russia in August 1941. In March 1943, the squadron moved to Abadan on the Persian Gulf, then Shaibah in Iraq and finally Egypt, which it reached in May, converting to Spitfires three months later. It left behind about ten of its Hurricanes, which were quickly commandeered by the Persian Air Force.

In August 1943, another neutral but friendly state received the Hawker fighters, when fifteen IICs went to Portugal. By the end of the year, a further fifty IIBs or IICs had also arrived, providing equipment for three squadrons. In return for this improvement in the quality of its armed forces, on 8 October the Portuguese Government handed over to the Allies bases in the Azores, which proved invaluable in the fight against the U-boats, since from then onwards, air cover could be mounted over the whole of the North Atlantic.

Also in August 1943, the ground crews of a new Polish Hurricane squadron, No 318, reached the Middle East, being followed a few days later by the pilots and their IICs. 318 was detailed for co-operation exercises with the army, serving in this role at Muqueiliba near Nazareth, a hideous location in a malarial, scorpion-infested valley, where the heat was so great that it was agony to touch a metal part of the machines. It subsequently moved to Gaza in October, then, in December, to Egypt, where two pilots were killed in a collision during one of its innumerable manoeuvres. The squadron converted to Spitfires in March 1944.

The tasks of 318 Squadron were fairly typical of those of the Hurricane units during late 1943 – important but definitely monotonous. Again from August 1943, Hurricanes were used, as in Britain, for meteorological duties. By early 1944 four Flights were so employed, their unarmed IICs climbing twice a day to as much as 35,000 feet from various bases as far apart as Palestine, Syria, Iraq, East Africa and the Sudan; – a role in which they continued until after the end of the war.

Those Hurricanes that were armed flew mainly as guardians for Allied convoys. They saw little action for the Regia Aeronautica had ceased to be a foe after the Italian capitulation on 8 September[62] while the Luftwaffe – faced by the need to protect the German homeland against massive bombing raids, support the crumbling Russian front and prepare to meet the threat of invasion in north-west Europe – was very much on the defensive. However, the occasional Junkers Ju 88 provided combat experience. 213 reached an official 'score' of 200 victories, all while flying Hurricanes, in October 1943 – though no doubt human error had meant that somewhat fewer than that number of enemy aeroplanes had in fact fallen. Even in the early months of 1944, the Hurricanes were still downing the odd German reconnaissance machine.

As a change from these defensive sorties, the Hurricane squadrons raided Crete or nearby islands. 213 again featured prominently in such activities as did the Greek squadron, 336. Indeed, its first action was a ground-attack on various enemy installations in Crete on 23 July 1943, from which two pilots failed to return. The Greeks followed this with other intruder

---

[62]Italian representatives had signed a secret armistice with the Allies five days earlier. Previously, on 25 July, Mussolini had been overthrown by a coup led by Marshal Badoglio.

missions, which included, on 12 August, shooting-up a U-boat caught on the surface.

Operations after dark were also featured in the Hurricanes' repertoire, especially on the night of 9 July 1943, when Allied airborne troops landed in south-eastern Sicily, as a preliminary to the main invasion from the sea early next day. The assaulting forces were scattered over far too wide an area but they did seize the important Ponte Grande bridge near Syracuse, as well as causing much desirable confusion. They were supported by the Hurricane IICs of 73, now based in Malta, which flew over the coast attacking any searchlights that might try to pick out the gliders or the aircraft towing them.

Thereafter, 73 flew a series of night intruder sorties over Sicily, though at a high cost. However, by the end of the month it had retired to North Africa, where it converted to Spitfires. It finished the war credited with 320 victories, of which practically all had been gained with its Hurricanes – an immense achievement, even after making all allowance for exaggerations.

If there is thus a strong case for ranking 73 as the finest Hurricane fighter unit in the Middle East, No 6 surely has an equal claim to be considered the best of the ground-attack squadrons. The 'Winged Tin-openers' had returned to Egypt on the conclusion of the Tunisian campaign, to re-equip with rocket-firing Hurricane IVs. In February 1944, these flew to Grottaglie in southern Italy, whence, on 29 March, they raided a German headquarters at Durazzo in Albania. Naturally, this was well defended – No 6 losing one Hurricane with its pilot, while two more were damaged, though not enough to prevent their safe return.

From that time onwards, however, although No 6 did attack targets on land, ranging from radar stations to dockyard cranes, its chief victims were enemy ships, which it assaulted on moonlit nights as well as by day. In sorties prior to July, when the squadron moved north to Foggia, it was credited with having sunk or damaged more than fifty vessels of various types including freighters, schooners, barges, ferries, landing-craft, dredgers and tugs.

As later in Normandy, Hurricanes also served in Sicily and Italy as fast communication aircraft. At least one such was flown by American personnel in southern Italy, painted silver overall with US markings. A slightly different form of communications was provided by the Hurricanes of 680 Squadron – the old 2 PRU. This now had Lockheed Lightnings to take its pictures, but it was the task of the Hawker aircraft to carry these back to the appropriate headquarters from forward bases all over the Mediterranean, which they did from September 1943 to October 1944, covering more than a million miles in the process. When all allowance has been made for the skills of the pilots, or even for good luck, it must be a testimony to the Hurricanes' reliability that in all that time, not one aeroplane was damaged, not one single photograph was lost.

Among those areas covered by the photo-reconnaissance machines, one of the most important was the Balkans. In September 1943, the Italian collapse had led to a vast uprising throughout Yugoslavia centred upon the resistance forces of Marshal Tito. The Germans reacted violently by pouring in massive reinforcements. Slowly, at immense cost, these began to regain the initiative from the partisans, to whom the western Allies, pre-occupied by their problems in Italy, gave perhaps less help than might have been possible, though the RAF dropped steadily mounting quantities of supplies.

Also, in mid-October, the Royal Navy set up an advanced base in the Adriatic island of Vis, from which, in course of time, Hurricanes would operate.

By early June 1944, Tito had had to be evacuated by Dakota to Bari in Italy, where he asked for close air-support for his troops as well as help for his wounded, of whom about 11,000 were at various times flown out to RAF hospitals in Italy. To meet the first request, the Balkan Air Force was formed. This included Hurricanes, a detachment from No 6 moving to Vis in August, where it was joined a month later by 351, a squadron of Yugoslav airmen that had been formed in North Africa on 1 July 1944 with Hurricane IICs but was now also flying rocket-armed Hurricane IVs.

From Vis, the Hurricanes, carrying four rockets under one wing and a 45-gallon tank under the other to increase their range, supported the Yugoslav resistance by attacking troop concentrations, roads or railways. The latter were soon to be more important to the Germans as escape routes than as means of communication, for, by the time 351 flew its first sorties on 13 October, the Axis positions in the Balkans were crumbling into ruin.

On August 20 1944, after months of ominous preparation, the Soviet Army, whose triumphant advance had carried it to the borders of Rumania, burst into that unhappy country, which asked for an armistice three days later – declaring war on Germany two days after that. On 8 September, Bulgaria followed suit, whereupon a continued occupation of Greece became impossible for the Germans, who began a slow, highly skilful retreat into Yugoslavia.

As they moved out, British seaborne and airborne forces moved in. On 23 September, these captured Araxos airfield in the north of the Morea. Megara aerodrome twenty miles west of Athens was taken on 10 October. The capital was secured on the 14th. Meanwhile those members of 6 Squadron who had not gone to Vis, flew to Araxos, whence they used their rockets to good effect in harrying the retreating Germans.

During the remaining months of 1944, the Axis soldiers continued their withdrawal: from Albania; from Belgrade which fell to the Russians and partisans on 19 October; from the Yugoslavian provinces of Macedonia, Montenegro and Serbia; back to north-eastern Bosnia and Croatia. They were worried continually by resistance fighters, who were aided by the Hurricanes in Vis and Greece, the latter destroying the railway bridge at Spuz, a vital link on the enemy's line of retreat from Albania, while other targets ranged from transport of all sorts to more headquarters buildings.

The coming of winter did not check the Hurricanes' activities. On 1 December, for instance, three of No 6's machines flew from Araxos to Niksic in Montenegro, north of Spuz, whence two days later, although carrying only four rockets apiece, they attacked a group of the Germans' most dangerous tanks – Tigers armed with 88 mm guns – destroying one and badly damaging two others. Next day they were again in action, again all scoring hits to wreck another tank and two mobile guns. 351 was also getting into its stride – during December, it destroyed twelve motor transports, a power station, a wireless station and a command post.

The closing year of the war, 1945, saw no relaxation of the Hurricanes' efforts. Strikes against shipping were resumed but the majority of the Hawker fighters' early missions were directed against enemy troop concentrations in the Sarajevo area. In February, 351 went to Prkos in Yugoslavia, where it was joined, first by a few aircraft from No 6, then by the remainder of the squadron in April. Spurred on by their attacks, the Germans resumed their retreat. The Yugoslavs followed up. At the end of April, cutting off large numbers of the enemy, who were left with no alternative but to surrender, they broke through the defences to meet up with the advance guard of Eighth Army at Monfalcone, north-east of Trieste.

In these last weeks, the Hurricanes' main targets were various forms of transport including coastal shipping. Their campaign was now to reach a perfect culmination. On 1 May, a week before the surrender of Germany ended the fighting in Europe, 6 Squadron flew its ultimate mission over the Gulf of Trieste. Twenty-five vessels of different types, sixteen of them troopships, hoisted white flags on the Hurricanes' appearance. It was a final, symbolic capitulation to the aircraft that had been numerically the most important Allied fighter at the start of the war and was still striking at its country's foes when this ended.

Yet even now the Hurricane still had some vital services to give. Amid the rejoicing, Churchill again sounded notes of warning. Some related to the problems of peace, which would not be the Hurricane's concern, but the Prime Minister also issued this reminder: 'Beyond all lurks Japan, harassed and failing, but still a people of a hundred millions, for whose warriors death has few terrors.'

Chapter Thirteen

The Road to Mandalay

In the beginning, the Hurricane's 'road' back into Burma was literally that – the Red Road in Calcutta, running parallel to the city's main street, the Chowringhee. At one end, according to the description in the *RAF Official History*,

> a white marble building, erected to commemorate an empress already half forgotten, gleams and winks in the sultry sunshine, and about halfway along it, calm-faced Lord Lansdowne and inscrutable Lord Roberts, graven in stone, look down benignly from their pedestals. In the summer, autumn and winter of 1942, the statues of the pro-consul and his military colleague were the mute witnesses of a new and striking use for this impressive highway. It had become the main landing ground for the Hurricanes defending Calcutta.

The Hurricanes in question belonged to 17 Squadron, which had a less easy runway than might be supposed, because of the road's sharp camber. 79 Squadron was also detailed to Calcutta's defence at this time, while other Hurricane units played their part later in protecting the crowded population of the huge city.

In fact, they saw surprisingly little action. It should be remembered throughout, that the conflict in this area was very much a 'sideshow' for the Japanese. As Brigadier Smith puts it in *Battle for Burma*, they had conquered the country

> at a time when their army was fully committed elsewhere in the Pacific, as well as fighting a campaign on the mainland of China. Thereafter those fronts would not allow the dispatch of large-scale reinforcements to Burma.

The same situation existed in the air as on the ground. Always short of machines, the Japanese were compelled to transfer any reserves to more threatening scenes. Also, Japan's finest interceptor, the Zero, being a Navy fighter, took virtually no part in the struggle for Burma, though, perhaps in unconscious tribute to its excellence, reports of Allied army officers constantly refer to all Japanese fighters as Zeros. Thus, the most important Allied activity in late 1942, was the establishment of an air-route to China over the 'Hump', the mountain range separating that country from Assam, by the Dakotas of the American Ferry Command, whose young pilots carried thousands of tons of supplies over dangerous, little-known territory in often ghastly weather conditions. On their return trips, moreover, they ferried 13,000 Chinese troops to reinforce the command in northern Burma of the American General Stilwell, the famous – some would say infamous – 'Vinegar Joe'.

By December 1942, however, the Hurricanes were also scenting combat. A number of them, eventually totalling eight squadrons, moved up to the forward base of Chittagong in East Bengal. This was raided on the 15th but 79 Squadron, which had only arrived the previous day, got among the bombers, destroying three, damaging two. At the same time, the Japanese began attacking Calcutta after dark, causing minor damage but considerable panic among the civilian population. However, No 17 was soon in action against the night raiders, scoring its first victory on the 23rd.

Some Hurricane detachments went even closer to the front line, transferring to air-strips constructed by the hardy sappers in paddy fields. These were swamps during the monsoon, they became dust-bowls at the height of the dry season, but now, as they first dried out, they were suitable for aircraft as rugged as the Hurricanes, which indeed with their reliable versatility were ideal for the primitive jungle fighting.

They were there to support the first faltering British and Commonwealth counter-attack. Famine and political unrest in India, a very high rate of malaria among the troops, a lack of training, an absence of supplies, all combined to prevent Field-Marshal Wavell (as he became on 1 January 1943) from attempting more than a push by 14th Indian Division along the coast to clear the Mayu Peninsula, in preparation for a short amphibious crossing to the island of Akyab – where once Hurricanes had been based – by the few landing craft available in the Burmese theatre.

Thus, despite dramatic publicity, which portrayed it as a major attempt to clear the whole of Burma, the offensive had very limited aims. Starting from their advanced base at Cox's Bazar, close to the Burmese frontier, on 9 December 1942, the Allied soldiers struggled through the sharp, jungle-covered hills until 6 January 1943, when, only ten miles short of Foul Point as the tip of the peninsula was named, they ran into fortified positions at a place called Donbaik.

Although held only by a single Japanese company, this proved a formidable obstacle. For the first time the Allies encountered Japanese 'bunkers' – strong-points usually made from heavy logs, covered with earth, sited to support each other, almost impervious to shells and so cleverly camouflaged that they were rarely located until they had opened fire. Before the bunkers of Donbaik, the first campaign in the Arakan – this being the general name for the coastal area of Burma – died away in a succession of increasingly heavy but equally futile attacks, the last of which met with ruin on 18 March.

While Donbaik exerted its fatal fascination, the Japanese gathered reinforcements for a counter-offensive. By the 29th, their outflanking movements had forced the Allied troops into a retreat, which ended by mid-May back at Cox's Bazar – a failure made all the more bitter by the exaggerated hopes with which the campaign had been accompanied.

During the advance, and, to an even greater extent, the retreat, the Hurricanes kept up continual pressure against the enemy, on the ground as in the air, though it will cause little surprise to state that their services have received scant recognition. The usual story is that command of the air only came to the RAF with the arrival of Spitfires in early October 1943, months after the end of 'First Arakan'. In particular, it is claimed that the Hurricanes were unable to intercept the Mitsubishi 'Dinah' reconnaissance machines.

This specific charge must be admitted. The Dinah, a very speedy aircraft not unlike the RAF Mosquito, had a service ceiling of 40,600 feet, which put it almost out of reach of the Hurricane, whose rate of climb was not its finest point. Although the odd Dinah was brought down by the Hawker fighters, it was only after Spitfires reached the front that Japanese scouts began to be lost at a considerable rate – four within a month. However, it should be pointed out that while it was clearly important that the enemy should not obtain information of Allied dispositions, this was the only gap in the Hurricanes' command of the air; in several far more important aspects they achieved almost complete domination.

For a start, if the Hurricanes could not stop the Dinahs, neither could these prevent the reconnaissance flights of the Hurricanes, which, incidentally, were much more valuable if only because they were more frequent. At the time of 'First Arakan', long-range recce duties were carried out by 681 Squadron, which formed on 25 January 1943 from the old No 3 PRU, flying Hurricane IICs with drop-tanks and a forward-facing camera, while the army co-operation role was entrusted to the Tac R IICs of 28 Squadron. As will be seen, however, many more Hurricanes would serve on such tasks in the near future.

Equally unstoppable were the ground-attack Hurricanes, which strafed all manner of targets, including sampans, river steamers, trains and bridges. Even Christmas Day 1942 saw 615 Squadron in action, shooting up Japanese machine-gun posts. In the dense jungle, strikes against enemy personnel were less effective but they were not without their value especially during the Allied retreat. It is interesting that Colonel Tanahashi, most capable of the enemy commanders, felt the need to exhort his men to have no fear of aircraft.

When the Japanese air force attempted to interfere, or raided Chittagong or the forward strips, the Hurricanes reverted to the role of interceptors for which they had originally been designed. At this time a majority of the squadrons were flying IIBs while a number of those with IICs restricted their armament to only two cannon, as in the Western Desert, in order to increase their range. Yet they gained a succession of victories, which, if less dramatic than those achieved on other fronts, was equally effective in view of the far smaller number of enemy aircraft encountered.

Thus, 615, for instance, shot down eight enemy machines during the First Arakan campaign, as well as claiming eighteen 'probables', some of which undoubtedly did not return to base. Its best effort was on 1 April 1943, when it broke up a raid of thirty Mitsubishi Sally bombers, destroying three, damaging nine more, though in the fight one of its best pilots, the Belgian Flying Officer Ortmans, lost his life.

No 136 Squadron gained its first victories on 23 January, when it shot down two Nakajima Oscar fighters. It then took part in a series of combats, including one on 5 April, in which Sergeant Cross had the extraordinary experience of damaging no fewer than six Oscars without being able to claim any confirmed 'kills'. Perhaps the squadron's fiercest battle came on

29 May, which saw the destruction of at least one Sally and two Oscars. The finest individual success of the period, however, was gained by a pilot of 135, the Australian Flight Lieutenant Storey, who, on 5 March, won a DFC by downing three Oscars over Akyab.

Of course, it goes without saying that the Hurricane pilots had their losses as well, the saddest perhaps being that of Flight Lieutenant Bowes of 79, who had gained at least nine victories before he fell in action on 21 May; but by June when the monsoon greatly reduced their activities, they were already clearly supreme in the skies over the Arakan – several months, be it noted, before any Spitfire fired its guns on this front. Lieutenant Colonel Frank Owen, in his book *The Campaign in Burma*, best sums up their attainments. 'They gripped the Japanese air-power,' he reports, 'and repelled its offensive, offsetting the reverses on the ground.'

This dominance was equally vital for another operation that took place in early 1943. On the night of 14 February 3,200 men, under Colonel Orde Wingate, crossed the Chindwin River to penetrate deep into Japanese-held territory. The 'Chindits', who took their name from the mythical beast, half-lion, half-griffin, that guarded the Burmese pagodas, depended for their supplies on parachute drops from Allied Dakotas. These were accompanied by strong escorts of Hurricanes but, in practice, the fighters saw hardly any combats. Even allowing for the Japanese Air Force's perennial shortage of machines, its lack of aggression must be counted as a high compliment to the Hurricanes.

Although the immediate achievements of Wingate's raiders were minor, only two-thirds of them returned to India in mid-April and a large proportion of the survivors were so ravaged by disease that they were judged unfit for further service. Yet, in the long-term, their mission was justified. Like wildfire, the grand news spread through the Allied troops: for the first time British and Indian soldiers had outfought and out-witted the

Japanese in the jungle. Morale, the essential basis for any army's success, rose dramatically.

In addition, Wingate had given an object lesson for future operations, by showing that the Allied supply lines could be transferred successfully to the care of transport aircraft, which would not be endangered by the swift Japanese out-flanking movements. Equally, since artillery was in short supply, it would in future be ground-attack aeroplanes that would have to play the crucial part in breaking the assaults of the enemy army.

During the remainder of 1943, the number of Hurricanes steadily increased – and not only in the RAF. In August, No 6 Squadron Indian Air Force received Hurricane IICs, as did seven other Indian squadrons during the next few months. About 300 Hurricanes eventually served with the IAF, on patrols over that traditional trouble-spot the North-West Frontier, on unexciting, though important, training duties, but chiefly in the Tac R role. Fitted with a forward-facing camera inboard of the starboard cannon and frequently carrying drop-tanks, they were used for photographic work, artillery spotting, or the delivery of important messages.

Though, like their companions in North Africa, these 'recce' Hurricanes, which normally operated in pairs, seldom enjoyed much fame, they faced some daunting dangers. Admittedly, the Japanese Air Force did not prove a fraction of the menace that the Luftwaffe had been in the Desert, but the hostile terrain, the often-violent weather, the inadequacy of the landing grounds, especially when the aircraft had to return after dark, and the absence of any emergency ones, all made such flights hazardous – while highly skilled observation was needed to spot the almost perfectly camouflaged enemy positions. Nor did the Indian pilots hesitate to carry out strafing attacks whenever they had the opportunity, despite extremely accurate Japanese fire. During the remainder of the campaign in Burma, they flew more than 16,000 missions, more than fifty of them being lost in action. Their services received proper recognition in the award of forty-four decorations for gallantry.

Of the RAF Hurricane squadrons, 136, 607 and 615 converted to Spitfires in the last months of 1943, but that still left twenty-one units in the Far East equipped with the Hawker fighters. Among these may be mentioned 273, which provided protection for Ceylon after the three squadrons that had fought Nagumo's airmen – 30, 258 and 261 – had all moved off to the Burma front. Also, 176, which flew Hurricane IICs as well as Beaufighters in the defence of Calcutta by night.

The Hurricanes still carried out day-fighter duties also, but combats were now few and successes still more rare. On 20 October 1943, 261 intercepted a raid on Chittagong, destroying or damaging four enemy aircraft but losing three Hurricanes, two of whose pilots were killed. On 5 December, the Japanese bombed Calcutta's docks causing minor damage but major panic among the civilian labour force, and in addition the Hurricanes of 67 and 146, sent off too late and attacked from above by the escort, lost five of their number, though only one pilot died. A second raid an hour later was opposed only by five Hurricanes from 176, three of which were shot down, two pilots being killed. Fortunately, the enemy never repeated this attack, though the Hurricanes were still able to claim the occasional Sally over the Arakan.

In the main, however, the Hurricanes were preparing for close-support tasks. By January 1944, all the squadrons except three were re-equipping with IICs though several still had some IIBs on strength. Apart from those used by the reconnaissance squadrons, all such Hurricanes were now accustomed to carry two 250 lb bombs apiece. The three squadrons not so armed were No 42, which in addition to IICs also flew Mark IV fighter-bombers; and 5 and 20, manning IIDs with 40 mm anti-tank guns. The splendid potentialities of these varied types of ground-attack aircraft would soon be realised beyond all possible expectations.

–

The leadership of the Far Eastern theatre was also reorganised. In October 1943, Wavell was appointed Viceroy of India, a land fast becoming one gigantic Allied base. A month later, Admiral Lord Louis Mountbatten, at the age of forty-three, took up his post as Supreme Commander of a new, unified South East Asia Command, entrusted with the task of recovering Burma.

Mountbatten's land forces, who were destined to fulfil this aim, were grouped in three areas. In the centre, IV Corps of Lieutenant-General William Slim's Fourteenth Army, joined in April 1944 by XXXIII Corps, guarded Assam and the remote, beautiful little frontier state of Manipur, the capital of which is Imphal. XV Corps of Fourteenth Army, under Lieutenant-General Christison, held the southern front in the Arakan. All these were under the overall control of General Sir George Giffard, but Stilwell's Chinese troops in the north were not, though the American commander did agree to take orders from Slim. To complicate an already absurd situation, Stilwell was also Deputy Supreme Commander, thereby outranking everyone except Mountbatten. In practice, the whole question was academic for the irascible 'Vinegar Joe' acted throughout completely independently, an attitude that may be excused by the remoteness of the area entrusted to his care.

The opposing Japanese troops, under Lieutenant-General Kawabe, were divided into three so-called 'armies' – though in fact each was equal only to a weak British Corps. These were the 15th on the central front, the 28th in the Arakan and the 33rd in the north. They were commanded by Lieutenant-Generals Mutaguchi, Sakurai and Honda respectively.

In view of their superior numbers, the Allies' proposals for 1944 seem somewhat limited – no doubt the memory of First Arakan still cast its shadow. Apart from another push along the coast by Christison's command, supported by a minor advance in the central zone, the main move planned was the conquest of northern Burma by Stilwell in order to secure adequate communications with China.

To assist these operations, the Allied Air Forces, led by Air Chief Marshal Sir Richard Peirse, under whom the American Major-General George Stratemeyer was responsible for missions over Burma, began their preliminary raids. Hurri-bombers ranged everywhere, their targets including bridges, troop positions, and transports on roads, railways or rivers. The anti-tank guns of the Hurricane IIDs were also put to good use: No 5 Squadron destroyed a total of twenty-seven lorries and fourteen bullock-carts on 8 January 1944 alone, while No 20, which had gone into action a few days earlier, was credited at the end of six months of operations with having hit, among other targets, 348 Japanese strong-points and 501 sampans.

However, the Japanese, with reckless daring, were also planning an offensive. Alarmed by Wingate's raid, which indicated that the Chindwin was not the secure barrier that they had believed, they had decided to gain a better defensive position by a purely 'preventative' seizure of the Imphal plain – although the invading forces were to be accompanied by the 'Indian National Army' of Subhas Chandra Bose, a former president of the Congress Party, they never had any intention of making a 'march on Delhi' as both sides would proclaim for propaganda purposes at different times. As a preliminary, moreover, the enemy determined on a diversionary attack in the Arakan.

Thus, early in the morning of 4 February, Allied forces, now again advancing down the Mayu peninsula, were engaged from flank and rear. Major-General Messervy's 7th Indian Division was completely cut off. It pulled back into the 'Admin Box' at Sinzweya, where, although assaulted from all sides, it was able to hold out until relieved, since its supplies were dropped from a swarm of Dakotas, which were so well escorted by Hurricanes that only one was lost. At the same time, other Hurricanes made strafing attacks on the enemy positions.

By 11 February the Japanese gamble had failed. Their troops, not being in receipt of supplies from the air, were desperately short not only of ammunition but even of food. At first refusing to admit that their plans had been disrupted, a folly they would shortly repeat on a greater scale, the enemy commanders finally pulled back on the 24th. Since they had forced the Allies to commit many of their reserves, perhaps the Japanese cannot be blamed for deciding to continue with their main advance on Imphal. Yet the course of the fighting at Sinzweya strongly suggests that they would have been far better advised if they had remained on the defensive when they would have been very hard to defeat.

This seems further confirmed by the fighting on the northern front, where, during the early months of 1944, Stilwell's Chinese, spear-headed by the American troops of Brigadier-General Merrill, had been making slow but steady progress. To assist them, on 5 March, the Chindits were flown behind the enemy lines; they enjoyed early successes despite the death of Wingate, now a Major-General, on the 24th. The finest feat of all, however, must be credited to the Americans. On 17 May, after a hazardous forced-march, 'Merrill's Marauders' seized the airfield near the key town of Myitkyina[63] thereby opening a new, far less perilous air-route to China.

After that, everything went wrong. As exhaustion began to take its toll of the attackers, hopes that the fortified town would fall equally swiftly withered away in the face of a stubborn Japanese defence. Not until 3 August was it taken. Its commander, Major-General Mizukami, committed suicide, presumably to preserve his honour – as if that could ever be questioned of the man who had held Myitkyina for seventy-nine days against odds of ten to one.

The Japanese stand was the more remarkable in that throughout they received no aid from the air. This was due in the main to Colonel Cochran's United States 1st Air Commando, the exploits of which in support of the Chindits have rarely received adequate praise.

---

[63]Pronounced 'Mitchina'.

The Americans' most valuable contributions took the form of low-level strikes on Japanese aerodromes by Mitchells dropping fragmentation bombs. These were followed by RAF Hurricanes, which strafed any machines that still survived. Such sorties destroyed well over 100 enemy aircraft, about a third of those available, thereby, as Brigadier Smith remarks, dealing 'a catastrophic blow to Japanese air power in Burma' – one which would have immense consequences, not least to the three divisions of General Mutaguchi's 15th Army, which on the night of 7/8 March, commenced the offensive against Imphal.

Considering that General Slim had a preponderance in manpower, artillery and armour, that his tanks could out-gun the few that the Japanese possessed, that he had total command of the air, and that he was aware of the impending enemy move, it might seem that Mutaguchi's assault was the wildest folly. Yet such was the bravery of the Japanese soldiers that they almost snatched victory in the face of impossible odds.

In the first place, the Japanese speed of movement through difficult terrain caught Slim by surprise. Covered by strikes from Hurribombers, three divisions retired in some disarray to Imphal, where they were joined by the bulk of a fourth, carried by air from the Arakan. Although heavily outnumbered, on 29 March Mutaguchi's 15th and 33rd Divisions, nominally aided by a division of Bose's Indian National Army[64], cut the road to Fourteenth Army's major base at Dimapur on the Bengal-Assam railway at a point about thirty miles north of Imphal. As Liddell Hart rather sourly remarks: 'Japanese nimbleness and thrustfulness had once again thrown their numerically superior opponents off balance and put them in an awkward plight.'

Fortunately, at 1800 on 29 March, the Allies were saved from being placed in a still worse plight by the sharp eyes of Squadron Leader Aijan Singh of 1 Squadron Indian Air Force,

---

[64]In reality, its members deserted whenever occasion offered.

who sighted a complete Japanese battalion fewer than ten miles north-east of Imphal airfield. With dusk already gathering, there was no time to be lost, and thirty-three Hurricanes – the rest of 1 IAF; the Tac R aircraft of 28 Squadron; the fighter-bombers of 34 and 42, hastily armed by a motley array of helpers – were soon heading for the danger-zone. At first, they could locate no targets, but flying very low, they were able to pick up the Japanese column in their landing lights – and routed it, killing, as was confirmed later from captured enemy documents, fourteen officers and more than 200 men. Most of the Hurricanes were able to get back to base before nightfall but one at least made an excellent landing in the dark without the aid of a flare-path.

Mutaguchi's remaining division, the 31st, was making for the most vulnerable point on the Imphal-Dimapur road, where it crossed the Naga hills at an altitude of some 4,700 feet at the village of Kohima. Since Slim believed that this was a target for only minor enemy forces, Colonel Richards, the officer in charge, had only 1,500 combat troops under his command. He abandoned the village to fall back on the Kohima Ridge, where the Japanese attacked him on 4 April.

However, it is extraordinary that they did not push on another forty-five miles to take Dimapur. Mutaguchi wished to do so but permission was refused by the overall Japanese commander in Burma, General Kawabe, whose plans forbade any advance deeper into India. By this typical inflexibility, he lost the campaign, for the capture of Dimapur would have provided vitally needed food and equipment, disrupted the supply routes not only of Fourteenth Army but of Stilwell's divisions, and deprived Slim of the base from which his counter-offensive would develop.

Thus, the fighting took the form of 'twin battles': the one as the Japanese assaulted Imphal from all directions; the other as they besieged the garrison on Kohima Ridge, while other contingents moved forward to block the advance from Dimapur of British and Commonwealth forces, which included further units flown from the Arakan.

Over the whole front, the struggle became one of stubborn attrition: one in which it has been claimed that more casualties were suffered in proportion to the numbers of troops engaged, than in any other action of the war; one in which participants on both sides have compared to the brutal battles of World War I. These comments say very little for the standard of generalship displayed on either side, but they do emphasise the extraordinary tenacity of the soldiers, whether from Britain, India or Japan, who, with Bose's followers dishonourably excepted, by common consent were worthy of each other.

What turned the battles in favour of the British and Commonwealth forces was the use of air power. The work of the transport machines in bringing up reinforcements has already been mentioned. In addition, they provided many of the needs of the troops advancing to the aid of Kohima and all the needs of those in the Imphal plain who could not be supplied by any other means. Equally important was the presence of 221 Group to give close support. This was under the command of Air Commodore, later Air Vice-Marshal, Vincent, who had once led the Hurricanes' gallant but vain fight against the Japanese in the East Indies. Now he would direct an equally brave but more fortunate campaign, in which the Hawker fighters again equipped a majority of the squadrons in his charge.

At the start of the struggle, there were six Hurricane units at Imphal or the nearby airfields at Palel and Tulihal: the Tac R machines of No 28 and of Squadron Leader Arjan Singh's 1 IAF; the fighter-bombers of 11, 34, 42 and 113. Later 60 and 123 with Hurribombers, 5, which having converted from IIDs to IICs due to a shortage of 40 mm shells, was used mainly for escorting transports, though it also carried out strafing attacks, 20, which still possessed IIDs, and two more IAF reconnaissance units, 6 and 9, also took part in the fighting.

By an odd quirk of fate, the Hurricanes saw least action on their interceptor duties, for the Japanese did not dare to attack a single one of the vital, vulnerable Dakotas that were guarded by them. Indeed, so feeble was enemy activity in the air, that in the entire campaign, only three unescorted transports were shot down.

The only time the Hurricanes encountered the Japanese Air Force came on 28 April, when a dozen Oscars made an early-morning fighter-bomber raid on the Allied supply base at Zubza on the Dimapur-Kohima road. They did some minor damage but then Hurricanes appeared, whereupon, to the disgust of the Japanese soldiers watching from the neighbouring hills, their airmen 'ran away without fighting'.

However, the main attention of the Hurricanes was directed against the enemy army. The extent and variety of their efforts can be seen from this summary by Frank Owen:

> Hurribombers became part of the pattern of the assault, strafing bunkers, foxholes and gun positions. Hurricanes spotted for the artillery and reconnoitred for the infantry. They shot up and bombed enemy concentrations, dumps, transport, bridges, river craft and locomotives... Hurricanes specialized in picking out Japanese road transport at night by its headlamps. When the Japanese 'blacked out' and travelled by moonlight the Hurricane pilots tracked them by the shadows the vehicles threw across the road. The monsoon in no way diminished their activity.

The Hurricane pilots kept up their raids against the Japanese lines of communication even in the most difficult weather conditions. Though it was often impossible for them to assess the exact details of the damage they had inflicted, the overall effect would become clear later in enemy reports of the 'frightfulness' of Hurricane attacks, which had left 'burned-out vehicles' with dead drivers 'crashed down into the valleys'. Since their available motor transport was already extremely limited, the Japanese were soon forced to use more primitive means of bringing up supplies. These included elephants and it is rather sad to have to relate that at least three of them were killed in a strafing attack by 5 Squadron. Yet the most important Hurricane missions of all were those flown in close support of the Allied land forces.

The 250 lb bombs that the Hurricane fighter-bombers normally dropped – though in some instances they used 500 lb weapons – were considered so effective that five Wellingtons were withdrawn from their own operations to ferry these to the front line. In April 1944, more than a million pounds of them rained down on the hapless Japanese infantrymen. In the area of Kohima alone, the Hurribombers of 11, 34, 42 and 113 squadrons flew more than 2,200 sorties, giving a tremendous boost to the spirits of the trapped Allied soldiers.

With such aid and encouragement, Richards and his staunch command held out, until, on 18 April, supported by a fierce Hurribomber raid, the first troops of the British 2nd Division, advancing from Dimapur, broke through to them. Next day, the Hurribombers concentrated on enemy units threatening the division's communications in the Zubza valley, helping to drive these back from their advanced positions. On the 20th, they returned to strike the Japanese troops in the vicinity of Kohima so effectively that these gave minimum interference to the main relief forces, which now also reached the besieged garrison.

At Imphal the crucial moment had come somewhat earlier. On 6 April, the Japanese seized the hill of Nungshigum, which rose abruptly from the plain little more than six miles to the north – the nearest they came to their objective. It was clear that if they could build up their numbers, they might well be able to break through to the vital Imphal airfield and the supplies of food, petrol and ammunition on which the defenders depended.

Next day therefore, the Hurribombers fell on Nungshigum, causing such casualties that an Allied counter-attack re-took it with remarkably little difficulty. However, the Japanese were not prepared to let the matter end there. By early morning on the 11th, they had again captured the hill, which they held despite further harassment by the fighter-bombers. It was apparent that a major effort would be needed to drive them out. Once again, Hurribombers were called on to provide close support.

Just after 1030 on 13 April, the Hurribombers blasted the Japanese positions, causing great destruction. Observers noted that 'clouds of earth, shattered branches of trees and mangled bodies were flung into the air on the summit.' Then, 'diving almost into the tree-tops', they strafed the hill, interspersing their strikes with 'dummy runs' to keep the defenders' heads down. Below them the attacking infantry, aided by tanks, supported by artillery fire, moved forward. They took Nung-shigum, albeit with heavy casualties, and it was never lost again.

Two days earlier, on 11 April, counter-attacks by British and Commonwealth forces in the hills around Shenam, to the south-east, had prevented the enemy breaking through to another crucial aerodrome at Palel. This was the base of the fighter-bomber Hurricane IVs of 42 Squadron, which, since Japanese troops were only two or three miles distant, were concealed in bays cut into the near-by hillside. Pilots and ground-crew were at readiness from dawn to dusk. Yet during April the squadron made 506 sorties, dropping more than 100 tons of bombs. Not the least important of its actions came during the crucial fighting on the 11th, when an essential ingredient of the Allied success was an attack by 42 Squadron on enemy defences, one bunker being seen to be blown up by a direct hit.

Yet if the Kohima Ridge had been relieved, the Japanese still held neighbouring high ground against heavy pressure; if Imphal had survived the initial assaults, the Japanese continued to launch further ones. The weary battles of attrition dragged on through May, with the Hurricanes operating at maximum intensity.

New developments were now being brought into the Hurricanes' repertoire. One such was later known as 'Cab Rank', in which they flew standing patrols over particularly vital areas, attacking targets of opportunity. Another was 'Earthquake – though this term also did not come into general use until later – in which, as at Nungshigum, they first bombed a position on which the ground troops were advancing, then strafed it with their cannon, then made mock-attacks, which kept the enemy under cover while avoiding any danger of hitting their own infantry.

Such new techniques were made possible by increasing inter-service co-operation. It became the practice that after 221 Group had drawn up the programme for each day's operations, in consultation with Army leaders, any squadrons, including Tac R ones, to which no specific tasks had been assigned, were allotted to a unit known as an Army Air Support Control, to which the forward ground troops could radio requests for assistance. The senior RAF officer at the Control – usually a wing commander – made the final decision, without reference to Group HQ, as to the employment of the aircraft entrusted to him, which were thus, for all practical purposes, operated by Fourteenth Army.

The pilots of the Hurribomber squadrons also took a personal interest in the soldiers' problems. It became quite usual for them to visit a brigade about to mount an attack, consult the commanders of the units designated to participate in this, and study from advanced observation posts the situation of the enemy positions that they would shortly have to bomb or strafe.

In consequence, strikes by Hurribombers soon reached a degree of accuracy that enabled them to raid targets only 200 yards from the front-line, or even less on occasions – as at Tengnoupal, near Palel, on 10 May, when twelve Mark IVs from 42 attacked so close to the advancing Allied soldiers that these clearly saw fragments of bunkers and of bodies blown upwards by the blast of the bombs. Such actions caused Air Chief Marshal Peirse to report that: 'The enemy's efforts to deploy in the Imphal plain during May 1944 were decisively defeated by Hurricane attacks at short intervals on any concentrations reported by ground troops through our Army Air Support Control operating at a high standard of efficiency.'

Equal praise was extended to the Hurribombers by the men of Fourteenth Army, from Slim in his reminiscences, to private soldiers in their letters home. There was good reason for their admiration, as fighter-bomber strikes became a regular, indispensable prelude to Allied ground assaults, almost always greatly reducing their cost.

On 18th May, the enemy's misery reached new heights when the monsoon broke. This did not stop the flights by the Hurricanes, which stood up to the weather's most furious bufferings, but it did wreck the already perilous Japanese supply lines. Yet Mutaguchi, in the face of his subordinates' protests, still ordered the troops at Kohima to hold their ground, while those in the Imphal area mounted attack after futile attack in pursuit of a plan that had long since become impossible to carry out.

By refusing to withdraw while there was yet time, he doomed his valiant 15th Army. On 31 May, outnumbered by at least five to one, desperately short of food, ammunition and medical supplies, the Japanese 31st Division retreated from Kohima apart from a small rearguard. This also pulled back on the night of 6/7 June. The British and Commonwealth forces moved southward, with the Hurricanes helping to clear the way for them by strikes on any targets that appeared. It was during these last few weeks of the fighting, that the IIDs of 20 Squadron, which previously had had to be content with lesser victims, at last sighted enemy tanks – three during May; twelve during June. They made the maximum use of their opportunities, putting every one of these out of action, though on 7

June, they suffered their only loss of the campaign during such an attack, when Flying Officer Brittain was killed by AA fire. On 22 June the troops advancing from Kohima made contact with their comrades in Imphal. Early in July, even Mutaguchi gave up and ordered a general retirement.

The campaign had been decisive. In addition to ruinous losses of equipment, it had cost Mutaguchi 53,000 casualties. Of these, 30,000 were killed, while a large proportion of the wounded died from starvation or disease during the retreat. Astonishingly, only about 600 prisoners were taken, of whom all but a handful were either badly wounded or in the last stages of exhaustion. Such losses could never be made good. Thereafter, the Japanese could not hope to do more than delay the British re-conquest of Burma. Mutaguchi had already removed all his three divisional commanders. Before the year was out, he also was dismissed, as was his superior, Kawabe.

By contrast, the British and Commonwealth forces had suffered fewer than 17,000 casualties out of a much larger army. They could thus follow up the retiring enemy despite the constant rains. In the early stages of the pursuit, the Hurricanes devoted most of their attention to the Japanese lines of communication, dropping delayed-action bombs at river crossings, hitting bridges – including the vital ones at Yanan, near Tamu, and at Falam, to the south of Tiddim – transport and supply dumps. However, they were soon back providing close support, even managing to find some new ways of doing this.

Fourteenth Army headed for the Chindwin River by two routes. The northern one led from Palel to Tamu, then south through the Kabaw Valley. Japanese resistance was slight until the crucial town of Kalemyo was neared, but the march passed through the territory of a more deadly foe. The Kabaw Valley with its sinister nickname of 'the Valley of Death' was reputed to be the most highly malarial spot on earth.

Yet the Hurricanes could deal with the mosquitoes as well. In between fighter-bomber raids, they carried DDT in their smoke-laying containers, with which they sprayed the road together with its approaches on either side – an action which, to the delight of the High Command, reduced the casualties caused by disease to a level lower than anyone had dared to hope. This very important task was entrusted on this or later occasions, to the Hurricanes of 5, 20, 34, 42, 60 and 113 Squadrons.

The other prong of the advance took the Tiddim road. This did meet with heavy opposition, for although the Japanese 15th and 31st Divisions had virtually disintegrated, the crack 33rd Division, now led by the intrepid Lieutenant-General Tanaka, was here falling back slowly in good order. Here also the Hurribombers again came into their own; harrying every pocket of resistance; helping the ground forces as they climbed the 'Chocolate Staircase' with its earth surface and its thirty-eight hairpin bends, to Tiddim. This they seized on 18 October, after the fighter-bombers had dived through the early morning mist for a devastating attack, which furthermore masked the noise of tanks moving up to the assault. 'Vital Corner', the most strongly held position of all, which had been blasted out of the solid rock of a precipice, was captured on 2 November, after a bombardment of artillery, aided by four squadrons of Hurribombers. Kennedy Peak, where the Hurribombers had destroyed three bunkers, fell on 4 November; Fort White fell on the 8th. Finally, linking up with the men pushing through the Kabaw Valley, the troops took Kalemyo on 13 November.

It should be mentioned, moreover, that these support missions took place in monsoon weather during which 175 inches of rain fell in Northern Burma, while no fewer than 500 inches fell in Assam. In September, this had slowed down even the Hurricanes' activities, but in the somewhat improved conditions of October, their efforts reached a maximum. 34 Squadron made 861 sorties during the month; 60 made 602; 113 flew 1,049 operational hours. The number of pounds of bombs launched by Hurribombers during the campaign rose to five-and-a-half million. Everywhere the Hurricane was seen – everywhere triumphant.

On 2 November the first RAF squadron re-entered Burma when No 11 landed at the airfield at Tamu. No 11 was a Hurricane squadron – of course!

–

Since the Hurricanes had borne such a vital burden in the decisive phase of the conflict, it seems only just that they should win further laurels during the triumphant reconquest of Burma – though by the end of 1944, many Hurricane squadrons were re-equipping with later types, chiefly the powerful seven-ton Republic Thunderbolts. The Hawker fighters even carried new weapons, for, in November 1944, 20 Squadron received Hurricane IVs armed not only with bombs and anti-tank guns but now with rockets as well.

It also appeared appropriate that on 23 February 1945, Air Marshal Sir Keith Park took over the post of Allied Air Commander-in-Chief. The man who had directed the Hurricanes in their greatest moments in the Battle of Britain was again on hand to guide them to their final victories.

As an accompaniment to the main advance of Fourteenth Army, it was rightly decided by the new commander of the Allied land forces, Lieutenant-General Sir Oliver Leese, who had succeeded Giffard on 12 November 1944, that XV Corps must continue its progress along the Burmese coast. Its ultimate objective was the capture of the islands of Akyab and Ramree as bases from which air transports could keep Slim's men supplied, once the capture of Mandalay had enabled them to push south towards Rangoon.

The Hurricanes were already making those reconnaissance flights that are the necessary preludes to any offensive. The day after Leese took up his appointment, such a mission was flown by one of Bomber Command's most famous pilots, Group

Captain Leonard Cheshire, who had won perhaps the hardest-earned of all VCs, not for one sublime act but for four years of continuous bravery well above the average. He had never been in a Hurricane before but doubtless so reliable an aeroplane presented few problems to such an airman. His sortie was over Akyab, which, he reported, appeared to be deserted.

The new Japanese Commander-in-Chief, Lieutenant-General Kimura, who had replaced the discredited Kawabe, was indeed pulling back from his advanced positions. In consequence, XV Corps moved forward with surprising speed. Donbaik, the staunch defence of which had ruined 'First Arakan', fell on 23 December 1944. On 2 January 1945, two Hurricanes from 20 Squadron, meeting no resistance while flying low over Akyab, dropped a message in Urdu and Burmese to the inhabitants of a village, asking them to sit on the ground if there were Japanese on the island, but if they had left, to raise both hands high above their heads. Ten minutes later the pilots were delighted to see groups of villagers with uplifted arms.

The island was occupied accordingly without fighting. From it, on 12 January, was launched an amphibious landing at Myebon on the Burmese mainland, some forty miles to the south-east. This proved successful only after a grim struggle in which the attackers were aided by low-level strikes by American Mitchell bombers and by Hurricanes, while other Hurricanes laid down a protective smoke-screen. Ramree Island was taken nine days later.

Elsewhere also the Japanese were in retreat. On the northern front, Lieutenant-General Sultan, Stilwell's successor, made steady progress, culminating in the capture of Lashio on 7 March 1945, which reopened the Burma Road to China. In the centre, Fourteenth Army, supported by Hurricane fighter-bombers, crossed the Chindwin on 4 December 1944, whereupon, leaving behind only a few stubborn strongholds, the enemy fell back to the far side of the mighty Irrawaddy River, while the Hurricanes, flying by night as well as by day, continued to harry all forms of transport.

Indeed, it may be wondered whether the Japanese would not have been better advised to abandon Burma altogether. They were heavily outnumbered but their High Command, facing a far greater menace in the Pacific could not spare reinforcements – on the contrary, the luckless Kimura was deprived of one division and the bulk of his supporting naval and air forces, so that he could call on fewer than a hundred aircraft, most of which were obsolescent.

It was in fact the progress of the war in the Pacific that made the Japanese actions in Burma so useless. On 20 June 1944, two days before the relief of Imphal, the Japanese Navy retired from the Battle of the Philippine Sea having suffered losses to its naval air strength so crippling that they could never be replaced. This defeat sealed the fate of the Mariana Islands, which the Americans had secured by mid–August, enabling them to bring air-raids on the Japanese cities with which the defences were quite unable to cope. These disasters also brought down the government of General Tojo, which fell on 18 July.

In October, while the Hurricanes were making their greatest endeavours in the fight for the Tiddim road, came the next major American advance. On the 20th, they landed on Leyte Island in the Philippines, the capture of which would cut off the Japanese from the vital resources to the south, for which they had gone to war in the first place. In particular, if their Navy remained in home waters it would lose its supplies of fuel – yet if it went south, it would be unable to receive ammunition. Since it would therefore by pointless for the Japanese to preserve their fleet but lose the Philippines, it was committed in one last desperate attempt to turn the tide of war.

The result was the titanic Battle of Leyte Gulf, the largest naval battle in history, which lasted from 23 to 26 October. This brought about the destruction of twenty-six major Japanese warships and the end of their fleet as an effective fighting force. Admiral Yonai, the Japanese Navy Minister, would state at the conclusion of hostilities that this defeat was 'tantamount to the loss of the Philippines'. 'When you took the Philippines', he added, that was the end of our resources.' The conquest of strategic positions in the islands quickly followed. Luzon, the main one, was invaded on 10 January 1945; the capital Manila fell on 4 March. The United States Navy, soon to be joined by the British Pacific Fleet, moved on to Iwo Jima and Okinawa, the final steps before the assault on the Japanese homeland.

In such circumstances, it is difficult to disagree with the verdict of Brigadier Smith:

> Courage alone kept Kimura's soldiers fighting during 1945 but although we must admire such devotion to duty, in retrospect it seemed a pointless and futile gesture by the Japanese High Command to continue the struggle. After the summer of 1944, the High Command was no longer able to reinforce Burma Area Army and although a more flexible defensive policy was adopted, the rump of their forces was left to combat numerically superior ground forces, virtually unsupported from the air, with an administrative machine in tatters. A complete withdrawal from Burma would have saved thousands of lives without radically changing the outcome of the war in South East Asia.

Since the Japanese did not withdraw from southern Burma, the British prepared to take it. The port of Rangoon would be a valuable acquisition to speed up the delivery of supplies, including those to China, but it is difficult to escape the conclusion that the main reason for this expenditure of money, materials, and, most important, men's lives, was the desire to preserve face by liberating a former colony. It would be interesting to know what the feelings of those taking part would have been if they could have foreseen the future: how the reconquest of Burma by British and Commonwealth troops would be followed by that country not only ceasing to be a British possession but leaving the Commonwealth; how the magnificent Indian Army would be broken up by the partition of the sub-continent; while the members of Bose's Indian National Army would return to be hailed as patriotic heroes by the Congress Party.

The irony of it all seems heightened, if anything, by the brilliant tactics used. When it was clear that the Japanese were retreating, Slim at once re-cast his plans. His new scheme bears an interesting, though seemingly unnoticed, affinity to the one laid down by Montgomery in Normandy; so much so that one wonders whether it was consciously or unconsciously influenced by the success of that campaign. In Normandy, constant pressure was exerted on the Allied left, threatening the vital target of Caen, in order to tie down the enemy forces, particularly the armour, while the decisive thrust was made on the right. This in essence was also the basis of the Battle of the Irrawaddy Shore.

In more detail, it was proposed that while Lieutenant-General Stopford's XXXIII Corps exerted constant pressure on the Allied left, threatening the vital target of Mandalay, IV Corps under the dynamic Messervy, now also a Lieutenant-General, would make the decisive thrust on the right by seizing the supply centre at Meiktila, the loss of which would cripple Kimura's army. Messervy's approach march, which was made in strictest secrecy, was blocked by a Japanese stronghold at Gangaw. Not wishing to use too many troops in case this aroused suspicion, he summoned aid from above. On 10 January 1945 American Mitchells, supported by Thunderbolts, launched a heavy assault on the enemy defences, after which the Hurribombers of 34, 42, 60 and 113 made precision raids on any positions left untouched, hitting five out of six bunkers. The subsequent infantry attack captured the stronghold with only two men wounded, leading to a delighted signal of thanks from Messervy and the statement by Slim that 'Gangaw was taken by the Air Force'.

The advance of XXXIII Corps to the Irrawaddy was similarly assisted by the Hurricanes. The main obstacle in Stopford's path was at Monywa, the fight for which, commencing on 14 January, was extremely bitter, for the defenders were a regiment of 33rd Division, whose resolution was as great as ever. In consequence, on the 20th, seven pilots from 20 Squadron flying Hurricane IVs armed with rockets, were called in to strike the defences – which they did with 'great accuracy'. Two days later, Monywa was in British hands.

All was now ready for the passage of the huge river by XXXIII Corps, backed by the close support of Mitchells, Thunderbolts and Hurricanes, the latter operating from airstrips hastily built in the jungle only eight miles behind the front line by the Royal Engineers, whose work throughout the campaign was beyond praise. As darkness fell on 12 February, 20th Division crossed at Myinmu, posing a direct threat to Mandalay.

Against 20th Division therefore, Kimura ordered his reserves, including his few remaining tanks. To continue the comparison with Normandy, this miserable handful would have seemed laughable to the men advancing on Caen in the face of numbers that varied from 530 to 725, but for the Japanese each one was doubly precious, simply because they had so few of them.

They would soon have even less. Slim has recorded that rocket-firing Hurricanes 'proved our most successful anti-tank weapon' – but, in fact, when Flight Lieutenants Farquharson and Ballard took off in the early hours of 19 February, at the start of what turned out to be 20 Squadron's greatest mission, they were flying aircraft armed with 40 mm cannon. Army units had reported the presence of hostile tanks but at first the pilots could spot no more than their tracks. Then, shortly after 1000, they sighted what appeared to be a small shed. The only odd fact about it was that its roof was covered with boughs of trees, though there were none near at hand. Suspicious, the pilots fired at the roof, which ripped away to reveal that the 'building' was a cleverly camouflaged tank. This they promptly set on fire. Soon afterwards, they found a second tank, which they also destroyed – then others. Thereupon they called to their aid the rest of the squadron, which, with a hail of anti-tank shells and rockets, reduced eleven more of the enemy's vital armoured vehicles to blazing wrecks.

Since this episode effectively shattered the Japanese counter-attack, it must be regarded as one of the Hurricanes' most telling interventions. Its importance was well summed up by Stopford who signalled: 'Destruction of enemy armour of major import in our battle for the bridgehead. Well done.' Major-General Gracey, commanding 20th Division, was even more enthusi-astic, if less formal. 'Nippon Hardware Corporation has gone bust,' he recorded gleefully. 'Nice work 20 Squadron. Tanks, repeat tanks, a million.'

While Japanese attention was directed upon Stopford, Messervy also crossed the Irrawaddy at Nyaungu, more than a hundred miles south of Mandalay. This was the widest opposed river crossing of the war, but since most of the defenders were members of the Indian National Army who were only too eager to surrender, it was not nearly so dangerous as many a shorter one. Kimura, still regarding the bridgehead at Myinmu as the main threat, dismissed this new move as a diversion – but he was grimly disillusioned when, on 21 February, Messervy's men, closely supported by fighter-bombers, made a dramatic dash for Meiktila.

It seems that all that saved the Japanese from being trapped in a massive 'killing ground' – a 'Falaise Pocket' – south of Mandalay, was that their administrative troops showed as much ferocious determination as those in the front line. Major-General Kasuya, the energetic commander of Meiktila, threw every available man, including patients from the hospital, into the fight. By the time all had been wiped out on 3 March, they had given Kimura the opportunity to send forces from the Mandalay area, which blocked any further advance north

– which posed indeed such a threat to the Allied garrison in Meiktila that, like that of Imphal, it had to be reinforced by air. In addition, the struggle for Mandalay continued unabated.

Also, as in the Imphal campaign, the ultimate Allied triumph was aided by constant pressure from the ground-attack Hurricanes. The Mark IVs of No 20 flew 566 sorties during March, but the Hurribombers, if less spectacular, were almost equally deadly and even more hard-working: 11 Squadron had 602 sorties to its credit in that same month; 60 Squadron no fewer than 722.

As a fitting culmination to the forays of the Allied Air Forces, Fort Dufferin – a great, rectangular, walled enclosure surrounded by a moat, which dominated Mandalay – was subjected, on 20 March, to a series of raids by Mitchells, Thunderbolts and the rocket-firing Hurricanes of No 20. The final blows were delivered by 42 Squadron, flying Hurricane IVs armed with 250 lb bombs, which made three separate attacks on the defenders, under cover of which Thunderbolts blasted a breach in the walls. Even the Japanese had now had enough – they made their escape through the drains. By the end of the day, the city was in Allied hands. By the end of the month, all enemy forces, whether in the Mandalay or Meiktila areas, were in disorderly retreat eastward to the Shan Hills, abandoning most of their equipment.

This left the way to Rangoon wide open to Fourteenth Army, particularly since the central Burmese plain was ideal for its tanks and armoured cars. On the other hand, with the monsoon imminent, time was short as Messervy led a spectacular race southward for the Burmese capital, while above him the Hurricanes, constantly moving up to advanced airfields as these were lost to the Japanese, fell on enemy transport with renewed zeal.

The most prominent unit during this final advance was undoubtedly 20 Squadron. On 13 April, its Hurricane IVs destroyed seventeen vehicles. On the 18th, they wrecked a lorry and eight bullock-carts, as well as damaging a steam-roller, of all things. On the 25th, the pilots first caused further heavy losses of enemy transport, then joined with the Hurribombers of 60 Squadron to strike at hostile troops in a village. Such raids reached a peak on 30 April, when No 20 was credited with the largest total of vehicles destroyed by any squadron in a single day – no fewer than forty-six of them.

Aided by such actions, Messervy reached Pegu, fewer than fifty miles from Rangoon, by 29 April, but here IV Corps was robbed of its reward by a last-ditch Japanese defence, since, on 3 May, before this could be overcome, a seaborne assault occupied the capital – which by a final irony was found abandoned. It was a somewhat miserable prize in any case, as the Japanese had wrecked the electricity supply before leaving, looters had been more than busy and filth of every kind clogged the streets.

It only remained for the Allied armies – under the leadership of Stopford, Slim having returned to England on leave – with the help of their Air Force colleagues, to wipe out the enemy forces cut off by Messervy's advance as they attempted to reach the safety of Thailand. This action was still in progress when, on 6 August, an atomic bomb fell on Hiroshima, to be followed by another on Nagasaki three days later. On the 14th, the Japanese accepted the terms of surrender. This was officially announced the next day, though the formal document was signed on board the American battleship *Missouri* in Tokyo Bay only on 2 September, six long years since Britain had declared war on Germany in 1939.

At the time of the capitulation, most of the RAF Hurricanes had been replaced by more modern aircraft, chiefly Thunder-bolts, but 20 Squadron, with its Mark IVs, and 28, with its Tac R IICs, retained the Hawker fighters, while six Photographic Reconnaissance IICs also remained on the strength of 681. In addition, the trusty IICs were still being flown by the pilots of eight Indian Air Force squadrons.

Thus, the Hurricane, which had been numerically the most important Allied fighter at the start of the Second World War, which had made the bulk of the interceptions in the Battle of Britain, remained on front-line service to the end of the great conflict. Of its exploits in the Burmese theatre, the finest summary must be the following from Francis Mason's *The Hawker Hurricane*:

> It had been called on to perform an unprecedented variety of duties, by day and by night, in fair weather and in the foulest. It had covered the armies in their retreat and it had helped them forward again. It had started with eight machine-guns and it finished with the equivalent of a warship's broadside. From the dusty strip inside Calcutta to the rough surfaces hacked in the dense jungle came this fighter.

Came also this fighter-bomber, this 'tank-buster', this reconnaissance machine. As indeed it had come earlier in a score of other theatres, from Murmansk to Madagascar, from Gibraltar to Java. No other aircraft had fought in so many campaigns, in so many varied conditions, in so many different roles, so successfully. That much misused word 'unique' is inadequate to describe its achievements.

Epilogue

But Not Forgotten

Since the Hurricane's exploits in war had been so outstanding, the fact that, with victory safely won, it should vanish from the scene with startling rapidity, seemed almost suitable – though the fate of many individual aeroplanes cannot but appear tragic in retrospect. Typical was that of the six PR Hurricanes still being flown by 681 Squadron. These machines, which prior to their service in Burma had performed similar duties in the Middle East, were simply abandoned on Meiktila airfield, where, after being stripped of everything valuable by the local inhabitants, they were allowed to rot away.

The other RAF Hurricanes lasted little longer. In the Far East, 20 and 28 Squadrons converted to Spitfires at the end of September 1945, though a few of the Hawker fighters remained on secondary duties, such as meteorological flights, for a further year. In north-western Europe and the Mediterranean area, even those Hurricane units engaged on similar minor tasks had disbanded or converted by the end of 1945, while the aircraft and personnel of 351 Squadron were transferred to the Yugoslav Air Force – no longer Royal – on 15 June.

Thus, by the beginning of 1946 the Mark IVs of 6 Squadron formed the RAF's only force of Hurricanes. In July 1945 they had returned to Palestine, where they co-operated with the Army on internal security duties until September 1946, when they moved to Nicosia, Cyprus. However, at the end of the year it was decreed that No 6 should re-equip with Tempests and, on 15 January 1947, the last Hurricanes left RAF squadron service.

Those Hurricanes flying with other air forces would also soon disappear. A small number that had fought with the SAAF squadrons in the Middle East went to South Africa on the close of hostilities, joining about eighty more already there as trainers – but within a few months all had been scrapped. Only two Indian Air Force squadrons, 2 and 7, still manned Hurricanes by early 1946. These guarded the North-West Frontier for another year, though some individual machines remained airborne for a little longer with training units. The Royal Egyptian and Turkish Air Forces flew Hurricanes throughout 1946. The Irish Air Corps at Baldonnel retained its Hurricanes until 1947.

The only countries that increased their stocks of Hurricanes after the war were Portugal and Iran. Fifty IICs, many of them provided with 45-gallon drop-tanks, were sent to Portugal during 1945-46; they served as both day and night fighters until 1951. Sixteen IICs, with their cannon removed and Merlin 22 engines in place of the usual XXs, joined the Persian Air Force as trainers in 1946.

Iran also ordered a pair of two-seat Hurricane trainers. Since Hawkers had prepared drawings of such a version for the RAF back in 1940 – though the project never materialised – work was quickly put in hand. On 27 September 1946, the first two-seater, adapted from a Mark IIC KZ232, took off with test pilot Bill Humble at the controls. The second cockpit was incorporated immediately behind the existing one, thereby necessitating little modification to the basic structure. Both cockpits were left uncovered originally, the first position being shielded by a windscreen and frame, the rear one only by a simple transparent fairing. Tests, however, showed that the latter was uncomfortably draughty, so a sliding hood, adapted from that in use on the Tempest, was provided.

This aircraft, now numbered 2-31, was delivered to the Persian Air Force early in 1947 and was followed by No 2-32, a similarly modified IIC, shortly afterwards. Their Merlin 22s gave them a top speed of 320 mph and their service ceiling was 26,000 feet.

Another Hurricane modification of this period was provided by Z3687, a Mark IIB that carried an Armstrong-Whitworth laminar-flow wing of reduced thickness designed to lessen drag. Painted white and highly polished, this aircraft was tested repeatedly at Farnborough during the years 1946-48 and was able to provide much valuable information.

Mention might also be made here of what might be called a Hurricane Variant, or at least a Hurricane-Inspired Variant. This was a five-eighths scale replica of the Hurricane – created by an American admirer, Mr Fred Sindlinger of Puyallup, Washington State – powered by a 150 hp Lycoming engine driving a Hartzell constant-speed two-blade propeller. Purists will rightly point out that this machine was not really a Hurricane at all, but as one of the Hurricane's outstanding traits was its versatility, it seems only fair to record it as an honorary member of the Hurricane type.

Of the true Hurricanes, by 1955 almost all had disappeared – mostly into oblivion. A few, though, went into museums and as the years passed the numbers of these increased slowly but steadily, as did the list of countries in which the museums were to be found: Britain, the United States, Russia, Canada, India, South Africa, Malta, Belgium, Finland and Serbia. Examples of Hurricanes that were still flying in 1955 and for some time thereafter, however, were limited to two, only one of which was in service with the RAF.

LF363 was a Hurricane Mark IIC that was among the last batch of Hurricanes – Mark IIBs and IVs but mainly IICs – built by Hawkers. It was delivered to 63 Squadron on 20 March 1944, subsequently flying with 309 Squadron, 63 again and, finally, 26, a squadron equipped with both Hurricanes and Mustangs, which, in addition to reconnaissance duties, specialised in artillery spotting. It also served at a number of Operational Training

Units. It has been widely reported that it was the last Hurricane to be delivered to the RAF, but apparently this is not the case. It would, however, become the last remaining Hurricane in the RAF not long after the war, being allocated to the Station Flights of a number of successive RAF bases, from which it regularly emerged to lead the annual fly-past over London on Battle of Britain Sunday. In 1948 and 1949 its pilot for these displays was Air Vice-Marshal Sir Stanley Vincent.

In 1957, LF363 was at Biggin Hill, where it was joined on 11 July by three Spitfires to form the Battle of Britain Flight. This would later move in sequence to North Weald, Martlesham Heath, Horsham St Faith in Norfolk, Coltishall in Norfolk and Coningsby in Lincolnshire, becoming the Battle of Britain Memorial Flight in 1973 after receiving a Lancaster, not, of course, an aircraft that had fought in the Battle. Sadly, in 1959, the Air Ministry decided to ban veteran aircraft from the Battle of Britain fly-past following an accident, not to LF363 but to its accompanying Spitfire, which was written off.

Indeed, while Spitfires came and went in the Memorial Flight, the faithful LF363 remained a constant member, appearing in a variety of different markings to commemorate famous Hurricane squadrons or pilots: its first one in 1969 was Douglas Bader's 242 Squadron. In the process it won the hearts of everyone connected with the Flight. So much was this the case that when it had a horrific crash and caught fire on 11 September 1991 – the pilot escaped with a broken ankle and minor burns – there was no question of it not being repaired even though the funds for this had to be raised by the sale of one of the Flight's Spitfires. On 29 September 1998, LF363 flew again in the markings of 56 Squadron, one of the first to receive Hurricanes and appropriately displays a phoenix on its squadron badge.

PZ865, also a IIC, was the last Hurricane built. It was not issued to the RAF for, as was mentioned earlier, it was purchased by Hawker Aircraft Limited. Later, with a Merlin 22 engine installed and its guns removed, it was placed on the civil register as G-AMAU. Painted a splendid royal-blue and gold, it was entered by Hawkers in numerous air races until 1960, when it was restored to its original military camouflage as well as its military registration number though, uniquely, it also retained its civil registration. In 1960-61 its particular speed range led to PZ865 being chosen to fly as 'chase plane' during the trials of the Hawker P1127 vertical take-off aircraft – predecessor to the famous Harrier.

PZ865 remained the property of Hawker Aircraft Limited for many years after this, giving displays at air shows where it was flown by Hawker test pilots. It was apparently also their favourite 'mount' if they needed to travel from its home base – first at Langley, later at Dunsfold – for any reason. It was not wholly unconnected with the RAF, however, for in 1954 and 1956, when LF363 was undergoing servicing, PZ865 led the RAF Battle of Britain fly-past. This foreshadowed the eventual decision by Hawkers to present PZ865 to the Battle of Britain Flight. After a short test flight at Dunsfold on 21 March 1972, it was duly delivered to the Flight's then base at Coltishall, thereby earning the indisputable claim of being the last Hurricane received by the RAF. Its civil registration was cancelled and, like LF363, it would appear in various markings, the first that of 'Bob' Stanford Tuck's 257 Squadron.

Meanwhile, the year of 1966 had witnessed an event, minor in itself but major in its consequences. A Canadian pilot and engineer, Bob Diemert, had acquired an old Hurricane Mark XII – the equivalent of the RAF's Mark IIB but powered by a Packard-Merlin 29 engine – which had served with the Royal Canadian Air Force from November 1942 to the end of the war. He rebuilt and restored this aircraft, usually believed to be RCAF No 5585 but apparently really No 5377, and early in 1966 flew it with the civil registration of CF-SMI. Next year it was transported to Britain where, in the company of LF363 and PZ865, it embarked on a temporary film career.

In 1969, this Hurricane was purchased by the Strath-allan Aircraft Collection of Auchterarder, Perthshire, Scotland, whose engineers devoted more than 5,000 man-hours to its restoration. Re-registered G-AWLW and now powered by a Merlin 25, it flew again from the Strathallan airfield on 28 June 1973 under the control of Duncan Simpson, chief test pilot of Hawkers – Hawker-Siddeley Aviation as the company had by then become – who is reported to have found this a pleasing change from his normal work with Harriers. In 1984 it was returned to Canada where it received another civil registration, C-GCWH. Sad to relate, however, it was destroyed in a hangar fire on 15 February 1993.

Bob Diemert's restoration of his Hurricane had ended in an unhappy anti-climax, but its influence lived on, for it provided an inspiration to other Hurricane enthusiasts not only in Canada but in the United States, where the Hurricane has often been more highly regarded than in the Britain it had done so much to save. A similar surge of interest followed the end of the Communist regime in Russia. For political reasons, this had kept very quiet about the Hurricanes it had received but, when it fell, attention was directed to them with the strong support of surviving pilots who remembered them with gratitude and affection.

In consequence, battered Hurricanes or parts of Hurricanes emerged from various sites in Canada in the 1970s and 1980s, and in Russia in the 1990s. They were entrusted to several different organisations and groups, among which special mention must be made of Hawker Restorations Limited, and during the above periods and into the first two decades of the twenty-first century, these set about repairing and restoring Hurricanes or newly creating them from the remains of more than one wrecked machine. Some were added to the static exhibits in museums including, disappointingly, one aircraft in Canada that had been restored to full airworthy condition, but others were able to take to the skies and reinforce LF363 and PZ865.

A total of fifteen of these would appear at different times, though the revived airborne careers of three of them would end sadly – in one case fatally. This Hurricane, a Mark XII, had first flown again in January 1996, but on 15 September 2007 it had a ghastly crash, the pilot being killed and the aircraft becoming a total loss. Of the remaining two, one was damaged in a landing accident and the other landed safely, only to be effectively disabled by another aircraft that struck it while it was stationary on the runway. It is probable that the former, and possible that the latter, will yet be restored to full flying standards and it is also known that other Hurricane projects are still patiently proceeding with the intention of seeing the aircraft fly again, or at least be able to taxi. Considering how long only LF363 and PZ865 were airborne, this is surely a most encouraging situation.

Of those Hurricanes successfully restored to a new life in the skies, there is space here for mentioning just two that seem to be of particular interest. Hurricane R4118, a Mark I, was returned from India where it had been used as a 'ground instructional' model for pilots of the Indian Air Force. Earlier in its career, however, it had served with 605 'County of Warwick' Squadron and is the only Hurricane still flying that had seen combat in the Battle of Britain. This it did in September and October 1940, being credited with at least a share in the destruction of five enemy aircraft.

Hurricane P3351 also began its career as a Mark I. At the end of May 1940, it joined 73 Squadron in France and took part in the last days of the fighting there: the only Hurricane still flying of which this can be said. In July 1940, an accident in a night-fighter sortie resulted in its being out of action until 6 September. The Battle of Britain was still at its height but 32 Squadron, to which P3351 was then assigned, had been pulled back to Acklington, Northumberland, and it saw no combat either with 32 or later with 71 'Eagle' Squadron of American volunteers. In late 1941, it was converted to a Mark IIA Series

2, and given a Merlin XX engine and a new serial number: DB393. It was sent to Russia on Convoy P016 in May 1942, serving there until late in 1943. It was returned to Britain in 1992.

It might also be mentioned that the restored Hurricanes have added to the type's extraordinary record by appearing in skies where this had rarely or never flown previously. One went to Australia, one, rather ironically, to Germany, a whole batch to various parts of the United States. P3351/DB393, however, travelled the farthest from the Hurricane's homeland, making its first flight on 12 January 2000 in the South Island of New Zealand. It continued to fly at air shows in New Zealand until 2013, when it moved, perhaps appropriately, to France.

Sir Sydney Camm – he had belatedly been honoured with a Knighthood in 1953 – who had been responsible for the very existence of all Hurricanes, would no doubt have been delighted to learn of the increasing airborne numbers of probably his favourite and certainly his most famous fighter. It was not to be for he died suddenly of a heart attack on 12 March 1966 at the age of seventy-three. The passing of Britain's finest aircraft designer received surprisingly little notice at the time – but everyone mentioned that he had created the Hurricane, and that alone was perhaps a sufficient epitaph for any man.

Acknowledgements

My thanks are due to the following for making available to me information some of which has not been published:

Mr Chaz Bowyer, Mr J.D. Brown: Naval Historical Branch, Ministry of Defence.

Mr J.P. MacDonald: Air Historical Branch, Ministry of Defence.

Captain Donald Macintyre.

Air Commodore H.A. Probert: Air Historical Branch, Ministry of Defence.

Mr P.E. Richardson: Strathallan Aircraft Collection.

Mr Christopher Shores.

My thanks are also due to the following who have kindly given me permission to quote from material within their copyright control: Messrs Eyre Methuen (*The Second World War* by Maj-Gen J.F.C. Fuller); Winant, Towers Ltd (*Narvik* by Donald Macintyre), The Controller of Her Majesty's Stationery Office (*Royal Air Force* 1939-45 by Denis Richards and Hilary St George Saunders); Macdonald Futura *Hawker Hurricane* by Francis Mason); Beaumont Aviation Literature (*43 Squadron* by J. Beedle) Collins Publishers (*The Navy at War* by Stephen Roskill); Batsford (*The Battle for Burma* by Brigadier Smith and *The North African Campaign* by General Sir William Jackson) Weidenfeld (Publishers) Limited (*Duel of Eagles* by Group Captain Peter Townsend); Doubleday (*Singapore, The Battle that*

Changed the World by James Leasor; A.M. Heath & Co Ltd (*Phoenix into Ashes* by Wing Commander Roland Beamont); Wing Commander Paul Richey, DFC, (*Fighter Pilot*).

Bibliography

Overall Strategic Background

Churchill, Winston S. *The Second World War*, Cassell, 1948-54.
Fuller, Major-General J.F.C. *The Second World War 1939-1945*.
Eyre & Spottiswoode. 1948, Revised ed 1954.
Liddell Hart, Captain B.H. *History of the Second World War*.
Cassell. 1970.
Purnell's *History of the Second World War*.

The Hurricane

Allward, Maurice. *Hurricane Special*. Ian Allan, 1975.
Bader, Group Captain Douglas. *Fight for the Sky: The story of the Spitfire & Hurricane*, Collins, 1973.
Green, William. *Aircraft of the Battle of Britain*. Macdonald/Pan, 1969.
Green, William *War Planes of the Second World War* (Vol 2 Fighters: Great Britain & Italy). Macdonald, 1961.
Mason, Francis K. *The Hawker Hurricane*. Macdonald, 1962.
Mason, Francis K. *The Hawker Hurricane I*. Profile Publications Ltd, 1966
Mason, Francis K. *The Hawker Hurricane IIC*. Profile Publications Ltd, 1965.
Mason, Francis K. *Hawker Hurricane Described*. Kookaburra Technical Publications, Australia, 1970.
Mason, Francis K. *Hawker Aircraft since 1920*. Putnams, 1961.

Moyes, Philip J.R. *Hawker Hurricane I.* Visual Art Productions Ltd., Oxford, 1978.

Robertson, Bruce & Scarborough Gerald. *Hawker Hurricane.* Airfix Products Ltd, 1974.

Shores, Christopher. *Hawker Hurricane Mk I/IV in Royal Air Force & Foreign Service.* Osprey Publishing Ltd, 1971.

The Battle of Britain

Bishop, Edward. *The Battle of Britain.* George Allen & Unwin, 1960.

Collier, Basil. *The Battle of Britain.* Batsford, 1962.

Collier, Richard. *Eagle Day.* Hodder & Stoughton, and J.M. Dent, 1966.

Lee, Wing Commander Asher. *Blitz on Britain.* Four Square, 1960.

Mason, Francis K. *Battle over Britain.* McWhirter Twins, 1969.

Middleton, Drew. *The Sky Suspended: The Battle of Britain May 1940 – April 1941.* Seeker & Warburg. 1960.

Townsend, Group Captain Peter. *Duel of Eagles.* Weidenfeld & Nicholson, 1970.

Wright, Robert. *Dowding & The Battle of Britain.* Macdonald, 1969.

Wood, Derek & Dempster, Derek. *The Narrow Margin.* Hutchinson, 1961, Revised ed. 1969.

Other Aspects of War in the Air

Barker, Ralph. *The Hurricats.* Pelham Books, 1978.

Bowman, Gerald. *War in the Air.* Evans, 1956.

Brookes, Andrew J. *Photo Reconnaissance.* Ian Allan, 1975.

Brown, David. *Carrier Fighters.* Macdonald & Janes, 1975.

Franks, Norman. *The Greatest Air Battle: Dieppe 19th August 1942,* Kimber, 1979.

Jackson, Robert. *Air War over France May-June 1940*. Ian Allan, 1974.

Jackson, Robert. *Strike from the Sea: A History of British Naval Air Power*. Arthur Barker Ltd, 1970.

Killen, John. *A History of Marine Aviation 1911-68*. Frederick Muller, 1969.

Mason, Francis K. *The Gloster Gladiator*. Macdonald, 1964.

Ministry of Information. *The Air Battle of Malta*. HMSO, 1944.

Owen, Roderic. *The Desert Air Force*. Hutchinson, 1948.

Poolman, Kenneth. *Faith Hope & Charity*. Kimber, 1954.

Richards, Denis & Saunders, Hilary St G. *Royal Air Force 1939-1945*. HMSO, 1953-4.

Shores, Christopher & Ring, Hans. *Fighters over the Desert: The Air Battles in the Western Desert June 1940 – December 1942*. Neville Spearman, 1969.

Shores, Christopher, Ring, Hans & Hess, William N. *Fighter Over Tunisia*. Neville Spearman, 1975.

Shores, Christopher. *Ground Attack Aircraft of World War II*. Macdonald & Janes, 1977.

Sims, Charles. *The Royal Air Force: The First Fifty Years*. Adam & Charles Black, 1968.

Wykeham, Air Vice-Marshal Peter. *Fighter Command*. Putnams, 1960.

The Pilots (Biography, Autobiography and Collections of Memoirs)

Baker, E.C.R. Pattle; *Supreme Fighter in the Air*. Kimber, 1965.

Baker, E.C.R. *The Fighter Aces of the RAF*. Kimber, 1962.

Beamont, Roland. *Phoenix into Ashes*. Kimber, 1968.

Bickers, R.T. *Ginger Lacey, Fighter Pilot*. Robert Hale, 1962.

Bowyer, Chaz. *Hurricane at War*. Ian Allan, 1974.

Brickhill, Paul. *Reach for the Sky: The Story of Douglas Bader*. Collins, 1954.

Forrester, Larry. *Fly for your Life*. Frederick Muller. (Wing Commander Stanford Tuck), 1956.

Johnson, Group Captain J.E. *Wing Leader.* Chatto & Windus, 1956.

Kelly, Terence. *Hurricane over the Jungle.* Kimber, 1977.

Masters, David. *So Few.* Eyre & Spottiswoode, 1941 Revised ed. 1943.

Ministry of Information. *Over to You.* HMSO, 1943.

Richey, Wing Commander Paul. *Fighter Pilot.* Hutchinson, 1941. Revised ed: 1955 & 1980 (Janes).

Shores, Christopher & Williams, Clive. *Aces High: The Fighter Aces of the British & Commonwealth Air Forces in World War II.* Neville Spearman, 1966.

Turner, John Frayn. *VCs of the Air.* Harrap, 1960.

Vincent, Air Vice-Marshal S.F. *Flying Fever.* Jarrolds, 1972.

Squadrons and Stations

Beedle, J. *43 Squadron.* Beaumont Aviation Literature, 1966.

Hunt, Leslie. *Twenty-One Squadrons: The History of the Royal Auxiliary Air Force 1925-1957.* Garnstone Press Ltd, 1972.

Kinsey, Gordon. *Martlesham Heath: The Story of the Royal Air Force Station 1917-1973.* Terence Dalton Ltd, 1975.

Lewis, Peter. *Squadron Histories RFC RNAS & RAF since 1912.* Putnams, 1959. Revised ed. 1968.

Moulson, Tom. *The Flying Sword: The Story of 601 Squadron.* Macdonald, 1964.

Moyes, Philip J.R. *Bomber Squadrons of the RAF & their Aircraft.* Macdonald. (Includes the Fighter–Bomber Hurricanes), 1964.

Polish Air Force Association. *Destiny Can Wait: The Polish Air Force in the Second World War.* Heinemann, 1949.

Rawlings, John D.R. *Fighter Squadrons of the RAF & their Aircraft.* Macdonald, 1969.

Shaw, Michael. *Twice Vertical: The History of No 1 Squadron, Royal Air Force.* Macdonald. 1971.

Tidy, Douglas. *I Fear No Man: The History of No 74 Squadron, Royal Air Force*. Macdonald, 1972.

Wallace, Graham. *RAF Biggin Hill*. Putnams, 1957.

Background to Campaigns

Ash, Bernard. *Norway 1940*. Cassell, 1964.

Barker, A.J. *Eritrea 1941*. Faber, 1966.

Blaxland, Gregory. *Destination Dunkirk: The Story of Gort's Army*. Kimber, 1973.

Bond, Brian. *France & Belgium 1939-1940*. Davis-Poynter, 1975.

Buckley, Christopher. *Norway, the Commandos, Dieppe*. HMSO, 1952.

Cameron, Ian. *Red Duster, White Ensign: The Story of the Malta Convoys*. Frederick Muller, 1959.

Campbell, Vice-Admiral Sir Ian & Macintyre, Captain Donald, *The Kola Run*. Frederick Muller, 1958.

Carew, Tim. *The Longest Retreat: The Burma Campaign 1942*. Hamish Hamilton Ltd, 1969.

Carver, Field-Marshal Sir Michael. *El Alamein*. Batsford, 1962.

Carver, Field-Marshal Sir Michael. *Tobruk*. Batsford, 1964.

Clark, Alan. *The Fall of Crete*. Anthony Blond, 1962.

Divine, David. *The Nine Days of Dunkirk*. Faber, 1959.

Evans, General Sir Geoffrey & Brett-James, Anthony. *Imphal: A Flower on Lofty Heights*. Macmillan, 1962.

Falk, Stanley L. *Seventy Days to Singapore: The Malayan Campaign 1941-1942*. Robert Hale, 1975.

Jackson, General Sir W.G.F. *The North African Campaign 1940-43*. Batsford, 1975.

Leasor, James. *Singapore: The Battle that Changed the World*. Hodder & Stoughton, 1968.

Lucas Phillips, Brigadier C.E. *Alamein*. Heinemann, 1962.

Lucas Phillips, Brigadier C.E. *Springboard to Victory*. Heinemann. (Kohima), 1966.

Macintyre, Captain Donald. *Narvik*. Evans, 1959.

Macintyre, Captain Donald. *The Battle for the Mediterranean*. Batsford, 1964.

Macintyre, Captain Donald. *The Naval War Against Hitler*. Batsford, 1971.

Ministry of Information. *The Abyssinian Campaigns*. HMSO, 1942.

Moorehead, Alan. *The Desert War: The North African Campaign 1940-1942*. Hamish Hamilton, 1965.

Owen, Col Frank. *The Campaign in Burma*. HMSO, 1946.

Owen, Col Frank. *The Fall of Singapore*. Michael Joseph, 1960.

Perowne, Stewart. *The Siege Within the Walls: Malta 1940-1943*. Hodder & Stoughton, 1970.

Robertson, Terence. *Channel Dash*. Evans, 1958.

Roskill, Captain S.W. *The Navy at War 1939-1945*. Collins, 1960.

Roskill, Captain S.W. *The War at Sea*. HMSO, 1954-61.

Schofield, Vice-Admiral B.B. *The Russian Convoys*. Batsford, 1964.

Shankland, Peter & Hunter, Anthony. *Malta Convoy*. Collins, 1961.

Slim, Field-Marshal Sir William. *Defeat into Victory*. Cassell, 1956.

Smith, Brigadier E.D. *Battle for Burma*. Batsford, 1979.

Smith, Peter C. *Arctic Victory: The Story of Convoy PQ18*. Kimber, 1975.

Smith, Peter C. *Pedestal: The Malta Convoy of August 1942*. Kimber, 1970.

Spencer, John Hall. *Battle for Crete*. Heinemann, 1962.

Swinson, Arthur. *Kohima*. Hutchinson, 1966.

Thomas, David A. *Japan's War at Sea: Pearl Harbor to the Coral Sea*. Andre Deutsch, 1978.

Tomlinson, Michael. *The Most Dangerous Moment*. Kimber. (Japanese attack on Ceylon), 1976.

Tsuji, Colonel Masanobu. *Singapore: The Japanese Version*. Mayflower-Dell, 1960 (UK edition 1966).

Warner, Geoffrey. *Iraq & Syria 1941*. Davis-Poynter, 1974.